Agrobacterium Protocols

Methods in Molecular Biology™

John M. Walker, SERIES EDITOR

44. *Agrobacterium* Protocols, edited by *Kevan M. A. Gartland and Michael R. Davey, 1995*
43. In Vitro Toxicity Testing Protocols, edited by *Sheila O'Hare and Chris K. Atterwill, 1995*
42. ELISA: *Theory and Practice*, by *John R. Crowther, 1995*
41. Signal Transduction Protocols, edited by *David A. Kendall and Stephen J. Hill, 1995*
40. Protein Stability and Folding, edited by *Bret A. Shirley, 1995*
39. Baculovirus Expression Protocols, edited by *Christopher D. Richardson, 1995*
38. Cryopreservation and Freeze-Drying Protocols, edited by *John G. Day and Mark R. McLellan, 1995*
37. In Vitro Transcription and Translation Protocols, edited by *Martin J. Tymms, 1995*
36. Peptide Analysis Protocols, edited by *Michael W. Pennington and Ben M. Dunn, 1994*
35. Peptide Synthesis Protocols, edited by *Ben M. Dunn and Michael W. Pennington, 1994*
34. Immunocytochemical Methods and Protocols, edited by *Lorette C. Javois, 1994*
33. *In Situ* Hybridization Protocols, edited by *K. H. Andy Choo, 1994*
32. Basic Protein and Peptide Protocols, edited by *John M. Walker, 1994*
31. Protocols for Gene Analysis, edited by *Adrian J. Harwood, 1994*
30. DNA–Protein Interactions, edited by *G. Geoff Kneale, 1994*
29. Chromosome Analysis Protocols, edited by *John R. Gosden, 1994*
28. Protocols for Nucleic Acid Analysis by Nonradioactive Probes, edited b *Peter G. Isaac, 1994*
27. Biomembrane Protocols: *II. Architecture and Function*, edited by *John M. Graham and Joan A. Higgins, 1994*
26. Protocols for Oligonucleotide Conjugates, edited by *Sudhir Agrawal, 19*
25. Computer Analysis of Sequence Data: *Part II*, edited by *Annette M. Griffin and Hugh G. Griffin, 1994*
24. Computer Analysis of Sequence Data: *Part I*, edited by *Annette M. Griffin and Hugh G. Griffin, 1994*
23. DNA Sequencing Protocols, edited by *Hugh G. Griffin and Annette M. Griffin, 1993*
22. Optical Spectroscopy, Microscopy, and Macroscopic Techniques, edited by *Christopher Jones, Barbara Mulloy, and Adrian H. Thomas, 1994*
21. Protocols in Molecular Parasitology, edited by *John E. Hyde, 1993*
20. Protocols for Oligonucleotides and Analogs, edited by *Sudhir Agrawal, 1993*

Earlier volumes are still available. Contact Humana for details.

Methods in Molecular Biology™ • 44

Agrobacterium Protocols

Edited by

Kevan M. A. Gartland

Department of Molecular and Life Sciences,
University of Abertay Dundee, Scotland, UK

and

Michael R. Davey

Department of Life Science, Plant Genetic Manipulation
Group, University of Nottingham, UK

Humana Press Totowa, New Jersey

Library of Congress Cataloging in Publication Data

Main entry under title:

Methods in molecular biology™.

Agrobacterium protocols / edited by Kevan M. A. Gartland and Michael R. Davey.
    p.   cm. — (Methods in molecular biology™ ; 44)
    Includes index.
    ISBN 0-89603-302-3
    1. *Agrobacterium.* 2. Agricultural biotechnology. I. Gartland, K. M. A. II. Davey, M. R. (Michael Raymond) III. Series: Methods in molecular biology™ (Totowa, NJ) ; 44.
QR82.R45A37   1995
589.9'5—dc20
                                         95-15031
                                         CIP

Preface

Since the identification, at the beginning of the present century, of *Agrobacterium* as the causative agent of crown gall disease of many dicotyledonous plants, the unique interaction of this Gram-negative soil bacterium with higher plants has fascinated biologists. However, it was not until the 1970s that advances in molecular techniques enabled evidence to be obtained that crown gall induction was associated with gene transfer from *A. tumefaciens* to higher plant cells. Likewise, the induction of "hairy roots" on susceptible plants was also confirmed to be associated with the transfer of bacterial genes from *A. rhizogenes* to plants and the integration of such genes into the genome of recipient species. In both cases, the expression of *Agrobacterium* genes in higher plants has a profound effect on plant development. Although many of the intricate details of the mechanisms associated with this naturally occurring genetic engineering system are still unresolved, the information that has been obtained has been exploited to transfer a range of genes into plants, including many of our most important dicotyledonous crops. The enigma—which remains—concerns the apparent inability of *Agrobacterium* to transform many monocotyledons, although current evidence indicates that genetically engineered "supervirulent" strains of *Agrobacterium* may well transform a greater diversity of host plants, including cereals.

Agrobacterium *Protocols* contains contributions that reflect the diversity of interests in *Agrobacterium*—higher plant interactions. The topics range from the maintenance of bacterial culture collections to aspects of the metabolism and physiology of transformed tissues and transgenic plants. Methods for introducing specific DNA sequences into target species are described in detail, and the environmental implications of genetically engineered plants are discussed. Although some chapters are review-oriented, the majority contain laboratory protocols,

v

in line with other volumes in this series. Such detailed procedures should facilitate rapid transfer of these techniques to other laboratories, together with exploitation of this technology in fundamental and applied biology. The editors take this opportunity of thanking all colleagues who have devoted their time and effort to provide the information presented in this Humana Press publication.

Kevan M. A. Gartland
Michael R. Davey

Contents

Preface .... *xx*

Contributors .... *xx*

CH. 1. Growth and Storage of *Agrobacterium*,
          Fraukje M. van Asma *1*

CH. 2. Simple Cultural Tests for Identification of *Agrobacterium* Biovars,
          Hacène Bouzar, Jeffrey B. Jones, and Andrew L. Bishop *9*

CH. 3. *Agrobacterium* Virulence,
          Jill S. Gartland *15*

CH. 4. *Agrobacterium tumefaciens* Chemotaxis Protocols,
          Charles H. Shaw *29*

CH. 5. Effect of Acetosyringone on Growth and Oncogenic Potential
          of *Agrobacterium tumefaciens*,
          Patrice Dion, Christian Bélanger, Dong Xu,
          and Mahmood Mohammadi *37*

CH. 6. Binary Ti Plasmid Vectors,
          Gynheung An *47*

CH. 7. Leaf Disk Transformation,
          Ian S. Curtis, Michael R. Davey, and J. Brian Power *59*

CH. 8. *Agrobacterium*-Mediated Transformation of Protoplasts
          from Oilseed Rape (*Brassica napus* L.),
          Jürgen E. Thomzik *71*

CH. 9. *Agrobacterium*-Mediated Transformation of Stem Disks
          from Oilseed Rape (*Brassica napus* L.),
          Jürgen E. Thomzik *79*

CH. 10. Peanut Transformation,
          Elisabeth Mansur, Cristiano Lacorte, and William R. Krul *87*

CH. 11. Transformation of Soybean (*Glycine max*) via *Agrobacterium*
          tumefaciens and Analysis of Transformed Plants,
          Paula P. Chee and Jerry L. Slightom *101*

CH. 12. *Agrobacterium*-Mediated Transformation of Potato Genotypes,
          Amar Kumar *121*

vii

CH. 13.   *Agrobacterium*-Mediated Transformation of Soft Fruit *Rubus, Ribes, and Fragaria,*
        Julie Graham, Ronnie J. McNicol, and Amar Kumar *129*

CH. 14.   High-Frequency and Efficient *Agrobacterium*-Mediated Transformation of *Arabidopsis thaliana* Ecotypes "C24" and "Landsberg *erecta*" Using *Agrobacterium tumefaciens,*
        Mehdi Barghchi *135*

CH. 15.   Transformation Protocols for Broadleaved Trees,
        Trevor M. Fenning and Kevan M. A. Gartland *149*

CH. 16.   *Agrobacterium*-Mediated Antibiotic Resistance for Selection of Somatic Hybrids: *The Genus* Lycopersicon *as a Model System,*
        Michael R. Davey, Rajendra S. Patil, Kenneth C. Lowe, and John B. Power *167*

CH. 17.   Histochemical GUS Analysis,
        Stanislav Vitha, Karel Beneš, Julian P. Phillips, and Kevan M. A. Gartland *185*

CH. 18.   Fluorometric GUS Analysis for Transformed Plant Material,
        Kevan M. A. Gartland, Julian P. Phillips, Stanislav Vitha, and Karel Beneš *195*

CH. 19.   The Detection of Neomycin Phosphotransferase Activity in Plant Extracts,
        Ortrun Mittelsten Scheid and Gabriele Neuhaus-Url *201*

CH. 20.   The Plant Oncogenes *rol*A, B, and C from *Agrobacterium rhizogenes: Effects on Morphology, Development, and Hormone Metabolism,*
        Tony Michael and Angelo Spena *207*

CH. 21.   Quantifying Polyamines in *Agrobacterium rhizogenes* Strains and in Ri Plasmid Transformed Cells,
        Nello Bagni, Marisa Mengoli, and Shigeru Matsuzaki *223*

CH. 22.   IAA Analysis in Transgenic Plants,
        John F. Hall, Sarah J. Brown, and Kevan M. A. Gartland *237*

CH. 23.   Quantifying Phytohormones in Transformed Plants,
        Els Prinsen, Pascale Redig, Miroslav Strnad, Ivan Galís, Walter Van Dongen, and Henri Van Onckelen *245*

CH. 24.   Manipulating Photosynthesis in Transgenic Plants,
        Jacqueline S. Knight, Francisco Madueño, Simon A. Barnes, and John C. Gray *263*

CH. 25.   Gene Activation by T-DNA Tagging,
        Klaus Fritze and Richard Walden *281*

CH. 26.   Assessing Cadmium Partitioning in Transgenic Plants,
        George J. Wagner *295*

CH. 27.   Overexpression of Chloroplastic Cu/Zn Superoxide Dismutase in Plants,
        Randy D. Allen *309*

Contents

Cн. 28. Agroinfection,
Nigel Grimsley 325

Cн. 29. T-DNA Transfer to Maize Plants,
Wen-Hui Shen, Jesús Escudero, and Barbara Hohn 343

Cн. 30. Use of Cosmid Libraries in Plant Transformations,
Hong Ma 351

Cн. 31. Regulation of *Agrobacterium* Gene Manipulation,
Frank Dewhurst 369

Cн. 32. Transgenic Oilseed Rape: *How to Assess the Risk of Outcrossing to Wild Relatives,*
Alain Baranger, Marie-Claire Kerlan, Anne-Marie Chèvre,
Frédérique Eber, Patrick Vallée, and Michel Renard 393

Cн. 33. Electroporation Protocols for *Agrobacterium,*
Garry D. Main, Stephen Reynolds, and Jill S. Gartland 405

Index 413

Contributors

RANDY D. ALLEN • *Department of Biological Sciences, Texas Tech University, Lubbock, TX*

GYNHEUNG AN • *Institute of Biological Chemistry, Washington State University, Pullman, WA*

FRAUKJE M. VAN ASMA • *Phabagen Collection, Netherlands Culture Collection of Microorganisms, Utrecht, The Netherlands*

NELLO BAGNI • *Dipartimento di Biologia, Universita degli Studi di Bologna, Italy*

ALAIN BARANGER • *Station d'Amelioration des Plantes, Institut National de la Recherche Agronomique, Le Rheu, France*

MEHDI BARGHCHI • *Department of Applied Biology and Biotechnology, De Montfort University, Scraptoft, Leicester, UK*

SIMON A. BARNES • *Department of Plant Sciences, University of Cambridge, UK*

CHRISTIAN BÉLANGER • *Département de Phytologie, Faculté des Sciences de l'Agriculture et de l'Alimentation, Université Laval, Sainte-Foy, Québec, Canada*

KAREL BENEŠ • *Faculty of Biology, University of South Bohemia, České Budějovice, Czech Republic*

ANDREW L. BISHOP • *Institute of Food and Agricultural Sciences, Gulf Coast Research and Education Center, University of Florida, Bradenton, FL*

HACÈNE BOUZAR • *Institute of Food and Agricultural Sciences, Gulf Coast Research and Education Center, University of Florida, Bradenton, FL*

SARAH J. BROWN • *Department of Applied Biology and Biotechnology, De Montfort University, Scraptoft, Leicester, UK*

PAULA P. CHEE • *Molecular Biology–Unit 7242, The Upjohn Company, Kalamazoo, MI*

ANNE-MARIE CHÈVRE • *Station d'Amelioration des Plantes, Institut National de la Recherche Agronomique, Le Rheu, France*

IAN S. CURTIS • *Plant Genetic Manipulation Group, Department of Life Science, University of Nottingham, UK*

MICHAEL R. DAVEY • *Plant Genetic Manipulation Group, Department of Life Science, University of Nottingham, UK*

FRANK DEWHURST • *Department of Applied Biology and Biotechnology, De Montfort University, Scraptoft, Leicester, UK*

PATRICE DION • *Département de Phytologie, Faculté des Sciences de l'Agriculture et de l'Alimentation, Université Laval, Sainte-Foy, Québec, Canada*

FRÉDÉRIQUE EBER • *Station d'Amelioration des Plantes, Institut National de la Recherche Agronomique, Le Rheu, France*

JESÚS ESCUDERO • *Friedrich Miescher-Institut, Basel, Switzerland*

TREVOR M. FENNING • *Department of Plant Breeding, Horticulture Research International, Wellesbourne, Warks., UK*

KLAUS FRITZE • *Max Planck Institut fur Zuchtungsforschung, Koln, Germany*

IVAN GALÍS • *Department of Plant Transgenesis, Institute of Plant Molecular Biology, Czech Academy of Sciences, České Budějovice, Czech Republic*

JILL S. GARTLAND • *Department of Molecular and Life Sciences, University of Abertay Dundee, Scotland*

KEVAN M. A. GARTLAND • *Department of Molecular and Life Sciences, University of Abertay Dundee, Scotland*

JULIE GRAHAM • *Department of Soft Fruit Genetics and Cell and Molecular Genetics, Scottish Crop Research Institute, Invergowrie, Dundee, Scotland*

JOHN C. GRAY • *Department of Plant Sciences, University of Cambridge, UK*

NIGEL GRIMSLEY • *Laboratoire de Biologie Moleculaire des Relations Plantes-Microorganismes, CNRS-INRA, Castanet-Tolosan, France*

JOHN F. HALL • *Department of Applied Biology and Biotechnology, De Montfort University, Scraptoft, Leicester, UK*

BARBARA HOHN • *Friedrich Miescher-Institut, Basel, Switzerland*

JEFFREY B. JONES • *Institute of Food and Agricultural Sciences, University of Florida, Gulf Coast Research and Education Center, Bradenton, FL*

MARIE-CLAIRE KERLAN • *Station d'Amelioration des Plantes, Institut National de la Recherche Agronomique, Le Rheu, France*

JACQUELINE S. KNIGHT • *Department of Plant Sciences, University of Cambridge, UK*

WILLIAM R. KRUL • *Department of Plant Sciences, University of Rhode Island, Kingston, RI*

AMAR KUMAR • *Department of Soft Fruit Genetics and Cell and Molecular Genetics, Scottish Crop Research Institute, Invergowrie, Dundee, Scotland*

CRISTIANO LACORTE • *Universidade Federal do Rio de Janeiro, Brazil*
KENNETH C. LOWE • *Plant Genetic Manipulation Group, Department of Life Science, University of Nottingham, UK*
HONG MA • *Cold Spring Harbor Laboratory, Cold Spring Harbor, NY*
FRANCISCO MADUEÑO • *Department of Plant Sciences, University of Cambridge, UK*
GARRY D. MAIN • *Department of Molecular and Life Sciences, University of Abertay Dundee, Scotland*
ELISABETH MANSUR • *Universidade do Estado do Rio de Janeiro, Brazil*
SHIGERU MATSUZAKI • *Dipartimento di Biologia, Universita degli Studi di Bologna, Bologna, Italy*
RONNIE J. MCNICOL • *Department of Soft Fruit Genetics and Cell and Molecular Genetics, Scottish Crop Research Institute, Invergowrie, Dundee, Scotland*
MARISA MENGOLI • *Dipartimento di Biologia, Universita degli Studi di Bologna, Bologna, Italy*
TONY MICHAEL • *AFRC Institute of Food Research, Norwich Research Park, Colney, Norwich, UK*
MAHMOOD MOHAMMADI • *Département de Phytologie, Faculté des Sciences de l'Agriculture et de l'Alimentation, Université Laval, Sainte-Foy, Québec, Canada*
GABRIELE NEUHAUS-URL • *Institute for Plant Sciences, Federal Institute of Technology, Zurich, Switzerland*
RAJENDRA S. PATIL • *Plant Genetic Manipulation Group, Department of Life Science, University of Nottingham, UK*
JULIAN P. PHILLIPS • *Gatersleben Research Institute, Gatersleben, Germany*
JOHN BRIAN POWER • *Plant Genetic Manipulation Group, Department of Life Science, University of Nottingham, UK*
ELS PRINSEN • *Department of Biology, University of Antwerp, Wilrijk, Belgium*
PASCALE REDIG • *Department of Biology, University of Antwerp, Wilrijk, Belgium*
MICHEL RENARD • *Station d'Amelioration des Plantes, Institut National de la Recherche Agronomique, Le Rheu, France*
STEPHEN REYNOLDS • *Department of Electrical and Electronic Engineering, University of Abertay Dundee, Scotland*
ORTRUN MITTELSTEN SCHEID • *Friedrich Miescher-Institut, Basel, Switzerland*
CHARLES H. SHAW • *Department of Biological Sciences, University of Durham, UK*

WEN-HUI SHEN • *Institut de Biologie Moléculaire des Plantes du CNRS, Université Louis Pasteur, Strasbourge, France*

JERRY L. SLIGHTOM • *Molecular Biology–Unit 7242, The Upjohn Company, Kalamazoo, MI*

ANGELO SPENA • *Max Planck Institüt für Züchtungsforschung, Köln, Germany*

MIROSLAV STRNAD • *Department of Plant Biotechnology, Institute of Experimental Botany, Czech Academy of Sciences, Olomouc, Czech Republic*

JÜRGEN E. THOMZIK • *Agrochemicals Division, Research/Biotechnology, Pflanzenschutzzentrum Monheim, Bayer AG, Leverkusen (Bayerwerk), Germany*

PATRICK VALLÉE • *Station d'Amelioration des Plantes, Institut National de la Recherche Agronomique, Le Rheu, France*

WALTER VAN DONGEN • *Department of Biology, University of Antwerp, Wilrijk, Belgium*

HENRI VAN ONCKELEN • *Department of Biology, University of Antwerp, Wilrijk, Belgium*

STANISLAV VITHA • *Faculty of Biology, University of South Bohemia, České Budějovice, Czech Republic*

GEORGE J. WAGNER • *Department of Agronomy, University of Kentucky, Lexington, KY*

RICHARD WALDEN • *Max Planck Institut fur Zuchtungsforschung, Koln, Germany*

DONG XU • *Departement de Phytologie, Faculté des Sciences de l'Agriculture et de l'Alimentation, Université Laval, Sainte-Foy, Quebec, Canada*

CHAPTER 1

Growth and Storage of *Agrobacterium*

Fraukje M. van Asma

1. Introduction

Proper preservation of strains is of great importance to all microbiologists. Working with contaminated or genetically changed cultures, or facing the loss of a strain are annoyances that often may be easily prevented. A reliable preservation method should fulfill the following criteria. First, the risk of introducing contaminants must be minimized. A method that invokes frequent subculturing is therefore considered less convenient. Such a method would also increase the risk of errors, including mislabeling and enrichment for mutants. Second, the cultures should be genetically stable during storage. Finally, loss of viability during preparation, storage, and reviving of the culture should be as low as possible. This guarantees a long shelf life and reduces the risk of selection for bacteria that are able to resist the exposure to stress conditions that occur during either processing or storage of cultures better.

Genetic stability and viability levels during storage can be improved by the reduction of the metabolic activity of the bacterial cells. This can be accomplished by deprivation of oxygen, storage in a medium with minimal nutrients, lowering the storage temperature, or near-complete removal of water. Depriving cultures of oxygen can be achieved simply by overlaying the culture with sterile liquid paraffin *(1,2)*. Since this method is seldom completely effective, strains can be kept in this way for a few years at the maximum. Plant pathogenic bacteria have been reported to survive storage in sterile distilled water *(3)*; after 20 yr of storage at 10°C, 27 of 30 *Agrobacterium tumefaciens* isolates were still

From: *Methods in Molecular Biology, Vol. 44:* Agrobacterium *Protocols*
Edited by: K. M. A. Gartland and M. R. Davey Humana Press Inc., Totowa, NJ

alive *(4)*. Storage and preservation of cultures by these methods can be applied in any laboratory, as no sophisticated facilities are needed (*see* Notes 1–4). These, and other simple preservation methods, have been described by Malik *(2)*. However, storage at low temperature (at least –50°C) and freeze-drying are the most reliable techniques, especially for the long term. These methods are discussed in the following text.

2. Materials

1. Tryptone-Yeast extract medium (TY): 5 g/L tryptone, 3 g/L yeast extract, for semi-solid media add 1.8% (w/v) agar (Baltimore Biological Laboratories, Becton Dickinson, Microbiology Systems, Cockeysville, MD) *(5)*.
2. Minimal medium (MM): 2.05 g/L K_2HPO_4, 1.45 g/L KH_2PO_4, 0.5 g/L $(NH_4)NO_3$, 0.5 g/L $MgSO_4 \cdot 7H_2O$, 0.15 g/L NaCl, 0.01 g/L $CaCl_2$, 2.5 mg/L $FeSO_4 \cdot 7H_2O$, 2.0 g/L glucose. *Agrobacterium* strains of biovar 1 require the addition of trace elements in the following amounts when grown in liquid culture: 0.5 mg/L $ZnSO_4 \cdot 7H_2O$, 0.5 mg/L $CuSO_4 \cdot 5H_2O$, 0.5 mg/L H_3BO_3, 0.5 mg/L $MnSO_4 \cdot H_2O$, and 0.5 mg/L $Na_2MoO_4 \cdot 2H_2O$ *(6)*.
3. Yeast Mannitol Broth medium (YMB): 0.5 g/L K_2HPO_4, 0.2 g/L $MgSO_4 \cdot 7H_2O$, 0.1 g/L NaCl, 0.4 g/L yeast extract, 10 g/L mannitol, pH 7.0. Sterile mannitol should be added immediately before use. For semi-solid media add 1.8% (w/v) agar *(7)*.
4. *Rhizobium* Minimal Medium (RMM): MM medium supplemented with trace elements, 50 µg/L thiamine, 50 µg/L biotine, 50 µg/L calcium pantothenate.
5. Dimethyl sulfoxide (DMSO): Filter sterilized or alternatively, autoclaved glycerol (*see* Note 5).
6. –80°C Freezer (*see* Notes 6 and 7).
7. Cryotubes (Nunc cryotubes, InterMed, Roskilde, Denmark).
8. Hornibrook medium: Dissolve in 50 mL distilled water 0.135 g tri-potassium citrate $C_6H_5O_7K_3 \cdot H_2O$, 0.245 g *tri*-sodium citrate dihydrate $C_6H_5Na_3O_7 \cdot 2H_2O$, 0.06 g K_2HPO_4, 0.06 g $MgCl_2 \cdot 6H_2O$, 0.1 g $K_2CO_3 \cdot 1.5H_2O$, and 5.75 g lactose. Dissolve the components separately and do not heat the solution. Add an equal volume of a $CaCl_2$ solution (0.133 g/50 mL). Adjust the pH to 7.0 with 10% (v/v) formic acid. Sterilize the solution by filtration, check the sterility by incubating a few drops in a rich medium like Luria Bertani (LB) *(8)* at 24, 30, and 37°C. Store the Hornibrook medium at –20°C.
9. Ampules (type A.M. 8-160, J. de Groot, de Mern, the Netherlands).
10. Freeze-dryer: VirTis BT 3 SL (Gardiner, New York) consisting of a vacuum pump, condenser, and manifold (*see* Notes 8–10).
11. Gas-air hand torch.
12. High frequency vacuum tester.

3. Methods

3.1. *Growth of* Agrobacterium

In general, *Agrobacterium* strains grow well on media containing yeast extract and a suitable carbohydrate source. For biovar 1 strains (encompassing strains of *A. tumefaciens* and *A. radiobacter*), TY medium *(5)* is a suitable rich medium. The majority of biovar 1 stains grow well on minimal medium containing nitrate or ammonium salts as the nitrogen source, such as MM *(6)*, SM *(9,10)*, or AB *(10)* media.

Agrobacterium strains of biovar 2 (encompassing strains of *A. tumefaciens, A. rhizogenes, and A radiobacter*) and of biovar 3 *(A. vitis)* grow well on YMB medium *(7,11)*. For liquid cultures, TY-medium can be used *(11)*. Biovar 2 strains grow on minimal media with nitrate when biotin and, in some cases also, L-glutamic acid is supplied *(10,12)*. RMM medium *(7,11)* can be used as a minimal medium for many of these strains.

Growth of *Agrobacterium* strains on all these media is moderate; on plates, it requires 2 d before single colonies appear. The recommended growth temperature for *Agrobacterium* strains is 25–30°C. Growth should occur under aerobic conditions. Strains can be kept for 2 mo on agar medium at 4°C in the dark.

3.2. Freezing of *Agrobacterium* Cultures

1. Add to an overnight broth culture 15% (v/v) sterile glycerol or 5% (v/v) DMSO.
2. Transfer 1 mL to a sterile cryotube and store at –80°C.
3. To subculture cells from this stock culture, open a vial and scrape off some frozen cells with a sterile loop. Use this sample to inoculate either a liquid broth culture, or to streak an agar plate. Transport the stock culture on dry ice.
4. Return the stock culture back into the freezer as soon as possible.

3.3. *Freeze-Drying of Cultures*

1. Grow the cells overnight on a 6.5-mL agar slope.
2. Turn on the refrigeration unit of the freeze-dryer. Allow the condenser to cool until it reaches a temperature of –40°C. Turn on the vacuum pump. It should reach 100 μm of Hg in less than 20 min; a vacuum running at 20–55 μm of Hg is sufficient.
3. Add 16 drops of Hornibrook medium to the cells.
4. Scrape off the cells into the Hornibrook medium using a sterile loop. Avoid gouging the agar with the loop.
5. Suspend the cells in the Hornibrook medium with a Pasteur pipet to form a thick, homogeneous cell suspension.

6. Distribute this cell suspension over 8 sterile ampules. The suspension should be placed at the bottom of the ampule. Try not to contaminate the sidewalls of the ampule and avoid frothing. Before replacing the cotton, flame the top of the ampule thoroughly.
7. Freeze the suspension, e.g., at −80C° or in a −35°C alcohol bath. This frozen suspension can be kept in the −80°C freezer until the time of freeze-drying.
8. Attach the ampules containing the frozen cell suspension to the manifold. Open the Quickseal valves directly after placing an ampule in order to prevent the suspension from thawing. Some freeze-dryers are equipped with a trough that can be filled with a mixture of isopropanol, dry-ice/ethanol or ice/sodium chloride to keep the cell suspension in the frozen state.
9. After 4 h the ampules can be flame-sealed, while still attached to the manifold under vacuum. For this, move the flame of the handtorch continuously around the constricted part of the ampule, until the glass begins to glow. Carefully pull away the ampule from the freeze-dryer, while not changing the relative position of the flame-sealer with respect to the ampule. Keep the tip of the ampule in the flame for a few seconds to blunt the end.
10. Open one Quickseal valve to break the vacuum, then turn off the vacuum pump and the condenser refrigeration.
11. Check the flame-sealed ampules with a high frequency vacuum tester.
12. Check one ampule of each lyophilized strain for purity and survival (*see* Notes 11 and 12).

3.4. Revival of Lyophilized Bacteria

1. Wipe off the ampule with an ethanol-soaked tissue.
2. Score the ampule with a glass cutter and break it at the scored mark. Do not allow traces of alcohol to enter the ampule; this will kill the dried cells immediately.
3. Flame the top of the ampule and add a few drops of an appropriate medium or saline, using a sterile Pasteur pipet.
4. Resuspend the cells gently in the medium and transfer one drop to a plate and streak for single colonies. Transfer the rest of the medium into a small bottle with medium.
5. Incubate at the appropriate temperature.

4. Notes

1. Cultures must always be labeled carefully and a record of the stored cultures must be kept separately.

2. The optimum growth temperature of biovar 1 strains is 32–33°C, but cells tend to lose their Ti plasmid at this temperature. Therefore, it is better to grow these strains at 28–29°C. Biovar 2 strains cannot grow at temperatures higher than 30°C.

3. Storage at temperatures higher than –30°C is not suitable for the long term, as the cells will be damaged by concentration of electrolytes and other soluble substances at these temperatures. Cells remain viable at this temperature for about 1 yr.

4. Freezing and thawing are known to have adverse effects on cells. At slow cooling rates of about 1°C/min, extracellular ice crystal growth occurs and passive dehydration of the cells takes place. This minimizes mechanical damage by intracellular formed ice crystals, but cells will be exposed to the damaging effects of increased electrolyte concentrations. When cells are frozen quickly, cell damage is more likely to occur by intracellularly formed ice crystals, rather than by increased concentrations of electrolytes and other soluble substances. Since bacteria are small, they dehydrate very quickly, with the result that rapid freezing is considered to be preferable. During storage at very low temperatures (e.g., –80°C or lower), viability counts remain constant after the rapid initial cell death. During thawing, cells again encounter the risk of being damaged by intracellular growth of ice crystals and increased electrolyte concentrations *(13)*. Therefore, the cells should be exposed as little as possible to freezing and thawing. The cells should be scraped from the vials while still frozen. They should be kept on dry ice and transferred back to the –80°C freezer as quickly as possible. If cells need to be subcultured regularly, store at least two stocks in the freezer and use one stock for subculture.

5. For storage in the vapor (–156°C) or liquid phase (–196°C) of nitrogen, the cells can be prepared in the same way as for storage in ultracold freezers (–80°C). Both DMSO and glycerol are suitable cryoprotective agents.

6. A disadvantage of storage in the freezer is that subculturing is necessary before transport.

7. To prevent loss of cultures because of power failure, consider storing a backup in another ultracold freezer, or having an alarm system installed on the freezer.

8. For successful lyophilization, use healthy cells. Liquid cell cultures can be used, but this requires a centrifugation step to harvest the cells. The best stage to harvest the cells is in the late logarithmic or early stationary phase *(1,14)*.

9. Skimmed milk (10% final concentration), sucrose or glucose solutions, horse-serum, and other suspending fluids *(1,15)* have also been used as suspending fluids.

10. Lyophilization can be performed in several ways. Besides the single-vial manifold procedure, methods for preparation of double-vial lyophiles *(16)* and centrifugal freeze-drying have been described *(14)*. Stoppered vials should be used only for storage periods shorter than 1 yr.
11. Lyophiles can be stored in the dark at room temperature. Experiences at Phabagen and LMD (Collection of Bacteria, Kluyver Laboratory of Biotechnology, Delft, the Netherlands) have shown that cultures remain viable for a period of at least 40 yr. However, their shelf life may be extended by storage at lower temperatures *(1)*.
12. Double vials can be opened by heating the tip of the outer vial in the flame of a Bunsen burner and squirting a few drops of water on the hot tip to crack the glass. Remove the glass by tapping with a forceps or file, and take out the inner vial.

References

1. Gherna, R. L. (1981) Preservation, in *Manual of Methods for General Bacteriology* (Gerhardt, P. and Costilow, R. N., ed.), American Society for Microbiology, Washington, DC, pp. 208–217.
2. Malik, K. A. (1991) Maintenance of microorganisms by simple methods, in *Maintenance of Microorganisms and Cultured Cells* (Kirsop, B. E. and Doyle, A., eds.), Academic, London, pp. 121–132.
3. DeVay, J. E. and Schnathorst, W. C. (1963) Single-cell isolation and preservation of bacterial cultures. *Nature* **199,** 775–777.
4. Iacobellis, N. S. and DeVay, J. E. (1986) Long-term storage of plant-pathogenic bacteria in sterile distilled water. *Appl. Environ. Microbiol.* **52,** 388,389.
5. Beringer, J. E. (1974) R-factor transfer in *Rhizobium leguminosarum. J. Gen. Microbiol.* **84,** 188–198.
6. Hooykaas, P. J. J., den Dulk-Ras, H., and Schilperoort, R. A. (1980) Molecular mechanism of Ti plasmid mobilization by R plasmids: isolation of Ti plasmids with transposon-insertions in *Agrobacterium tumefaciens. Plasmid* **4,** 64–75.
7. Hooykaas, P. J. J., Klapwijk, P. M., Nuti, M. P., Schilperoort, R. A., and Rörsch, A. (1977) Transfer of the *Agrobacterium tumefaciens* Ti plasmid to avirulent Agrobacteria and to *Rhizobium* ex planta. *J. Gen. Microbiol.* **98,** 477–484.
8. Bertani, G. (1951) Studies on lysogenesis. The mode of phage liberation by lysogenic *Escherichia coli. J. Bacteriol.* **62,** 293–300.
9. Klapwijk, P. M., Oudshoorn, M., and Schilperoort, R. A. (1977) Inducible permease involved in the uptake of octopine, lysopine, and octopinic acid by *Agrobacterium tumefaciens* strains carrying virulence-associated plasmids. *J. Gen. Microbiol.* **102,** 1–11.
10. Lichtenstein, C. and Draper, J. (1985) Genetic engineering of plants, in *DNA Cloning A Practical Approach* (Glover, D. M., ed.), IRL Press, Oxford, UK, pp. 67–119.
11. Costantino, P., Hooykaas, P. J. J., den Dulk-Ras, H., and Schilperoort, R. A. (1980) Tumor formation and rhizogenicity of *Agrobacterium rhizogenes* carrying Ti plasmids. *Gene* **11,** 79–87.

12. Smith, L. T., Smith, G. M. and Madkour, M. A. (1990) Osmoregulation in *Agrobacterium tumefaciens*: accumulation of a novel disaccharide is controlled by osmotic strength and glycine betaine. *J. Bacteriol.* **172,** 6849–6855.

13. Mackey, B. M. (1984) Lethal and sublethal effects of refrigeration, freezing and freeze-drying on micro-organisms, in *The Rivival of Injured Microbes* (Andres, M. H. E. and Russell, A. D., eds.), Academic, London, pp. 45–75.

14. Rudge, R. H. (1991) Maintenance of bacteria by freeze-drying, in *Maintenance of Microorganisms and Cultured Cells* (Kirsop, B. E. and Doyle, A., eds.), Academic, London, pp. 31–44.

15. Lapage, S. P., Shelton, J. E., and Mitchell, T. G. (1970) Media for the maintenance and preservation of bacteria, in *Methods in Microbiology* (Norris, J. R. and Ribbons, D. W., eds.), Academic, London, pp. 1–133.

16. Alexander, M., Daggett, P. M., Gherna, R., Jong, S., and Simione, F. (1980) Freeze-Drying methods, in *Laboratory Manual on Preservation, Freezing and Freeze-Drying as Applied to Algae, Bacteria, Fungi and Protozoa* (Hatt, H. D., ed.), American Type Culture Collection, Rockville, MD, pp. 19–21.

CHAPTER 2

Simple Cultural Tests
for Identification
of *Agrobacterium* Biovars

Hacène Bouzar, Jeffrey B. Jones,
and Andrew L. Bishop

1. Introduction

In *Bergey's Manual of Systematic Bacteriology (1)*, the genus *Agrobacterium* is divided into four species (*A. radiobacter, A. rhizogenes, A. rubi,* and *A. tumefaciens*) based on phytopathogenic characters. With the exception of *A. rubi*, each species is subdivided into three biovars based on physiological and biochemical tests *(1)*. In the past three decades, it has been shown that the true taxonomic structure of the genus is not based on pathogenicity but rather on chromosomal groups that are represented by the biovars *(1–4)*. As a result, there have been several proposed revisions of the species classification in *Agrobacterium*. However, to avoid confusion in this chapter, we refer to the strains by their biovar designation.

Presently, biovar designations are determined by the reaction patterns of strains on a gallery of biochemical and physiological tests *(1,5)*. Unfortunately, these tests are labor intensive and may require 3 wk before the results are obtained. The emergence of novel technologies resulted in the development of new methods that have been useful in discriminating among biovars. Polymerase chain reaction (PCR) amplification of DNA coding for 16S rRNA differentiated the three biovars *(6)*. A monoclonal antibody specific for biovar 3 strains has been developed and used for detection and diagnosis of grapevine crown gall *(7)*. The recent develop-

From: *Methods in Molecular Biology, Vol. 44:* Agrobacterium *Protocols*
Edited by: K. M. A. Gartland and M. R. Davey Humana Press Inc., Totowa, NJ

Table 1
Differential Characterization of *Agrobacterium* Biovars by Cultural Tests

Cultural test	Biovar 1	Biovar 2	Biovar 3
3-Ketolactose production	+	–	–
Acid clearing on PDA-CaCO$_3$	–	+	–
Motility at pH 7.0	+	+	–
Pectolytic activity at pH 4.5	–	–	+

ment of automated systems for bacterial identification, such as the MIDI Microbial Identification System for analysis of fatty acid profiles and the Biolog MicroPlate system for analysis of nutritional patterns, have proven effective for rapid and simple identification of the three biovars *(8)*. However, acquisition of these systems is relatively expensive and may be prohibitive for many laboratories.

Although inexpensive and simple physiological tests on agar plates have been developed, the confidence level for assays testing a single phenotypic trait is much lower than for complex automated tests that analyze many phenotypic traits. Nevertheless, these simple tests are very useful for rapid (48 h), presumptive identification. These include the 3-ketolactose test *(9)*, the differential acid production assay *(10)*, the motility assay *(11)*, and the polygalacturonic acid (PGA) test *(12,13)* *(see* Table 1).

2. Materials

1. Nutrient glucose agar (NGA): 0.3% (w/v) beef extract, 0.5% (w/v) peptone, 0.25% (w/v) glucose, and 1.5% (w/v) agar in distilled water. Autoclave the medium at 121°C for 15 min.
2. Lactose agar: 1% (w/v) α-lactose, 0.1% (w/v) yeast extract, and 2% (w/v) agar in distilled water. Autoclave the medium at 121°C for 15 min.
3. Benedict's reagent: Dissolve 17.3 g of sodium citrate and 10.0 g of Na$_2$CO$_3$ in 80 mL of distilled water by heating; filter the carbonate-citrate solution to remove insoluble particles. Dissolve in another container 1.73 g of CuSO$_4 \cdot$ 5H$_2$O in 10 mL of distilled water. Add the copper solution to the carbonate-citrate solution while stirring, then dilute the mixture to 100 mL.
4. Potato dextrose agar (PDA) calcium carbonate medium: 2% (w/v) PDA (Difco, Detroit, MI), 0.08% (w/v) calcium carbonate, and 0.75% (w/v) agar in distilled water. Add a magnetic bar prior to sterilizing the medium at 121°C for 45 min *(see* Note 1).

5. Motility medium: 0.1% (w/v) nutrient broth, 0.111% (w/v) K_2HPO_4, 0.049% (w/v) KH_2PO_4, 0.3% (w/v) agar, distilled water, adjust the pH to 7.0. Autoclave the medium at 121°C for 20 min. Cool the medium and pour it into 9-cm Y-sectored plates (Falcon No. 1004) 1 d prior to testing.
6. Pectate medium PGA: 50 mM potassium acetate (pH 4.5) containing 0.1% (w/v) polygalacturonic acid and 1% (w/v) agarose. After autoclaving at 121°C for 20 min, pour a thin layer (about 5 mL/plate) of PGA. Staining solutions: 0.05% (w/v) ruthenium red in distilled water or 1% (w/v) cetyltrimethylammonium bromide (CTAB) *(14)*.

3. Methods
3.1. Growth and Purification of Bacteria
1. Streak strains onto NGA and incubate the plates at 28°C until single colonies appear.
2. Transfer growth from one colony with a sterile loop to an NGA slant.
3. After 24 h at 28°C, transfer growth from the slant to the test media.

3.2. 3-Ketolactose Test
1. Place spot of growth, approx 5 mm in diameter, on a plate of lactose agar; four to six strains may be tested on the same plate.
2. Incubate the plate at 28°C for 2 d.
3. Flood the plate with a shallow layer of Benedict's reagent.

The presence of 3-ketolactose in the medium is indicated by the formation of a yellow ring (2–3 cm in diameter) of cuprous oxide around the growth of biovar 1 strains (*see* Note 2).

3.3. Differential Acid Production Assay
1. Inoculate a spot of bacterial growth on a plate of PDA-CaCO$_3$ (*see* Notes 1 and 3).
2. Incubate the plate at 28°C for 2–3 d.

The medium around biovar 1 and 3 strains remains opaque, whereas a clear zone develops around biovar 2 strains.

3.4. Test of Motility at pH 7
1. To perform this test, transfer a small quantity of bacterial growth to the center of the motility medium using the tip of a sterile toothpick.
2. Incubate the plate at 28°C for 48 h. Check size of colony (*see* Note 4).

3.5. Pectolytic Activity at pH 4.5
1. Inoculate a mass of bacterial growth as a spot on PGA medium.
2. Incubate the plate for 3–4 h at 28°C.

3. Wash the plate with running water to remove cell mass.
4. Flood the assay gel with staining solution for 20 min, then rinse gel with distilled water.

Biovar 3 strains digest the polygalacturonic acid at pH 4.5. The resulting 8–10-mm zone around the growth of biovar 3 strains is visualized with either ruthenium red or CTAB solution (*see* Note 5).

4. Notes

1. Maximum intensity of the yellow ring around 3-ketolactose positive strains is reached in about 1–2 h after flooding with Benedict's reagent. Biovar 1 strains have the unique ability to oxidize lactose into 3-ketolactose.
2. To obtain a homogeneously white PDA-CaCO$_3$ medium, cool the medium in a water bath to 55–60°C and thoroughly mix the calcium carbonate with a magnetic stirring bar prior to pouring the medium into plates.
3. The time required for the formation of a clear zone around biovar 2 colonies growing on the PDA-CaCO$_3$ medium is dependent on the CaCO$_3$ concentration. The zone is visible after 24–48 h when the calcium carbonate concentration in the medium is at 0.5–0.8 g/L. The diameter of the clearing increases as the colony ages.
4. Biovar 3 strains are nonmotile at pH 7.0 and form very small colonies (<0.5 mm in diameter), whereas biovar 1 and 2 strains remain motile and form large colonies (>2 mm in diameter).
5. Ruthenium red stains the nondegraded polygalacturonic acid present in the medium leaving the zone around biovar 3 strains unstained. CTAB precipitates the polygalacturonic acid in the medium; thus, the zone of pectolytic activity around biovar 3 strains appears clear on a white background. The medium around the growth of biovar 1 and 2 strains is stained red if ruthenium red is used, or white if CTAB is used.

References

1. Kersters, K. and De Ley, J. (1984) Genus III. *Agrobacterium* Conn 1942, in *Bergey's Manual of Systematic Bacteriology*, Vol. *1* (Krieg, N. R. and Holt, J. G., eds.), Williams and Wilkins, Baltimore, pp. 244–254.
2. Ophel, K. and Kerr, A. (1990) *Agrobacterium vitis* sp. nov. for strains of *Agrobacterium* biovar 3 from grapevines. *Int. J. Syst. Bacteriol.* **40,** 236–241.
3. Sawada, H., Ieki, H., Oyaizu, H., and Matsumoto, S. (1993) Proposal for rejection of *Agrobacterium tumefaciens* and revised descriptions for the genus *Agrobacterium* and for *Agrobacterium radiobacter* and *Agrobacterium rhizogenes. Int. J. Syst. Bacteriol.* **43,** 694–702.
4. Willems, A. and Collins, M. D. (1993) Phylogenetic analysis of rhizobia and agrobacteria based on 16S rRNA gene sequences. *Int. J. Syst. Bacteriol.* **43,** 305–313.

5. Moore, L. W., Kado, C. I., and Bouzar, H. (1988) *Agrobacterium,* in *Laboratory Guide for Identification of Plant Pathogenic Bacteria,* 2nd ed. (Schaad, N. W., ed.), APS Press, St. Paul, MN, pp. 16–36.
6. Ponsonnet, C. and Nesme, X. (1994) Identification of *Agrobacterium* strains by PCR-RFLP analysis of pTi and chromosomal regions. *Arch. Microbiol.* **161,** 300–309.
7. Bishop, A. L., Burr, T. J., Mittak, V. L., and Katz, B. H. (1989) A monoclonal antibody specific to *Agrobacterium tumefaciens* biovar 3 and its utilization for indexing grapevine propagation material. *Phytopathology* **79,** 995–998.
8. Bouzar, H., Jones, J. B., and Hodge, N. C. (1993) Differential characterization of *Agrobacterium* species using carbon-source utilization patterns and fatty acid profiles. *Phytopathology* **83,** 733–739.
9. Bernaerts, M. J. and De Ley, J. (1963) A biochemical test for crown gall bacteria. *Nature (London)* **197,** 406,407.
10. Bouzar, H. and Jones, J. B. (1992) Distinction of biovar 2 strains of *Agrobacterium* from other chromosomal groups by differential acid production. *Lett. Appl. Microbiol.* **15,** 83–85.
11. Bush, A. L. and Pueppke, S. G. (1991) A rapid and efficient new assay for determination of the three biotypes of *Agrobacterium tumefaciens. Lett. Appl. Microbiol.* **13,** 126–129.
12. Bishop, A. L., Katz, B. H., Burr, T. J., Kerr, A., and Ophel, K. (1988) Pectolytic activity of *Agrobacterium tumefaciens. Phytopathology* **78,** 1551.
13. Reid, J. L. and Collmer, A. (1985) An activity stain for the rapid characterization of pectic enzymes in isoelectric focusing and sodium dodecyl sulfate-polyacrylamide gels. *Appl. Environ. Microbiol.* **50,** 615–622.
14. McGuire, R. G., Rodriguez-Palenzuela, P., Collmer, A., and Burr, T. J. (1991) Polygalacturonase production by *Agrobacterium tumefaciens* biovar 3. *Appl. Environ. Microbiol.* **57,** 660–664.

CHAPTER 3

Agrobacterium Virulence

Jill S. Gartland

1. Introduction
1.1. Agrobacterium Biology

Agrobacteria are soil living Gram-negative rod-shaped bacteria. The genus includes two well-studied phytopathogenic bacteria, *Agrobacterium tumefaciens* and *A. rhizogenes*. Their ability to cause disease is associated with the presence of a large (about 200 kb) plasmid. The plasmid is named according to the disease caused by it. The *A. tumefaciens* plasmid is known as a Ti (tumor-inducing) plasmid and that of *A. rhizogenes* as a Ri (root-inducing) plasmid.

1.2. The Disease Process

The chromosomal genes *chv*A, *chv*B, *psc*A, and *att* are involved in the attachment of the bacteria to the cell walls of wounded susceptible plant species, which is a prerequisite for transformation. The disease is then brought about following the transfer and stable integration into the plant genome of a piece of the plasmid that carries a group of bacterial genes. These genes reside on a portion of the Ti or Ri plasmid called the T-DNA (transferred DNA). The Ti T-DNA contains plant growth regulator biosynthetic genes, as do some types of Ri T-DNA. Two genes encode a novel pathway for auxin biosynthesis, tryptophan monooxygenase (*tms*1 or *iaa*M), and indole-3-acetamide hydrolase (*tms*2 or *iaa*H). The gene *tmr* (or *ipt*) encodes an isopentenyl transferase involved in cytokinin biosynthesis *(1)*. Additional T-DNA genes (genes 5 and 6b) may encode functions which modify the plant cells' sensitivity to cytokinin and auxin,

From: *Methods in Molecular Biology, Vol. 44:* Agrobacterium *Protocols*
Edited by: K. M. A. Gartland and M. R. Davey Humana Press Inc., Totowa, NJ

respectively. *Rol* genes contained on the *A. rhizogenes* Ri plasmid T-DNA encode functions that are capable of modulating growth regulation (*see* Chapter 20). The transfer and expression of growth regulator genes within the plant cell brings about the characteristic tumor or hairy root that gives the diseases the respective names of crown gall and hairy root disease. This uncontrolled plant cell growth can occur in vitro in the absence of exogenously applied plant growth regulators.

1.3. T-DNA Structure

The T-DNA is delimited by directly repeated, conserved left and right border sequences. The right hand sequence is essential for efficient T-DNA transfer *(2)*.

There are many different forms of wild-type agrobacteria and these may be characterized by the nature of the T-DNA encoded opine (unusual amino acid sugar derivative), which they induce plant cells to produce. For example, *A. tumefaciens* C58 induces the production of nopaline and Bo542 agropine. The C58 T-DNA is a single uninterrupted region, whereas the T-DNAs of some other strains of bacteria, for example the Ri A4 plasmid, have two separate T-DNA regions (*see* Fig. 1).

1.4. Disarming Agrobacteria

In many instances the presence of the *Agrobacterium* growth regulator biosynthetic genes within transformed plant tissues means that it is not possible to regenerate plants of normal phenotype from these tissues. This can be overcome by a process called disarming, in which the genes can either be removed or be made nonfunctional by interrupting their sequence. In the absence of the bacterial-derived growth regulator genes, transformed cells can instead be identified by the inclusion of a selectable marker gene, such as resistance to the antibiotic kanamycin.

1.5. Binary Vectors

The transfer of the T-DNA into the plant cell is brought about by a set of genes called virulence (*vir*) genes, which is discussed in Section 1.8. The *vir* genes are situated outside the T-DNA and are not themselves transferred. It is therefore possible to remove the entire T-DNA from the Ti or Ri plasmid and to use the remaining *vir* genes to transfer a separate T-DNA, which is part of a genetically engineered smaller binary vector. A typical binary vector is shown in Fig. 2 (*see also* Chapter 6).

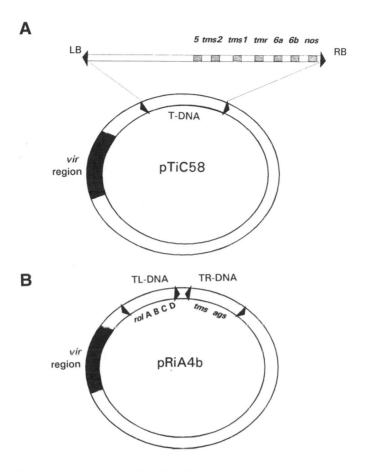

Fig. 1. (**A**) *A. tumefaciens* pTi C58 with a single T-DNA region. *tms*1 = tryptophan monooxygenase; *tms*2 = indole-3-acetamide hydrolase; *tmr* = isopentenyl transferase; and *nos* = nopaline synthase. (**B**) *A. rhizogenes* pA4b with split left (L) and right (R) T-DNA regions. *tms* = tryptophan monooxygenase and indole-3-acetamide hydrolase and *ags* = agropine synthase.

1.6. The Use of Reporter Genes

Not all genes encode products such as drug resistances, which are readily selectable. In contrast, gene products, which may be identified and quantified, are known as reporter genes. These genes are often also included on a binary vector. The most frequently used reporter gene, β-glucuronidase (or GUS) is discussed in subsequent chapters.

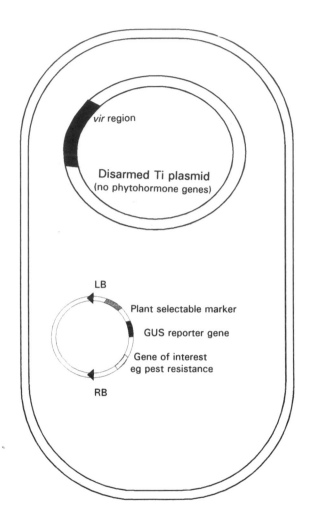

Fig. 2. A typical *A. tumefaciens* binary vector system.

1.7. Agrobacterium *as a Means of Plant Improvement*

Agrobacterium transformation technology has improved greatly since 1983 when Zambryski et al. *(3)* reported the first use of genetically modified *A. tumefaciens* to transform foreign genes into plant cells. Since then, many commercially important crop species have been successfully transformed with a variety of genes for qualities such as pest resistance, herbicide resistance, disease tolerance, and improved fruit storage capacity. Examples of such transformed plants are discussed in other chapters.

Transformation strategies vary depending on plant species and are described elsewhere in this volume. They all involve establishing plant culture systems that allow regeneration of intact plants from tissue explants that have been cocultivated with agrobacteria. This chapter concentrates on the *Agrobacterium* virulence genes, their relevance in the transformation process, and how to assess their induction.

1.8. Agrobacterium *Virulence Genes*

Extensive studies on *Agrobacterium* T-DNA transfer mechanisms have shown that transfer depends on the plasmid-based *vir* genes but also involves some chromosomal genes.

Most work to date has been undertaken on *A. tumefaciens*. Octopine strains possess *vir*A, B, C, D, E, F, G, and H (sometimes known as *pin*F) genes, nopaline strains harbor similar genes, but lack *vir*F and *vir*H and have in addition a *tzs* gene *(4)*.

With the exception of *vir*A and *vir*G, the *vir* genes are not expressed in free living bacteria but are transcriptionally activated by exudates, such as phenol-based compounds, e.g., acetophenones, chalcones, and cinnamic acid derivatives, along with opines and sugars produced by the host plant *(5)*. Other complex interactions involving pH, temperature, and inorganic phosphate levels affect *vir* gene expression.

The molecular mechanism of *vir* gene induction involves the two component *VirA/VirG* regulatory system and is homologous to other two-component regulatory systems that sense and respond to changes in the bacterial environment. VirA protein is a sensor protein and the plant signal is transmitted from VirA to VirG by a cascade of phosphorylation reactions. VirA contains a periplasmic amino terminal domain required for sensing monosaccharides and three cytoplasmic domains: a linker, a protein kinase, and a phosphoryl receiver *(6)*. The linker domain is required for sensing phenolic compounds and acidity. The phosphoryl receiver plays an inhibitor role in signal transduction which may be modulated by phosphorylation. VirG, the regulator component, is phosphorylated by VirA. Sequence-specific DNA binding of the VirG protein to the *vir* gene promoter regions then activates their transcription. All highly inducible *vir* genes have at least one binding site, a conserved 12 bp sequence (TNCAATTGAAAPy) known as a *vir* box *(7)*, at the same position relative to the inducible transcription start. Mutations involving elevated *Vir*G activity cause increased expression at other *vir* loci resulting in a super virulent phenotype *(8)*.

Chromosomal genes have also been shown to alter *vir* gene expression. The *ros* gene product is a repressor that interacts with a sequence in the promoter region that overlaps with the *vir* box of *vir*C and *vir*D *(9)*. The *chv*E gene (homologous to an *Escherichia coli* galactose binding protein) is required for *vir* induction by sugars *(10)*.

T-DNA transfer begins when a *vir*D encoded endonuclease cleaves within the right hand border sequence and proceeds in a unidirectional manner. The *vir*D operon has four open reading frames and encodes proteins that are essential for T-DNA processing *(11)*. VirD1 has topoisomerase activity *(12)*, and, following DNA cleavage, the VirD2 protein, tightly linked to the T-DNA, directs its passage into the plant cell nucleus by means of nuclear localization signals *(13)*.

T-DNA transfer is also thought to be aided by the coating of the DNA in VirE2 protein, a single-stranded DNA binding protein *(14)*, which may help protect the DNA during its passage from *Agrobacterium* to the plant cell nucleus.

The *A. tumefaciens vir*B operon has sequence similarities to the DNA transfer operons of broad host range conjugative plasmids and the 11 *Vir*B proteins are hypothesized to form at least part of a membrane localized T-DNA transport mechanism necessary for the transfer of T-DNA protein complexes from *Agrobacterium* into the plant cell *(15)*.

The four loci *vir*A, B, D, and G are essential for gene transfer, whereas *vir*C and *vir*E enhance the transfer efficiency *(16)*. Figure 3 simply summarizes the transformation process.

Not all *Agrobacterium* species have the same host range and there are many instances where plant species are highly susceptible to one strain but not to a range of others. Bacteria that are capable of transforming a large range of dicotyledonous plants are known as wide host range (WHR) others with only a narrow range, as limited host range (LHR). The reasons for this are complex and not fully understood. Wide host range bacteria probably respond to more chemical inducers than limited host range and are more sensitive to inducers such as acetosyringone *(17)*. These host range differences may be affected by differences in the structure of the VirA protein. VirA encoded by LHR and WHR strains are structurally related overall but are quite different in the amino termini, which are involved in signal recognition *(18)*.

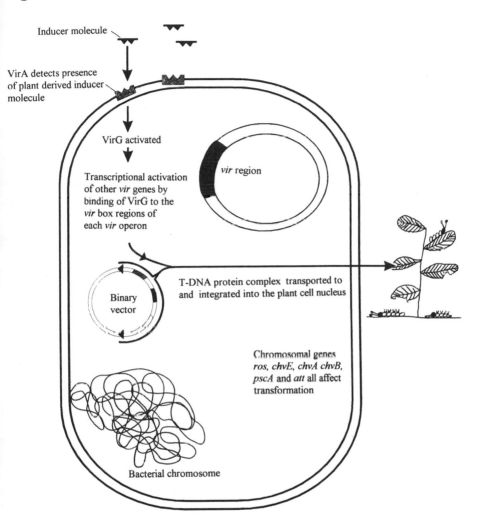

Fig. 3. Schematic representation of the transformation process.

Nopaline strains of *A. tumefaciens* are strongly attenuated in some plant species, such as *Nicotiana glauca*, compared with octopine strains. This has been shown to be owing to the absence of the *vir*F gene from nopaline Ti plasmids. Introduction of the VirF protein converts the nonhost *N. glauca* into a host for tumor formation *(19)*.

In some cases, the environment around the wounded plant cells is inhibitory to bacteria and VirH might be involved in detoxification *(20)*. Some plant exudates are highly acidic and it might be possible to improve the environment for agrobacteria merely by increasing the buffering capacity of the transformation media.

Some types of agrobacteria possess enzymes, such as β-glucosidase, which allow them to convert a normal plant metabolite such as the phenylpropanoid glucoside coniferin to coniferyl alcohol thought to be a *vir* inducer *(21)*.

Transformation efficiencies can often be greatly improved by varying the *Agrobacterium* host strain used together with differing chromosomal-plasmid combinations, tissue explant sources, and pretransformation treatment regimes. Even with amenable plant–bacteria interactions, difficulties are frequently encountered in optimizing the transformation process. Some species, for example monocotyledonous plants, however, may prove recalcitrant to infection with *Agrobacterium*.

By altering the transformation protocol, the pH (often important), temperature (<30°C), bacterial growth medium, type of sugar present, phosphate concentration, and, by inclusion of inducer molecules appropriate to the host bacteria/chromosome/plasmid combination being used, optimal *vir* induction conditions can be assessed.

Vir gene induction can be determined either by measuring virulence as a proportion of cells or explants that are transformed using a range of bacterial species and on one or more plant genotypes. Alternatively, induction of *vir* genes when agrobacteria are placed in a suitable medium in the presence of wounded plant extracts, is easily monitored by linking the *vir* genes as transcriptional fusions to a readily assayable reporter gene such as *lac*Z. In 1986, Stachel and Nester *(16)* generated a set of *vir::lac*Z fusions by randomly inserting the *lac*Z fusion transposon Tn3-HoHo1 into plasmid clones of the *vir* region of pTiA6. These were then recombined into the Ti plasmid of *Agrobacterium* (*see* Fig. 4). The promoter-less *lac*Z element acted as a reporter for transcription of the sequences into which it has inserted. Gene expression was monitored by assaying for β-galactosidase activity, via the *lac*Z gene product.

Due to the length of time (a few weeks to several months) that must be spent undertaking the transformation and plant regeneration process, it is important to begin to optimize transformation protocols as early in the experimental procedure as possible. In cases where difficulties are encountered, the use of *lac*Z fusion constructs in combination with the β-galactosidase assay, which is described in the following sections, allows the

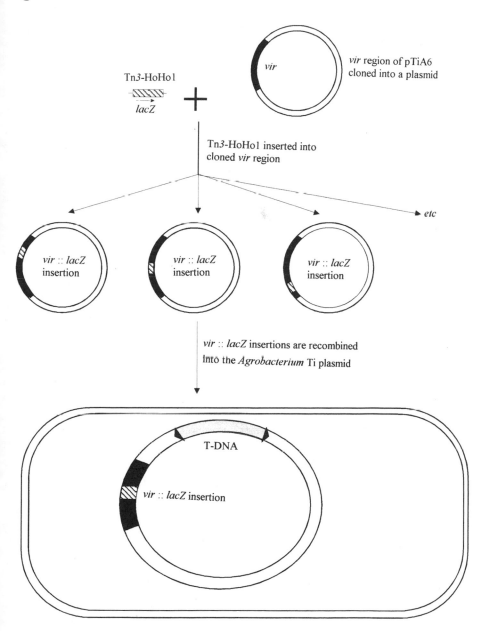

Fig. 4. Insertion of Tn*3*-HoHo1 into the *vir* region of pTiA6. β-galactosi-dase, the enzyme product of *lacZ*, is produced when expression of the *vir* gene carrying Tn*3*-HoHo1 is induced and can be detected using the protocol described in this chapter.

early assessment of a variety transformation protocols and *Agrobacterium* strains. Such experiments are also valuable in determining the tissue explant source that might be most suitable for transformation. Xu et al. *(22)*, for example, assessed different tissue extracts of the monocotyledon rice and found that *vir* gene expression-inducing factors were present but only in specific plant tissues and in particular developmental stages.

2. Materials

2.1. Growth of Agrobacterium *Strains (see Note 1)*

1. *A. tumefaciens* containing *lacZ* in one of the following fusions: *vir*A:: *lacZ*, *vir*B:: *lacZ*, *vir*C:: *lacZ*, *vir*D:: *lacZ*, *vir*E:: *lacZ*, *or vir*G:: *lacZ*, either as a merodiploid (the *vir::lacZ* is on a separate plasmid to the Ti plasmid and the bacterium is therefore diploid for part of the *vir* region) or integrated into the Ti plasmid.
2. AB medium (1 L): KH_2PO_4 (7.25 g), K_2HPO_4 (10.25 g), NaCl (0.15 g), $MgSO_4 \cdot 7H_2O$ (0.15 g), CaCl (0.01 g), glucose (5 g), and $FeSO_4$ (2.5 mg). Eight grams of agar may be added when solidified media for plates is required.
3. YEP medium (1 L): peptone (10 g), yeast extract (10 g), and NaCl (5 g), pH 7.0–7.2.
4. YMB (1 L): mannitol (10 g), yeast extract (0.4 g), NaCl (0.1 g), $MgSO_4 \cdot 7H_2O$ (0.2 g), and K_2HPO_4 (0.5 g), pH 7.0–7.2.
5. LB medium (Luria-Bertani medium): tryptone (10 g), yeast extract (5 g), and NaCl (5 g), pH 7.0–7.2.
6. 28°C Static incubator.
7. 28°C Shaking incubator.
8. Autoclave (to sterilize media at 15 lb/in. pressure for 20 min).

2.2. Preparation of Solutions to be Tested for vir *Gene Induction Activity*

1. AB minimal medium containing 0.5% (w/v) sucrose, YEP, LB, and YMB (*see* Section 2.1.).
2. Chemicals such as acetosyringone (3',5',-dimethoxy-4'-hydroxyacetophenone) to be assessed for *vir* gene induction properties (*see* Note 2).
3. Plant extracts to be examined for inducer molecule/inhibitor properties (*see* Note 3).
4. 1*M* MES (morpholinoethanesulfonic acid) pH 5.5.
5. 1*M* HEPES (*N*-2-hydroxyethylpiperazine-*N'*-ethanesulfonic acid) pH 8.5 stock solutions.
6. Mortars and pestles for grinding samples.

7. Centrifuge and containers to hold 20 mL bacteria.
8. −70°C Freezer (or −20°C if unavailable).

2.3. β-Galactosidase Assay (see Note 4)

1. Spectrophotometer and cuvets to measure absorbance at 420 and 600 nm.
2. Z-buffer: $Na_2HPO_4 \cdot 7H_2O$ (16.1 g), $NaH_2PO_4 \cdot H_2O$ (5.5 g), KCl (0.75 g), $MgSO_4 \cdot 7H_2O$ (0.246 g), and β-mercaptoethanol (2.7 mL). Adjust to pH 7.0 and bring to 1 L with H_2O (do not autoclave).
3. Toluene.
4. 4 mg/mL ONPG (orthonitrophenyl-β-D-galactoside) in $0.1M$ potassium phosphate buffer pH 7.0 (filter sterilized).
5. $1M$ Sodium carbonate.

3. Methods

3.1. Growth of Agrobacterium Strains

1. Streak *Agrobacterium* strain onto fresh AB minimal plates containing 0.5% (w/v) sucrose and incubate at 28°C for 2 d.
2. Remove single colonies and inoculate into a flask containing 20 mL of YEP medium with appropriate antibiotics dependent on the *lacZ* fusion used, and shake for 24 h at 28°C.
3. Inoculate 40 mL of AB sucrose medium with 200 µL of YEP bacterial culture. Grow cells at 28°C to an approximate density of 10^9 cells/mL.
4. Collect cells by centrifugation.
5. Wash cell pellet with 20 mL of distilled water.
6. Resuspend each pellet in 20 mL of the induction media (*see* Section 3.2.) to be tested at a cell density of approx 5×10^8 cells/mL. Remember to include control flasks (*see* Note 5).
7. Remove 5 mL of sample for each time point to be assessed in the β-galactosidase assay (*see* Section 3.3.).

3.2. Preparation of Solutions to be Tested for Inducer Activity

3.2.1. Plant Extract Preparation

Collect plant tissues for extraction and proceed as follows:

1. Chop plant tissue into pieces.
2. Freeze at −70°C.
3. Homogenize in a mortar and pestle.
4. Add AB minimal medium (5 mL/g plant material) and rehomogenize.
5. Extract for 12 h at 4°C.
6. Remove debris by centrifugation and filter sterilize.

7. Adjust pH within the range 5.0–8.5 as desired (buffering can be achieved where required by the addition of 50 mM MES or 50 mM HEPES).
8. Store at –70°C until required (*see* Note 6).

3.2.2. Testing Other Induction Agents

There are many different permutations of chemical, pH, and plant extract that can be assessed, as too can the presence of different sugars in addition to or instead of sucrose (*see* Note 2).

3.2.3. Use of Induction Agents

Dissolve each plant extract or chemical to be assessed for induction activity in AB medium (*see* Note 7) containing 0.5% (w/v) sucrose at the required concentration and pH. Antibiotics should be included as appropriate for the *vir::lacZ* constructs employed.

3.3. β-Galactosidase Assay

1. Wash the aliquot of cells from Section 3.1., step 7 with an equal volume of distilled water.
2. Resuspend in 0.5 mL Z-buffer.
3. Determine the bacterial density at 600 nm (1-cm path length).
4. Add 25 µL of toluene and mix vigorously.
5. Start the reaction by the addition of 0.1 mL of 4 mg/mL ONPG.
6. Incubate at 28°C for 30 min (a yellow color should develop).
7. Add 0.25 mL of 1M sodium carbonate to stop the reaction.
8. Carefully remove insoluble materials by centrifugation (*see* Note 8).
9. Measure optical density of the supernatant at 420 nm (1-cm path length).
10. Calculate specific units of β-galactosidase activity by using the formula:

$$(A_{420} \times 1000)/(A_{600} \times t)$$

where t is the incubation time in minutes, A_{420} = OD of supernatant at time t, and A_{600} = OD of bacterial cells. *See* Note 7 for comments on replication and suitable controls.

4. Notes

1. Before starting any transformation work you must ensure that you are complying with local plant pest and genetic manipulation regulations and guidelines.
2. Many other possible *vir* gene inducers can be tested, for example, synthetic phenolics, such as guaiacol and sinapinic acid; monosaccharides, such as fucose and arabinose, and the acidic sugars D-galacturonic acid and D-glucuronic acid.

3. Ideal samples for extract preparation are tissue explants (similar to those that will be used in subsequent transformation protocols), callus cultures, liquid-grown suspension cells or intact healthy plants.

4. You must be aware of how to correctly handle chemicals, for example, large volumes of β-mercaptoethanol and toluene should be used in a fume cupboard.

5. Remember to include control flasks grown under similar conditions (e.g., temperature, pH) but lacking the test inducer. For more reliable results each time point value should be derived from three independently induced cultures

6. Alternatively conditioned liquid media that contain actively metabolizing cells can be used to test for *vir* gene induction.

7. Alternative bacterial growth media such as LB, YEP, or YMB can be substituted for AB to suit different *Agrobacterium* strains.

8. Make sure that the debris is well pelleted, so that it does not interfere in subsequent OD readings.

References

1. Davey, M. R., Gartland, K. M. A., and Mulligan, B. J. (1986) Transformation of the genomic expression of plant cells, in *Plasticity in Plants* (Jennings, D. H. and Trewavas, A. J., eds.), Company of Biologists, Cambridge, UK, pp. 85–120.

2. Wang, K., Herrera-Estrella, M., Van Montagu, M., and Zambryski, P. (1984) Right 25 bp terminus sequences of the nopaline T-DNA is essential for, and determines the direction of DNA transfer from *Agrobacterium* to the plant genome. *Cell* **38,** 35–41.

3. Zambryski, P., Joos, H., Genetello, C., Leemans, J., van Montagu, M., and Schell, J. (1983) Ti plasmid vector for the introduction of DNA into plant cells without alteration of their normal regeneration capacity. *EMBO J.* **2,** 2143–2150.

4. John, M. C. and Amasino, R. M. (1988) Expression of an *Agrobacterium* Ti plasmid gene involved in cytokinin biosynthesis is regulated by virulence loci, and induced by plant phenolic compounds. *J. Bacteriol.* **170,** 790–795.

5. Stachel, S. E., Nester, E. W., and Zambryski, P. (1986) A plant cell factor induces *Agrobacterium vir* gene expression. *Proc. Natl. Acad. Sci. USA* **83,** 379–383.

6. Chang, C. H. and Winans, S. C. (1992) Functional roles assigned to the periplasmic, linker, and receiver domains of the *Agrobacterium tumefaciens* VirA protein. *J. Bacteriol.* **174(21),** 7033–7039.

7. Jin, S. G., Roitch, T., Christie, P. J., and Nester, E. W. (1990) The *Vir*G protein specifically binds to a cis-acting regulatory sequence involved in transcriptional activation of *Agrobacterium tumefaciens* virulence genes. *J. Bacteriol.* **172,** 531–537.

8. Jin, S., Song, Y.-N., Pan, S. Q., and Nester, E. W. (1993) Characterisation of a *virG* mutation that causes constitutive virulence gene expression in agrobacteria. *Mol. Microbiol.* **7(4),** 555–562.

9. D'Souza-Ault, M. R., Cooley, M. B., and Kado, C. I. (1993) Analysis of the Ros repressor of *Agrobacterium virC* and *virD* operons molecular intercommunication between plasmid and chromosomal genes. *J. Bacteriol.* **175(11),** 3486–3490.

10. Doty, S. L., Chang, M., and Nester, E. W. (1993) The chromosomal virulence gene, *chvE*, of *Agrobacterium tumefaciens* is regulated by a LysR family member. *J. Bacteriol.* **175(24)**, 7880–7886.

11. Jayaswal, R. K., Veluthambi, K., Gelvin, S. B., and Slightom, J. L. (1987) Double stranded cleavage of T-DNAs, and generation of single stranded T-DNA molecules in *Escherichia coli* by a *virD* encoded border-specific endonuclease from *Agrobacterium tumefaciens. J. Bacteriol.* **169**, 5035–5045.

12. Ghai, J. and Das, A. (1989) The operon of *Agrobacterium tumefaciens* encodes a DNA relaxing enzyme. *Proc. Natl. Acad. Sci. USA* **86**, 3109–3223.

13. Shurvington, C. E., Hodges, L., and Ream, W. (1993) A nuclear localisation signal and a C-terminal omega sequence in the *Agrobacterium VirD2* endonuclease are important for tumor formation. *Proc. Natl. Acad. Sci. USA* **89(24)**, 11,837–11,841.

14. Christie, P. J., Ward, J. E., Winnans, S. C., and Nester, E. W. (1988) The *Agrobacterium tumefaciens virE2* gene product is a single stranded DNA binding protein that associates with T-DNA. *J. Bacteriol.* **170**, 2659–2667.

15. Dale, E. M., Binns, A. N., and Ward, J. E., Jr. (1993) Construction and characterisation of Tn5virB a transposon that generates non polar mutations and its use to define *virB8* as an essential virulence gene in *Agrobacterium tumefaciens. J. Bacteriol.* **175(3)**, 887–891.

16. Stachel, S. E. and Nester, E. W. (1986) The genetic, and transcriptional organisation of the *vir* region of the A6 Ti plasmid of *Agrobacterium tumefaciens. EMBO J.* **5(7)**, 1445–1454.

17. Bolten, G. N., Nester, E. W., and Gordon, M. P. (1986) Plant phenolic compounds induce expression of the *Agrobacterium tumefaciens* loci needed for virulence. *Science* **232**, 983–985.

18. Leroux, B., Yanofsky, M. F., Winans, S. C., Ward, J. E., Ziegler, S. F., and Nester, E. W. (1987) Characterisation of the *virA* locus of *Agrobacterium* tumefaciens: a transcriptional regulator, and host range determinant. *EMBO J.* **6**, 849–856.

19. Regensburg-Tuink, A. J. G. and Hooykaas, P. J. J. (1993) Transgenic *N. glauca* plants expressing bacterial virulence gene *virF* are converted into hosts for the nopaline strains of *A. tumefaciens. Nature* **363(6424)**, 69–71.

20. Kanemoto, R. H., Powell, A. T., Akiyoshi, D. E., Regier, D. A., Kerstetter, R. A., and Nester, E. W. (1989) Nucleotide sequence and analysis of the plant inducible locus *pinF* from *Agrobacterium tumefaciens. J. Bacteriol.* **171**, 2506–2512.

21. Morris, J. W. and Morris, R. O. (1990) Identification of an *Agrobacterium tumefaciens* virulence gene inducer from the pinaceous gymnosperm *Pseudotsuga menziesii. Proc. Natl. Acad. Sci. USA* **87**, 3614–3618.

22. Xu, Y., Bu, W., and Li, B. (1993) Metabolic factors capable of inducing Agrobacterium vir gene expression are present in rice (Oryza sativa L.). *Plant Cell Rep.* **12**, 160–164.

CHAPTER 4

Agrobacterium tumefaciens Chemotaxis Protocols

Charles H. Shaw

1. Introduction

Directed movement in response to chemical attractants is of crucial importance to *Agrobacterium tumefaciens*. Motility is a determinant of rhizosphere competence *(1)*, and chemotaxis a conditional requirement for virulence *(2,3)*. *A. tumefaciens* is attracted toward a variety of sugars typical of plant exudates *(4)*. Responses to nanomolar amounts of phenolic wound exudates are chiefly determined by the Ti-plasmid encoded virulence genes *vir*A and *vir*G *(5,6)*. The sensitivity of these responses indicates that they play a role in the biology of *A. tumefaciens*. Thus, the attraction to sugars can partly explain the organism's prevalence in the rhizosphere *(7)* and the highly sensitive response to phenolics, may help to guide *A. tumefaciens* to wounded plant cells *(8,9)*.

An understanding of chemotaxis and its methodology is thus important to a full appreciation of *A. tumefaciens*. In this chapter the protocols currently in use for the study of motility and chemotaxis in this organism are described, together with some cautionary notes concerning possible pitfalls.

2. Materials

1. Culture media: *A. tumefaciens* are grown in LAB M nutrient broth number 2 (Amersham, Bucks, UK) or Min A + glucose *(10)*. Autoclave a 5X stock solution of MinA Salts. One liter contains 52.5 g K_2HPO_4, 22.5 g K_2HPO_4, 1 g $(NH_4)_2SO_4$, and 2.5 g Na citrate · $2H_2O$. Mix 100 mL 5X MinA Salts with 380 mL sterile water, 5 mL 20% glucose, and 0.5 mL $1M$ Mg_2SO_4. Swarm plates are prepared from nutrient broth supplemented with 0.17%

From: *Methods in Molecular Biology, Vol. 44:* Agrobacterium *Protocols*
Edited by: K. M. A. Gartland and M. R. Davey Humana Press Inc., Totowa, NJ

Bacteriological agar Number 1 (Oxoid, Gibco, Paisley, UK). Antibiotics (Sigma, St. Louis, MO) are used as previously described *(11)*.

2. Chemotaxis medium: 0.1 mM EDTA, 10 mM KH$_2$PO$_4$, pH 7.0 *(12)*.
3. Preparation of attractants: Stock solutions of phenolic attractants are prepared in decimal concentration decrements from 10^{-2} to $10^{-5}M$ using ethyl acetate:methanol (1:1) as a solvent. 10^{-5} to $10^{-8}M$ solutions are prepared for assays by dilution of these stocks with 1000 vol of chemotaxis medium. This ensures a constant and low concentration of solvent in all assays. Control solutions consist of chemotaxis medium with an equal concentration of solvent to that in assay solutions, but no attractant.
4. Dilution of bacteria for plating is done in 10 mM Mg$_2$SO$_4$.
5. 1-[Methyl-^{3}H]-methionine (Amersham).
6. Glycerol: 50% (v/v).
7. Formalin: 40% (v/v).
8. Glass capillaries.
9. Boyden chambers (*see* Section 3.5.).
10. Silicon grease.
11. Polycarbonate filter (13 mm) with 8-μm pores.
12. Isoton (Coulter, Luton, UK).
13. Tricarboxylic acid (TCA; Sigma).
14. Sodium dodecyl sulfate (SDS) sample buffer (Sigma).

3. Methods

3.1. Observation of Motility

As a matter of routine, *A. tumefaciens* cultures are monitored for motility. Ten microliters of bacteria are mixed with 50 μL of chemotaxis medium in the well of an indented glass slide, and covered by a coverslip. Observations are made using 40X or 100X phase-contrast, in a Nikon Optiphot microscope (Nikon, Tokyo, Japan). This can reveal the state of the culture, from the proportion of motile cells. Moreover, it is possible to tell whether the culture is contaminated, as *A. tumefaciens* is significantly smaller than *E. coli*, and has very characteristic motility *(4)*. Behavior and speed measurements are made during slow-motion playback of recorded motility, using a Hitachi (Tokyo, Japan) HV-720K CCTV camera coupled to a Ferguson 3V23 video recorder (Osaka, Japan), a Panasonic WJ-810 time-date generator, and a Hitachi VM-920K video monitor.

3.2. Swarm Plates

Swarm plates are sloppy agar plates made using a reduced quantity of agar (*see* Section 2.). When inoculated into the center of such plates, the bacteria utilize nutrients, thus creating a concentration gradient. Bacteria respond by steadily moving outward, consuming nutrients, and creating a gradient as they go. Swarm plates have a number of uses:

1. Routinely, cultures for chemotaxis assays are enriched for highly motile populations by three passages through swarm plates, taking bacteria from the outermost edge at each passage.
2. Spontaneous mutants can be selected by taking bacteria from the center of a swarm.
3. Mutants can be screened for movement defects after a mutagenesis protocol using the miniswarm assay. In our experiments, Tn5 is used to generate mutations, and the culture is diluted into swarm plates at a density of approx 100 cells/plate. As swarms develop, nonmotile mutants can be detected by the tight nature of their colonies. Nonchemotactic mutants produce small diffuse swarms, owing to random, undirected movements.
4. By incorporating potential attractants into minimal swarm plates, it is possible to perform crude tests of their attractiveness. This does usually require that the attractant also be metabolizable.

3.3. Preparation of A. tumefaciens for Chemotaxis Assays

1. Grow 50-mL cultures of the strains to be assayed for 16 h to exponential phase in MinA + glucose.
2. Pellet 40-mL aliquots at 7000g for 10 min in sterile tubes.
3. Resuspend and wash in 40 mL chemotaxis medium.
4. Repeat step 2, concentrating the pellets in 30, 20, 15, and finally, 10 mL of chemotaxis medium.
5. Measure cell density using a Coulter Electronics Multisizer II (Coulter), fitted with a 30-μm aperture.

3.4. Capillary Assays

Capillary assays *(12)* are one of the oldest and most widely used methods for measuring chemotaxis. They are cheap and sensitive. They are, however, very difficult to set up, and require large numbers of replicates for reproducibility.

1. Take a glass capillary bent into a U-shape (*see* Fig. 1) and attach it to a glass slide using silicon grease.
2. Fill the pool created with 300 μL of bacteria. Cover with a coverslip.
3. Suck 3 mL of attractant into a plastic capillary, and seal the opposite end with grease.
4. Place capillary into bacterial pool. Leave on bench for 60 min.

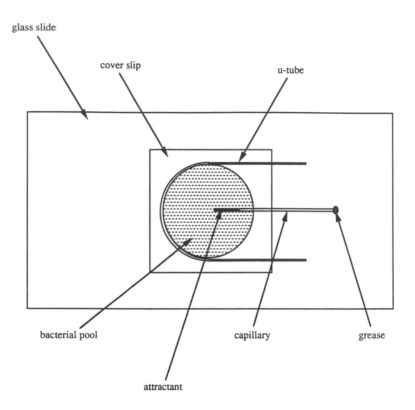

Fig. 1. The capillary assay.

5. Remove capillary and expel contents into diluent. Dilute and plate 100 μL with appropriate selection.
6. Count colonies after 2–5 d incubation at 28°C.
7. Perform assays in triplicate.

3.5. Blindwell Assays

Early experiments were performed using the capillary assay *(13,14)*. We have subsequently switched to the Blindwell assay *(5)* owing to its ease of use and greater reproducibility. These are performed in Boyden chambers *(15)*, which consist of a perspex base, drilled and threaded, into which a hollow Teflon plug is screwed *(see Fig. 2)*.

1. Lightly grease shelf above the bottom chamber.
2. Pipet 200 μL of bacteria into the bottom chamber.
3. Place on shelf a Nuclepore 13-mm diameter polycarbonate filter with 8-μm pores (Sterilin, Stone, Staffs, UK).
4. Screw in Teflon plug to hand tightness.

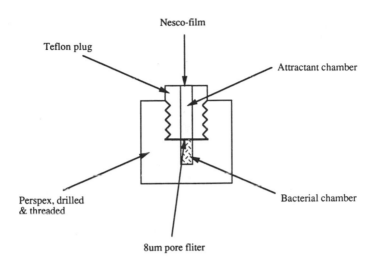

Fig. 2. The Blindwell assay.

5. Pipet 400 µL of attractant solution into the hollow center of the Teflon plug ensuring that no air bubbles are trapped
6. Tightly cover top of plug with Nesco-Film ensuring no air bubbles are trapped. Invert and tap to force any bubbles in the bacterial chamber to the apex of this chamber.
7. After 2 h inversion, remove Nesco-Film and sample 200 µL into 20 mL Isoton (Coulter). Count the number of cells passing through the membrane using a Coulter Electronics Multisizer II, fitted with a 30-µm aperture.
8. Perform assays in triplicate for each strain and attractant concentration. Three separate Multisizer readings are taken for each assay (*see* Notes 1 and 2).

3.6. MCP Labeling with L-[methyl-³H]-Methionine

Many bacteria respond to chemoattractants employing a set of membrane receptors, called methyl-accepting chemotaxis proteins (MCPs) *(16)*. These proteins are methylated during the course of chemotactic responses, the methyl groups being derived from methionine via S-adenosylmethionine *(17)*. By blocking protein synthesis with puromycin *(18)*, it is possible to label MCPs with methyl groups derived from L-[methyl-³H]-methionine.

1. Incubate 50 mL of *A. tumefaciens* overnight at 28°C.
2. Harvest cells by centrifugation and resuspend in 50 mL chemotaxis medium after three washes.

3. Put 2 mL bacteria into each small universal and add 10 mL 50% glycerol together with 16 μL of 12.5 mg/mL puromycin (final concentration 200 μg/mL).
4. Place in 30°C water bath and leave to equilibrate for 10 min.
5. Fifteen microliters of l-[methyl-^{3}H]-methionine (15 Ci/mmol) is added to each cell suspension in labeled Universal bottles (stock solution is 67 mM, making the final concentration 1 mM). Incubate 30 min at 30°C.
6. Chemoattractants are added to each Universal bottle to its optimum chemotactic concentration. Incubate for an appropriate time.
7. Transfer samples to labeled Eppendorf tubes and lyse cells with 62.5 μL of 40% formalin. Place on ice and precipitate proteins with 0.2 mL TCA. (Leave on ice at least 15 min.)
8. Centrifuge cells down and pour off supernatant.
9. Resuspend cells in 0.1 mL SDS sample buffer, vortex, and add 10 μL Tris-HCl pH 8.0 to neutralize the acid.
10. Boil samples for 2 min before analyzing by SDS-PAGE.
11. Impregnate gel with scintillant for fluorography.

4. Notes

1. Mathematical treatment of results: Results from Blindwell assays are presented in the form of the Chemotactic Index (CI), which is a measure of the proportion of cells in the bacterial population attracted toward the attractant. The CI corrects for differing initial cell density, and arises from the observation that for a given strain, the proportion of cells attracted to a particular attractant concentration is a constant, over a broad range of cell densities. To calculate the CI, Multisizer readings are converted to numbers of cells/mL. Control values are subtracted from each point to give a figure for number of cells attracted solely by the attractant. To correct for differences in original culture density, the data are then divided by the cell density of the suspension introduced into the chambers. Averaged figures are then converted to percentages of cells attracted by the compound (the CI) and graphically presented. The CI is thus a reproducible value, and permits comparison from one experiment to the next. In an average experiment, a CI of 2 represents approx 10^7 bacteria/mL in the attractant chamber.

    CI = {[(cells in upper chamber) – (cells in control assay upper chamber)]/ (cells initially introduced to lower chamber)} × 100%

2. Artifacts of chemotaxis assays: Typically, data from both capillary and Blindwell assays will generate graphs like those shown in Fig. 3. These usually depict a sharp peak of chemotaxis, implying that above this concentration, chemotaxis is reduced. This is in fact an artifact of the assays, owing

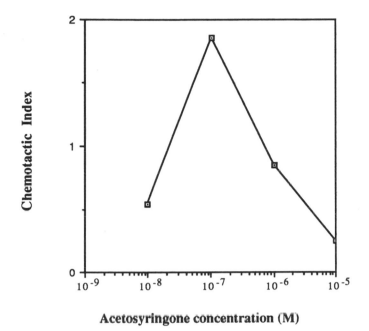

Fig. 3. An example of a typical graph showing data from both capillary and Blindwell assays.

to the fact that bacteria respond to concentration gradients. In the assays, the gradient is created by the attractant diffusing out of the capillary or through the membrane. At low concentrations, the gradient is thus between the attractant and bacterial chambers, and bacteria swim from one to the other. At higher concentrations sufficient gradient is created within the bacterial chamber/pool and the bacteria swim from one part of their chamber to another. Thus few bacteria get into the other chamber to be counted.

References

1. Shaw, C. H., Loake, G. J., Brown, A. P., Garrett, C. S., Deakin, W., Alton, G., Hall, M., Jones, S. A., O'Leary, M., and Primavesi, L. (1991) Isolation and characterisation of behavioural mutants and genes from *Agrobacterium tumefaciens. J. Gen. Microbiol.* **137,** 1929–1953.
2. Hawes, M. C., Smith, L. Y., and Howarth, A. J. (1988) *Agrobacterium tumefaciens* mutants deficient in chemotaxis to root exudates. *Mol. Plant–Microbe Int.* **1,** 182–186.
3. Hawes, M. C. and Smith, L. Y. (1989) Requirements for chemotaxis in pathogenicity of *Agrobacterium tumefaciens* on roots of soil grown pea plants. *J. Bacteriol.* **171,** 5668–5671.

4. Loake, G. J., Ashby, A. M., and Shaw, C. H. (1988) Attraction of *Agrobacterium tumefaciens* C58C1 towards sugars involves a highly sensitive chemotaxis system. *J. Gen. Microbiol.* **134**, 1427–1432.
5. Shaw, C. H., Ashby, A. M., Brown, A. P., Royal, C., Loake, G. J., and Shaw, C. H. (1988) *vir*A and G are the Ti-plasmid functions required for chemotaxis of *Agrobacterium tumefaciens* toward acetosyringone. *Mol. Microbiol.* **2**, 423–418.
6. Palmer, A. C. V. and Shaw, C. H. (1992) The role of *vir*A and G phosphorylation in chemotaxis towards acetosyringone by *Agrobacterium tumefaciens*. *J. Gen. Microbiol.* **138**, 2509–2514.
7. Kerr, A. (1974) Soil microbiological studies on *Agrobacterium radiobacter* and biological control of crown gall. *Soil Sci.* **118**, 168–172.
8. Shaw, C. H., Ashby, A. M., Loake, G. J., and Watson, M. D. (1988) One small step: the missing link in crown gall. *Oxford Surv. Plant Mol. Cell Biol.* **5**, 177–183.
9. Shaw, C. H. (1991) Swimming against the tide: chemotaxis in *Agrobacterium*. *BioEssays* **13**, 25–29.
10. Miller, J. H. (1972) *Experiments in Molecular Genetics*. Cold Spring Harbor Laboratory, Cold Spring Harbor, NY.
11. Leemans, J., Shaw, C. H., Deblaere, R., De Greve, H., Hernalsteens, J.-P., van Montagu, M., and Schell, J. (1981) Site-specific mutagenesis of *Agrobacterium* Ti-plasmids and transfer of genes to plant cells. *J. Mol. Appl. Gen.* **1**, 149–164.
12. Adler, J. (1973) A method for measuring chemotaxis and use of the method to determine optimum conditions for chemotaxis by *Escherichia coli*. *J. Gen. Microbiol.* **74**, 77–91.
13. Ashby, A. M., Watson, M. D., and Shaw, C. H. (1987) A Ti-plasmid determined function is responsible for chemotaxis of *Agrobacterium tumefaciens* towards the plant wound product acetosyringone. *FEMS Microbiol. Lett.* **41**, 189–192.
14. Ashby, A. M., Watson, M. D., Loake, G. J., and Shaw, C. H. (1988) Ti-plasmid specified chemotaxis of *Agrobacterium tumefaciens* C58C1 toward *vir*-inducing phenolic compounds and soluble factors from monocotyledonous and dicotyledonous plants. *J. Bacteriol.* **170**, 4181–4187.
15. Armitage, J. P., Josey, D. P., and Smith, D. G. (1977) A simple qualitative method for measuring chemotaxis and motility in bacteria. *J. Gen. Microbiol.* **102**, 199–202.
16. Springer, M. S., Goy, M. F., and Adler, J. (1979) Protein methylation in behavioural control mechanisms and in signal transduction. *Nature* **280**, 279–284.
17. Aswad, D. and Koshland, D. E. (1975) Evidence for an S-adenosyl-methionine requirement in the chemotactic behavior of *Salmonella typhimurium*. *J. Mol. Biol.* **97**, 207–223.
18. Alam, M., Lebert, M., Oesterhelt, D., and Hazelbauer, G. L. (1989) Methyl-accepting taxis proteins in *Halobacterium halobium*. *EMBO J.* **8**, 631–639.

CHAPTER 5

Effect of Acetosyringone on Growth and Oncogenic Potential of *Agrobacterium tumefaciens*

Patrice Dion, Christian Bélanger, Dong Xu, and Mahmood Mohammadi

1. Introduction

Expression of the virulence genes of *Agrobacterium tumefaciens* occurs in acidic media containing sugars and certain phenolics secreted at plant wounds. In some strains of *A. tumefaciens*, exposure to these inducing conditions leads to inhibition of bacterial growth, which can then be followed by the selection of avirulent mutants *(1)*. Observation of this effect of *vir* gene induction depends on the use of particular experimental conditions. This chapter first specifies experimental parameters allowing to reveal, in *A. tumefaciens*, the modifications that accompany *vir* gene induction. Other parts of the chapter then describe how the mutants that are generated in the presence of acetosyringone are recognized and characterized. These protocols may prove useful to those with an interest in *Agrobacterium* biology. In addition, acetosyringone enhances the ability of *A. tumefaciens* to transform recalcitrant host plants, and as such has been incorporated in medium used for co-cultivation of bacteria and plant tissue *(2–5)* or alternatively has been used to precondition explants prior to addition of bacteria *(6)*. Hence, the possibility of side effects of acetosyringone on *A. tumefaciens* bears direct relevance to experiments dealing with transformation of crop plants.

From: *Methods in Molecular Biology, Vol. 44:* Agrobacterium *Protocols*
Edited by: K. M. A. Gartland and M. R. Davey Humana Press Inc., Totowa, NJ

2. Materials

2.1. Measurement of Bacterial Growth

1. Spectrophotometer to accomodate 18-mm diameter test tubes.
2. Erlenmeyer flasks (50-mL vol) fitted with a 18-mm diameter test tube as a side arm. These test tubes should be calibrated for measurements of optical density at 600 nm (OD_{600}).
3. Incubator with rotary shaker.
4. Neutral ATx2 stock solution: Dissolve in 1 L of distilled water: 21.8 g KH_2PO_4, 0.32 g $MgSO_4 \cdot 7 H_2O$, 0.010 g $FeSO_4 \cdot 7 H_2O$, 0.022 g $CaCl_2 \cdot 2 H_2O$, and 0.004 g $MnCl_2 \cdot 4 H_2O$. Adjust to pH 7.0 with 6N KOH and autoclave. A slight precipitate will develop on autoclaving, which may be removed by filtration on Whatman (Maidstone, UK) no. 4 filter paper. The stock solution is then autoclaved a second time.
5. Acidic ATx2 stock solution: Dissolve in 1 L of distilled water: 30.0 g KH_2PO_4, 0.32 g $MgSO_4 \cdot 7 H_2O$, 0.010 g $FeSO_4 \cdot 7 H_2O$, 0.022 g $CaCl_2 \cdot 2 H_2O$, 0.004 g $MnCl_2 \cdot 4 H_2O$. Adjust to pH 5.8 with 6N KOH. Autoclave. The pH of this stock solution tends to be modified after autoclaving. The autoclaved solution is readjusted to the desired pH and autoclaved a second time.
6. Solid AT minimal medium, pH 7.0: 500 mL neutral ATx2 stock solution, 1 g ammonium sulfate, 2 g glucose, 15 g Bacto-Agar (Difco, Detroit, MI). Final vol, 1 L. Ammonium sulfate and glucose are prepared as 100X stock solutions and are autoclaved separately. Agar is also autoclaved separately in water.
7. Liquid AT minimal medium, pH 5.8: 500 mL acidic ATx2 stock solution, 1 g ammonium sulfate, 2 g glucose. Final vol, 1 L.
8. Acetosyringone (500 mM): Dissolve 98.1 mg acetosyringone in 1 mL dimethyl sulfoxide (DMSO). Keep at −20°C.

2.2. Recovery of Bacterial Clones Following Growth in the Presence of Acetosyringone and Recognition of Avirulent Mutants

1. Nopaline stock solution: Dissolve 800 mg nopaline (Sigma, St. Louis, MO) in 100 mL water. The addition of 1–2 drops of 6N KOH may be necessary to facilitate solubilization. Sterilize by filtration.

    Only nopaline-type strains have so far been found to be susceptible to acetosyringone-induced growth inhibition. Some other opines (e.g., octopine, octopinic acid, and mannopine) are also commercially available.
2. AT nopaline medium: 500 mL neutral ATx2 stock solution, 1 g ammonium sulfate, 100 mL nopaline stock solution, 15 g Noble Agar (Difco). Final vol, 1 L.

Strain C58 derivatives are unable to grow on arginine unless nopaline is also present in the medium as an inducer of the Ti plasmid-encoded catabolic genes. Thus, for this strain only, AT nopaline medium, which contains 800 mg/L of the opine, can be replaced with a cheaper medium composed, per liter, of 500 mL ATx2 stock solution, 1 g ammonium sulfate, 100 mg nopaline, 1 g arginine, and 15 g Noble Agar.

3. Indicator plants for testing virulence of *Agrobacterium*: These include kalanchoë plants and 2-wk-old tomato plantlets. To obtain a reliable response with kalanchoë, the youngest fully developed leaves must be selected for inoculation. Care must be taken that temperature in the greenhouse does not exceed 30°C for the 3 d following inoculation.
4. Saline water: 8.5 g NaCl/L water. Autoclave.

3. Methods

3.1. Measurement of Bacterial Growth

Several experimental parameters are crucial to observation of growth inhibition in the presence of acetosyringone. These include concentration of acetosyringone, pH, and sugar composition of the growth medium, density of bacterial inoculum, and the particular bacterial strain being studied. With respect to the last factor, a limited survey suggested that *A. tumefaciens* strains carrying a nopaline-type Ti plasmid, such as strains C58 and T37, exhibit retarded growth in the presence of acetosyringone, but that strains carrying an octopine-type plasmid, such as strains A348, B6S3, and Ach5, do not show this effect *(1)*. Furthermore, we found variability between strains labeled as "C58" obtained from different laboratories. The protocol presented here allows the observation of growth inhibition with all strain C58 derivatives that we have tested.

1. For long-term storage, bacteria should be kept as glycerol stocks at –80°C. For routine use, bacteria are maintained on solid AT minimal medium, pH 7.0. They should be transferred regularly from a single colony, ideally every 2 wk. Experiments are initiated from a single colony from a fresh culture (48–72 h) on solid AT minimal medium (*see* Note 1).
2. A single colony is transferred to a 10-mL liquid AT minimal medium, pH 5.8 in a 50-mL Erlenmeyer flask with calibrated side arm. Incubate with shaking (175 rpm at 27°C). At OD_{600} of 0.5 (after 30–48 h of incubation), the culture is diluted 100-fold in sterile saline water. This dilute suspension is then used immediately to inoculate the test and control media (*see* Note 2).
3. Test medium with 50 μM acetosyringone (liquid AT minimal medium, pH 5.8, containing 100 μL of acetosyringone stock solution/L) and control medium (similar to test medium but prepared by addition of DMSO instead of

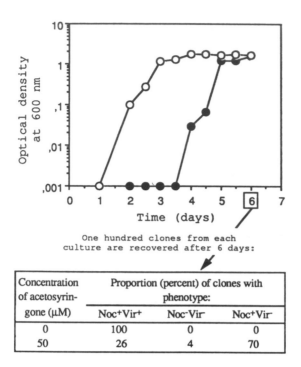

Fig. 1. Effect of acetosyringone on growth and virulence of *Agrobacterium tumefaciens* C58. A bacterial inoculum was prepared as described in this chapter and transferred to liquid AT minimal medium, pH 5.8. This medium contained 100 μL/L DMSO (○) or 50 μ*M* acetosyringone dissolved in DMSO (●); Noc, nopaline catabolism; Vir, virulence on tomato.

acetosyringone stock solution) are distributed in volumes of 20 mL in Erlenmeyer flasks with sidearm (*see* Note 3). The media are inoculated by addition of 100 μL of dilute bacterial suspension, or about 2.5×10^5 cfu (*see* step 2).

4. Ten milliliters of inoculated medium are taken to verify initial pH (*see* Note 4). The other 10 mL are left in the flask.

5. Cultures in Erlenmeyer flasks are incubated at 27°C with shaking for 6 d. OD_{600} is measured every 12 h. At the end of the experiment, the culture is centrifuged and pH of the supernatant is noted (*see* Note 5). A typical result obtained using this protocol is shown in Fig. 1. Exposure to acetosyringone results in a 2-d delay in onset of growth, as compared to the control culture. Following this delay, optical density in the acetosyringone-treated culture increases at a rate similar to what had been observed for the control culture (Fig. 1).

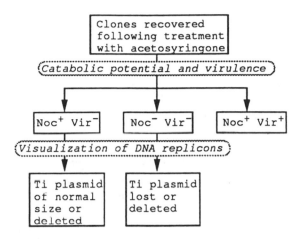

Fig. 2. Summary of procedure for initial screening of bacterial clones recovered following growth in the presence of acetosyringone. Clones are analyzed for catabolic potential, virulence, and presence of the Ti plasmid.

3.2. Recovery of Bacterial Clones Following Growth in the Presence of Acetosyringone and Recognition of Avirulent Mutants

After 6 d of incubation under the conditions just described, bacterial population in both control and acetosyringone-treated cultures is about $3-10 \times 10^9$ colony-forming units/mL. In the case of the acetosyringone-treated culture, this population is the outcome of initial inhibition of growth of those bacteria responding to the acetosyringone *vir*-inducing signal, followed by the selection of uninducible mutants. Genetic alterations in these mutants are the result of activity of IS elements *(7)*, and conceivably various types of mutants could be recovered following acetosyringone treatment. Analysis of these mutants is a two-step process, consisting first of recovery of clones and recognition of avirulent mutants, and then characterization of the genetic defect in the particular clones of interest. These clones are classified initially on the basis of two criteria, capacity for opine utilization and pathogenicity (Fig. 2). The first of these criteria is used as a convenient indicator of the presence of the Ti plasmid within the clones. Conclusions based on the opine utilization test are then confirmed and completed by visualization of the various DNA replicons carried by the clones on agarose gel.

1. Bacteria from the acetosyringone-treated and control cultures are serially diluted in saline water and the 10^6, 10^7, and 10^8 dilutions are plated on Nutrient Agar (Difco).
2. After 2–3 d of incubation at 27°C, 100 colonies from each culture are replica transferred with toothpicks to:
   a. Nutrient Agar;
   b. AT minimal medium, pH 7.0; and
   c. AT nopaline medium.
3. Nutrient Agar cultures are incubated for 2 d and used for inoculation of kalanchoë leaves or stems of tomato seedlings. Plants are scored for the presence of tumors 3–4 wk after inoculation.
4. Cultures on AT nopaline medium are incubated for 7 d and ability of each of the individual clones to grow on this medium is then monitored.
5. Clones of interest are recovered from the AT minimal medium, pH 7.0, and are stored as frozen glycerol stocks.
6. The presence of a Ti plasmid in the clones is investigated by agarose gel electrophoresis following in-well lysis. Numerous variations on the original protocol of Eckhardt *(8)* exist, and this basic protocol may have to be adapted to the particular strain being used and the conditions prevailing in the laboratory.

3.3. Analysis of the Nature of the Mutations in the Avirulent Clones

Clones of interest for further analysis are generally those that are still capable of nopaline utilization, but have become avirulent. The procedure followed to ascertain the nature of the genetic defect in these bacteria is outlined in Fig. 3. The various steps of this procedure involve standard methodology and therefore we simply provide the rationale for conducting those tests and a guide to interpretation of the results.

Since both chromosomal and Ti plasmid-encoded genes have been implicated in determination of oncogenic potential in *A. tumefaciens*, it is first necessary to determine whether the mutation resides on the Ti plasmid or elsewhere in the bacterial genome. This is done most conveniently by transferring the Ti plasmid from the avirulent mutants to a Ti plasmidless derivative of strain C58, such as C58C1. Protocols for rapid extraction and electroporation of the Ti plasmid are available (9,10; *see* Chapter 33). The electrotransformants are then tested for oncogenic potential to determine whether they carry mutant or wild-type Ti plasmids. In the latter case, the genetic defect in the original mutant is inferred to reside outside of the Ti plasmid.

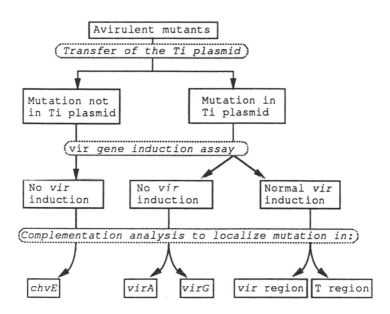

Fig. 3. Summary of procedure for analysis of the nature of the mutations in the avirulent clones. Clones are analyzed for presence of the mutation on the Ti plasmid or elsewhere in the genome. They are then tested for their ability to induce a *vir*B::*lac*Z fusion in the presence of acetosyringone and to be complemented by a variety of clones containing wild-type genes involved in oncogenicity.

Avirulent mutants recovered after treatment with acetosyringone are often altered in *vir*A and *vir*G, which are the two loci responsible for detection and transduction of the acetosyringone *vir*-gene inducing signal. As a consequence, these mutants are no longer capable of *vir* gene induction. The ability of the avirulent mutants to respond to acetosyringone by inducing the *vir* genes is assessed using standard methodology, involving the use of a promoterless reporter gene fused to a *vir* promoter *(11)*.

The mutants are finally subjected to virulence complementation analysis by transfer of clones carrying specific wild-type genes involved in oncogenicity. Clones listed in Table 1 allow to cover the entire range of cases encountered to date in our laboratory. It is conceivable that other types of mutants may be recovered in the future.

4. Notes

1. Bacteria subject to growth inhibition in the presence of acetosyringone also tend to be genetically unstable *(7)*. For this reason, care must be taken

Table 1
Clones Used in Virulence Complementation Analysis
of Avirulent Mutants Recovered Following Growth
in the Presence of Acetosyringone

Designation of clone	Operons or genes involved in tumorigenesis carried by clone	Refs.
pMWH146	*chv*E region from A136	*12*
pUCD2614	*vir*E, *vir*D, *vir*C, *vir*G, *vir*B, and *vir*A of pTiC58	*13*
pUCD2638	transcripts 1, 2, 4, and 5 of pTiC58 T-DNA	*13*
pVCK219	*vir*H, *vir*A, and part of *vir*B from pTiA6	*14*
pVCK221	*vir*B, *vir*G, and *vir*C from pTiA6	*14*
pVCK225	*vir*G, *vir*C, *vir*D, and *vir*E from pTiA6	*14*
pVCK257	*vir*H, *vir*A, *vir*B, *vir*G, and *vir*C from pTiA6	*14*

in maintenance of these strains, and particularly in avoidance of unfavorable incubation conditions.

2. Under most circumstances, inhibition of growth is not complete but is rather characterized by an increase in generation time. Thus, intensity of effect of acetosyringone (as measured by length of apparent lag phase) depends on the size of the inoculum.

3. We use 50-mL Erlenmeyer flasks with sidearm in our bacterial growth assays. We find that these provide much better and reliable results than test tubes. We clean these flasks in strong detergent solution, and rinse them carefully 10 times in tap water and then 10 times in distilled water. They are stoppered with aluminum foil.

4. It is of crucial importance that initial pH be controlled with precision. This pH may be varied between 5.3 and 5.8, with similar effects of acetosyringone being observed. However, growth retardation is more pronounced at pH 5.3–5.6 than at pH 5.8. No significant effect of acetosyringone on bacterial growth is observed above pH 5.8.

5. Bacteria produce acids from glucose, and the final pH value after 6 d should be about 5.3–5.4.

Acknowledgments

We thank C. I. Kado for the generous gift of plasmids. Experimental work from our laboratory has been supported by strategic grant 133859 from the Natural Sciences and Engineering Research Council of Canada (NSERC) to P. Dion. C. Bélanger gratefully acknowledges the receipt of a postgraduate fellowship from NSERC.

References

1. Fortin, C., Nester, E. W., and Dion, P. (1992) Growth inhibition and loss of virulence in cultures of *Agrobacterium tumefaciens* treated with acetosyringone. *J. Bacteriol.* **174,** 5676–5685.
2. Godwin, I., Todd, G., Ford-Lloyd, B., and Newbury, H. J. (1991) The effects of acetosyringone and pH on *Agrobacterium*-mediated transformation vary according to plant species. *Plant Cell Rep.* **9,** 671–675.
3. Godwin, I. D., Ford-Lloyd, B. V., and Newbury, H. J. (1992) *In vitro* approaches to extending the host-range of *Agrobacterium* for plant transformation. *Aust. J. Bot.* **40,** 751–763.
4. Holford, P., Hernandez, N., and Newbury, H. J. (1992) Factors influencing the efficiency of T-DNA transfer during co-cultivation of *Antirrhinum majus* with *Agrobacterium tumefaciens*. *Plant Cell Rep.* **11,** 196–199.
5. Owens, L. D. and Smigocki, A. C. (1988) Transformation of soybean cells using mixed strains of *Agrobacterium tumefaciens* and phenolic compounds. *Plant Physiol.* **88,** 570–573.
6. Guivarc'h, A., Caissard, J.-C., Brown, S., Marie, D., Dewitte, W., Van Onckelen, H., and Chriqui, D. (1993) Localization of target cells and improvement of *Agrobacterium*-mediated transformation efficiency by direct acetosyringone pretreatment of carrot root discs. *Protoplasma* **174,** 10–18.
7. Fortin, C., Marquis, C., Nester, E. W., and Dion, P. (1993) Dynamic structure of *Agrobacterium tumefaciens* Ti plasmids. *J. Bacteriol.* **175,** 4790–4799.
8. Eckhardt, T. (1978) A rapid method for the identification of plasmid deoxyribonucleic acid in bacteria. *Plasmid* **1,** 584–588.
9. Cangelosi, G. A., Best, E. A., Martinetti, G., and Nester, E. W. (1991) Genetic analysis of *Agrobacterium*. *Methods Enzymol.* **204,** 384–397.
10. Mozo, T. and Hooykaas, P. J. J. (1991) Electroporation of megaplasmids in *Agrobacterium*. *Plant Mol. Biol.* **16,** 917,918.
11. Stachel, S. E. and Nester, E. W. (1986) The genetic and transcriptional organization of the *vir* region of the A6 Ti plasmid of *Agrobacterium tumefaciens*. *EMBO J.* **5,** 1445–1454.
12. Huang, M. L. W., Cangelosi, G. A., Halperin, W., and Nester, E. W. (1990) A chromosomal *Agrobacterium tumefaciens* gene required for effective plant signal transduction. *J. Bacteriol.* **172,** 1814–1822.
13. Rogowsky, P. M., Powell, B. S., Shirasu, K., Lin, T.-S., Morel, P., Zyprian, E. M., Steck, T. R., and Kado, C. I. (1990) Molecular characterization of the *vir* regulon of *Agrobacterium tumefaciens*: complete nucleotide sequence and gene organization of the 28.63-kbp cloned as a single unit. *Plasmid* **23,** 85–106.
14. Stachel, S. E. and Zambryski, P. (1986) *virA* and *virG* control the plant-induced activation of the T-DNA transfer process of *A. tumefaciens*. *Cell* **46,** 325–333.

CHAPTER 6

Binary Ti Plasmid Vectors

Gynheung An

1. Introduction

Living organisms have been continuously evolving by assimilating new genetic material from the environment. However, this progress is very slow and often limited to transfer of genetic materials among closely related species. Recent developments in molecular biology and gene transfer techniques enable researchers to move genetic information among a variety of living organisms. Gene transfer techniques for higher plants can be divided into two major methods: direct DNA transfer and *Agrobacterium*-mediated tumor inducing (Ti)-plasmid-vector methods. Direct DNA transfer methods have been used for the transformation of a wide variety of species, especially those plant species that are recalcitrant to transformation with Ti plasmid vectors. However, the direct DNA transfer method requires more manipulation and transformation efficiency is generally much lower compared to the Ti-plasmid-vector system. Furthermore, it appears that stability of introduced genes is lower when DNA was directly transferred. Therefore, Ti vectors are commonly used for transformation of most dicot plant species that are, in general, more readily transformed using *Agrobacterium*-mediated DNA delivery.

2. Ti Plasmid Vectors
2.1. Disarmed Vectors

Ti plasmid vectors were developed in early 1980 by several laboratories based on observations that a specific region (called T-DNA) of the Ti plasmid is stably transferred from *Agrobacterium tumefaciens* into

From: *Methods in Molecular Biology, Vol. 44:* Agrobacterium *Protocols*
Edited by: K. M. A. Gartland and M. R. Davey Humana Press Inc., Totowa, NJ

the plant chromosome and any foreign DNA sequences inserted within the T-DNA are also cotransferred into plants *(1–3)*. Since only the 25-bp border sequences located at each end of the T-DNA are required for the DNA transfer mechanism, "disarmed" Ti plasmid vectors were developed. One of the most frequently used disarmed vectors is pGV3850 in which the T-DNA region was replaced with the linearized pBR322 sequence. A DNA fragment located in the pBR or related vectors can be cointegrated into the pGV3850 by homologous recombination *(4)*. The disarmed vector also contains various selectable markers that are necessary for identification of transformed plants. The most commonly used selectable marker is the chimeric kanamycin resistance gene, which was constructed from a bacterial transposable element Tn5 by replacing the 5' and 3' flanking sequences with various plant regulatory elements. Since some plants are resistant to a high level of kanamycin, other selectable markers, such as hygromycin and herbicide-resistant genes, have also been developed. The hygromycin resistant gene was obtained by placing the bacterial hygromycin phosphotransferase gene (*aph*IV) under the control of plant regulatory sequences *(5)*. Among several herbicide resistant genes the bialaphos resistant *(bar)* gene has been the most widely used for a plant selectable marker *(6)*. The *bar* gene that encodes for phosphinothricin acetyltransferase was originated from *Streptomyces hygroscopicus*. Plants carrying this selectable marker become resistant to the synthetic herbicide phosphinothricin (Basta®, Hoechst).

2.2. Binary Vectors

Although the disarmed vectors have been used widely in transferring foreign genes into the plant chromosome, this system is not easy to use because the Ti plasmid is large (at least 100 kb). To facilitate handling the transformation system, much smaller and simpler vectors were constructed. These "binary" Ti plasmid vectors were developed based on the finding that the T-DNA region does not have to be physically linked to the *vir* genes of a Ti plasmid *(7)*. The *vir* region contains seven operons, of which four (*vir*A, *vir*G, *vir*B, and *vir*D) code for a variety of proteins which are absolutely essential for the transfer process, whereas the remaining three operons (*vir*C, *vir*E, and *vir*F) are necessary only in certain host species *(8)*. The products of *vir*A and *vir*G genes form a two-component regulatory system that induces *vir* gene expression in response to plant wound signals *(9)*. The other *vir* genes produce proteins that are involved in the generation of single-stranded DNA, protection of the T-DNA, transport of the T-DNA across the bacterial membranes, and interaction with host chromosomes. In addition to the *vir* genes, several genes located on the *Agrobacterium* chromosome are also needed for the T-DNA transfer *(10)*. In the binary vector system, a Ti plasmid is divided into two portions. The larger fragment contains the entire plasmid minus the T-DNA region. Therefore, this "helper" plasmid can replicate and produces the factors necessary for T-DNA transfer. Since the smaller fragment containing the T-DNA region is unable to self-replicate, this fragment was inserted into a wide-host-replicon such as RK2 plasmid that multiplies in both *Agrobacterium* and *Escherichia coli*. Since the T-DNA genes are not needed for DNA transfer and expression of some of these genes generates tumorigenic tissues, this region was replaced with a plant-selectable marker. In addition, the binary vector contains a multiple cloning site for inserting a foreign gene and other useful features that is discussed later *(11)*.

The binary system is much easier to use since binary vectors are smaller (about 10–15 kb) and do not require cointegration. After insertion of a target gene into a unique site in the T-DNA region using *E. coli* as a host, the vector is transferred into *Agrobacterium* by either conjugation or DNA transformation. As the binary vector does not carry the mobilization function necessary for conjugal transfer, this function is provided by a mobilization plasmid carried in another *E. coli* strain. The advantage of the conjugal transfer method is a higher transformation frequency compared to the direct DNA transformation. However, the conjugal transfer method is slow and often results in rearrangement of the vector. To overcome these difficulties, the direct DNA transfer method had been developed. In this procedure, a vector DNA isolated from *E. coli* is transformed into *Agrobacterium* by the freeze–thaw method *(11)*. Since transformants can be obtained in 2 d and DNA rearrangement is a rare event, the direct DNA transfer method is widely used.

2.2.1. Characteristics of Binary Vectors

A large number of binary vectors has been constructed by many different researchers *(12)*. These vectors can be divided into several groups based on the experimental purpose. Since the structure and function are similar within the same group, the binary vectors developed by my laboratory are used as examples to describe characteristics and usage of each group.

The basic binary vector contains a wide host-range replicon, T-DNA borders, a selectable marker, and cloning sites. The replicons used for the vector construction are mostly RK2 derivatives such as pTJS75 or pRK252. These plasmids are not self-transmissible and are capable of replication in a wide range of gram-negative bacteria including *E. coli* and *Agrobacerium*. Unlike common cloning vectors such as pUC or pBR plasmids, the binary Ti plasmids are present in a low copy number and are not amplifiable by chloramphenicol. Both the selectable marker and cloning sites are located between the T-DNA borders so that an inserted DNA fragment is cotransferred with the selectable marker into plant chromosomes. Therefore, plants stably transformed with a foreign gene can be readily identified and maintained using the selectable marker.

Figure 1 shows maps of basic binary vector, pGA628, which is used in my laboratory. As a plant selectable marker, these vectors contain the kanamycin resistance gene that is under the control of the nopaline synthase *(nos)* promoter. The *nos* gene produces the most abundant mRNA in tumors induced by nopaline-type Ti plasmids *(13)*. Therefore, this promoter has been used by several laboratories for construction of chimeric selectable markers. However, it has been found that the *nos* promoter is not constitutively expressed in plant cells *(14)*. The promoter is highly active in roots but very weakly expressed in leaves. Furthermore, the *nos* promoter activity is developmentally and environmentally regulated. Therefore, selectable markers constructed with the *nos* promoter may not be suitable for certain plant species. A stronger and more constitutive promoter such as the CaMV 35S promoter *(15)* has been used in some binary vectors. Although all the kanamycin resistance genes were derived from the same neomycin phosphotransferase *(npt)* II gene of transposon Tn5, some binary vectors such as pBI121 carry a mutant *npt* gene that substantially reduces the enzyme activity *(16)*. Therefore, plants transformed with the mutant gene are less resistant to antibiotics than those transformed with the normal gene. In leaf disc transformation of tobacco, the binary vectors carrying the mutant *npt* gene were several-fold less resistant to kanamycin and G418. Selecting transformants with the mutant *npt* gene may be more difficult, especially when the plants are resistant to a high level of the antibiotics. All the pGA binary vectors reported in this chapter carry the wild-type *npt* gene.

The plasmids pGA628 and pGA810 contain different sets of multiple cloning sites which can be used for the cloning of various restriction

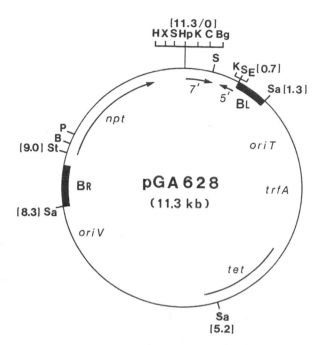

Fig. 1. A circular map of pGA628. This basic binary vector contains a wide host replicon (1.3–8.3), right border (BR), left border (BL), *npt* gene, and multiple cloning site. *ori*T, pRK2 origin of conjugative transfer; *ori*V, pRK2 origin of replication; *tet*, tetracycline resistant gene; *trf*A, transacting replication factor gene; 5', transcription termination region of the gene 5 of pTiA6; 7', transcription termination region of the gene 7 of pTiA6. Restriction enzyme sites: B, *Bam*HI; Bg, *Bgl*II; C, *Cla*I; *Eco*RI; H, *Hind*III; Hp, *Hpai*; K, *Kpn*I; P, *Pst*I; S, *Sst*I; Sa, *Sal*I; St, *Sst*II; X, *Xba*I.

fragments (Fig. 2). The DNA fragment carrying terminator regions of T5 and T7 genes of the octopine type Ti plasmid is placed between the multiple cloning sites and the left border in order to minimize the possible antisense transcription of the introduced gene when the T-DNA is inserted into the middle of an active gene. Integration of T-DNA into an active gene appears to be a common phenomenon as observed in *Arabidopsis* in which approx 20% of T-DNA insertions generate observable mutations *(17)*. Frequent observations of this phenomenon suggest that the T-DNA integrates preferentially into regions of the chromosome capable of being transcribed.

T-DNA transfer starts from the right border and generally ends at the left border. However, the transfer often ends before the left border when

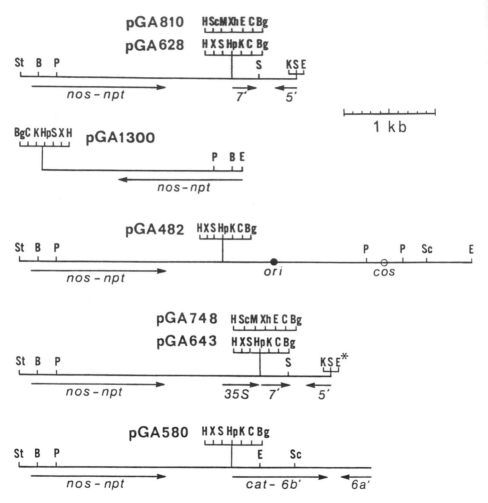

Fig. 2. Linear maps of T-DNA region. The T-DNA region between the right and left borders are shown. *cat*, Chloramphenicol acetyltransferase coding region; *cos*, cohesive end site of bacteriohpage λ; *nos-npt*, a chimeric gene fusion between the *nos* promoter and *npt* gene; *ori*, Col EI origin of replicon; 5', 7', 6a', and 6b', transcription termination region of the gene 5, 7, 6a, and 6b of pTiA6, respectively; 35S, promoter of cauliflower mosaic virus 35S transcript. Restriction enzyme sites: B, *Bam*HI; Bg, *Bgl*II; C, *Cla*I; E, *Eco*RI; H, *Hind*III; Hp, *Hpa*I; K, *Kpn*I; M, *Mlu*I; P, *Pst*I; S, *Sst*I; Sa, *Sal*I; Sc, *Sca*I; *Sst*II; X, *Xba*I; Xh, *Xho*I. E*, not present in pGA748.

there is a sequence similar to the T-DNA border, resulting in a partial transfer of the T-DNA. Since, in most binary vectors, a selectable marker is present immediately next to the right border and a foreign DNA is inserted between the selectable marker and left border, transformants often do not carry the foreign DNA. To ensure that transformants always carry the introduced gene, a selectable marker is placed next to the left border in some binary vectors, such as pGA1300 (Fig. 2).

2.2.2. Binary Vectors for Establishing Stable Transformation

Basic binary vectors have been modified for a variety of different experimental purposes. The vectors carrying an additional marker gene such as β-glucuronidase (GUS), chloramphenicol acetyltransferase *(cat)*, or luciferase gene under the control of a strong plant promoter are useful for establishing a stable transformation procedure. Expression of these reporters by a strong promoter can be assayed during various stages of transformation procedure in order to optimize the transformation frequency. However, one should be careful to avoid false positive data because some plant tissues may contain endogenous enzyme activity. In addition, bacterial contamination may also contribute to the enzyme activity. Reporter genes carrying an intron can be used to minimize the false signal from bacterial expression. The *cat* gene is a sensitive reporter used in many organisms, since there is no or little endogenous background *cat* activity in plant tissues *(18)*. The luciferase gene is a valuable reporter because the assay for this is nondestructive *(19)*. The luciferase gene isolated from the firefly, *Photinus pyralis*, encodes an enzyme that catalyzes the light-producing, ATP-dependent oxidation of luciferin. Unfortunately, the amount of light produced by the luciferin is extremely low, requiring an expensive luminometer for the analysis. The most frequently used reporter is the GUS gene *(20)*. The *E. coli* GUS gene encodes for an acid hydrolase that catalyzes the cleavage of a wide variety of β-glucuronides. Substrates for β-glucuronidase are generally water-soluble and many substrates are commercially available, including substrates for spectrophotometric, fluorometric, and histochemical analyses. The histochemical analysis is particularly useful since spatial and developmental expression pattern of the reporter gene can be easily monitored. The histochemical substrate that provides the most visually distinctive results is 5-bromo-4-chloro-3-indolyl β-D-glucuronide (X-gluc). This colorless substrate is hydrolyzed by β-glucuronidase to produce indolyl derivatives that are dimerized to form insoluble indigo dye. This analysis is usually performed after a tissue is fixed. However, expression of the GUS reporter gene can also be tested without destroying plant tissues by adding X-gluc directly onto the agar around the plantlets and detecting blue staining in the roots *(21)*. The X-gluc and its products do not seem to be lethal to the plants when exposed only for short periods. Therefore, the treated plants can be rescued by transferring to a new culture medium. The *npt* gene has also been used as reporters since their expression can be studied by very sensitive radioactive assays (*18, see* Chapter 19).

2.2.3. Vectors for Construction of Genomic Libraries

The plasmid pGA482 contains a *cos* site that is useful in constructing cosmid libraries of an entire plant genome (Fig. 2). *Arabidopsis thaliana* has a small genome size and therefore only a few thousand clones are needed to cover the entire genome. This library can be used for isolation of a gene by complementing a specific recessive mutant of *Arabidopsis* as well as other plant species. This method is an attractive way of obtaining useful genes, especially when well-characterized mutants are already available.

2.2.4. Binary Vectors for Expression of Foreign Genes

Binary vectors such as pGA643 and pGA748 can be used for expression of a foreign gene (Fig. 2). These vectors carry an expression cartridge, comprised of a strong promoter such as the 35S promoter, a multiple cloning site and a transcriptional termination region. Effects of the overproduction of a gene can be elucidated by inserting a cDNA or a structural portion of a foreign gene into the multiple cloning site and studying the effect of the gene expression in transformed plant cells. Cell type specific expression of a foreign gene may be achieved by using a binary vector containing a spatially regulated promoter. For example, a tapetum-specific promoter of tobacco has been used to express a foreign ribonuclease gene only in developing anthers *(22)*. Chimeric ribonuclease gene expression within the anther selectively destroyed the tapetal cell layer that surrounds the pollen sac, thus preventing pollen formation, and resulting in male sterility. These binary vectors are also useful in generating transgenic plants with antisense mRNA by inserting a foreign gene in an inverse orientation. Antisense mRNA binds to the corresponding sense mRNA, resulting in reduced level of expression of the gene *(23)*. Therefore, the antisense approach has been widely used to study the functional role of a gene and to eliminate or reduce undesirable characteristics from important crops. Transgenic tomato expressing antisense RNA of 1-aminocyclopropane-1-carboxylate (ACC) synthase, inhibits fruit ripening *(24)*. Ripening of a fruit can be induced by ethyl-

	HindIII	XbaI	SstI	HpaI	KpnI	ClaI	BglII
pGA580	A AGC TTC TAG AGC TCG TTA ACG GTA CCA TCG ATA GAT CT						

	HindIII	XbaI	SstI				BglII
pGA581	AA GC T TCT AGA GCT CG T TCG ATA GAT CT						

	HindIII	XbaI	SstI	HpaI	KpnI		BglII
pGA582	AAG CTT CTA GAG CTC GTT AAC GGT ACC ATC GCG ATA GAT CT						

	HindIII	ScaI	MluI	XhoI	EcoRI	ClaI	BglII
pGA809	AGC TTC AGT ACT ACG CGT CTC GAG AAT TCA TCG ATA GAT CT						

Fig. 3. The sequence of the polylinker regions in the expression vectors.

ene. The rate-limiting step in the biosynthetic pathway of ethylene is the conversion of S-adenosylmethionine to ACC, which is catalyzed by the ACC synthase. When antisense RNA of the ACC synthase gene was expressed by the 35S promoter, 3 of 34 independent transgenic tomato plants showed a marked inhibition in ethylene production and the onset of fruit ripening was delayed.

2.2.5. Binary Vectors for Promoter Analysis

Binary vectors can also be used for studying regulatory elements involved in transcription initiation. The promoter vectors, pGA580, pGA581, pGA582, and pGA809 (Figs. 2 and 3) were developed from a basic binary vector by inserting a promoterless *cat* reporter gene immediately downstream from the multiple cloning sites *(11)*. Insertion of a promoter fragment in front of the reporter gene results in transcription of the *cat* mRNA. If the fragment carries only the promoter region and does not contain the coding region, the *cat* mRNA will produce a native chloramphenical acetyltransferase (CAT) enzyme. If the promoter fragment also contains a portion of the coding region, it is essential to make an in-reading-frame translational fusion between the introduced protein and *cat* reporter so that the start codon belonging to the foreign fragment becomes the start codon of the hybrid fusion protein. Since the CAT assay is very sensitive and most plants do not have endogenous CAT activity, this reporter system has been widely used for studying various promoter elements. However, some plant species contain an inhibitor for the CAT enzyme activity that may be inactivated by brief heating of the plant extract at 55–65°C. A similar set of binary vectors for the GUS reporter gene have also been widely used for promoter analysis *(20)*. In both systems, addition of an extra peptide sequence at the amino terminal ends does not destroy the enzyme activities. However, the fusion proteins are often not as active as the native enzyme.

2.2.6. Binary Vectors for Direct Isolation of Promoter Elements

Ti plasmid vectors have been used for isolation of promoters directly from plant chromosomes. Plant promoters are normally obtained from genomic clones. However, this approach requires several intermediate steps that are tedious and time consuming. Using a binary vector could simplify this procedure because there is no prerequisite to know in advance the product of the gene chosen for study. For example, a binary vector pHTT88 was used to obtain organ-specific promoters from tobacco *(25)*. In this vector, a promoterless *npt* gene is placed immediately next to the T-DNA right border. Therefore, screening for kanamycin resistant transformants will select insertional activation of the silent *npt* gene when the T-DNA is placed downstream of plant transcription signals. The GUS marker is also excellent for this purpose since the expression pattern of a promoter can be easily monitored. For example, the pPRF120 binary vector was constructed to detect T-DNA insertions within transcriptionally active areas of the plant genome *(26)*. In this vector, a promoterless GUS gene was linked to the right border. Screening of leaf tissues from approx 1000 transgenic tobacco plants revealed that about 5% of the plants contained GUS activity.

2.2.7. Binary Vector for Mutagenesis

Insertion of a foreign DNA into a gene generates a mutation and serves as a vehicle for the isolation of the flanking plant DNA. Transposons have been used extensively to tag a variety of plant genes *(27)*. However, transposons have been studied extensively only in a few plants and, therefore, manipulation of a transposon is difficult in many species. Since T-DNA can be introduced into a variety of plants, *Agrobacterium*-mediated plant transformation is an alternative approach in tagging genes especially in *Arabidopsis (28)*. More than 1000 putative mutants were obtained from *Arabidopsis* using the T-DNA tagging method *(29, see* Chapter 25). From these mutants, several genes have been isolated and characterized.

3. Concluding Remarks

Binary vectors have been used to transform a variety of species including several agronomically important plants *(30)*. These are extremely valuable in studying gene regulation and introducing new genetic information into selected crop species. Since the *Agrobacterium*-mediated binary vector transformation method is efficient and easy to handle, this system is widely used for transformation of dicot plants. However, attempts on transformation of most monocot plants have been unsuccessful. It is hoped that a procedure for transformation of monocot plants can be established in the near future.

References

1. Ream, L. W. and Gordon, M. P. (1982) Crown gall disease and prospects for genetic manipulation of plants. *Science* **218,** 854–859.
2. Schell, J. (1987) Transgenic plants as a tools to study the molecular organization of plant genes. *Science* **237,** 1176–1183.
3. Hooykaas, P. J. J. and Schilperoort, R. A. (1992) *Agrobacterium* and plant genetic engineering. *Plant Mol. Biol.* **19,** 15–38.
4. Zambryski, P., Joos, H., Genetello, J., Leemans, J., Van Montagu, M., and Schell, J. (1983) Ti plasmid vector for the introduction of DNA into plant cells without alteration of their normal regeneration capacity. *EMBO J.* **2,** 2143–2150.
5. Waldron, C., Murphy, E. B., Boberts, J. L., Gustafson, G. D., Armour, S. L., and Malcolm, S. K. (1985) Resistance to hygromycin B. A new marker for plant transformation studies. *Plant Mol. Biol.* **5,** 103–108.
6. Holt, J. S., Powles, S. B., and Holtum, A. M. (1993) Mechanisms and agronomic aspect of herbicide resistance. *Annu. Rev. Plant Physiol. Plant Mol. Biol.* **44,** 203–229.
7. Hoekema, A., Hirsch, P. R., Hooykaas, P. J. J., and Schilperoort, R. A. (1983) A binary plant vector strategy based on separation of *vir*- and T-region of the *Agrobacterium tumefaciens* Ti-plasmid. *Nature* **303,** 179,180.
8. Stachel, S. E. and Nester, E. W. (1986) The genetic and transcriptional organization of the *vir* region of the A6 Ti plasmid of *Agrobacterium tumefaciens*. *EMBO J.* **5,** 1445–1454.
9. Stachel, S. E. and Zambryski, P. (1986) *vir*A and *vir*G control the plant-induced activation of the T-DNA transfer process of *Agrobacterium tumefaciens*. *Cell* **46,** 325–333.
10. Douglas, C. J., Halperin, W., and Nester, E. W. (1982) *Agrobacterium tumefaciens* mutants affected in attachment to plant cells. *J. Bacteriol.* **152,** 1265–1275.
11. An, G., Ebert, P. R., Mitra, A., and Ha, S. B. (1988) Binary vectors, in *Plant Molecular Biology Manual* (Gelvin, S. B. and Schilperoort, R. A., eds.), Kluwer, Dordrecht, Netherlands, pp. A3:1–19.
12. Lichtenstein, C. P. and Fuller, S. L. (1987) Vectors for genetic engineering of plants. *Genet. Eng.* **6,** 103–183.
13. Willmitzer, L., Dhaese, P., Schreier, P. H., Schmalenbach, W., Van Montagu, M., and Schell, J. (1983) Size, location and polarity of T-DNA-encoded transcripts in

nopaline crown gall tumors; common transcripts in octopine and nopaline tumors. *Cell* **32,** 1045–1056.

14. An, G., Costa, M. A., and Ha, S.-B. (1990) Nopaline synthase promoter is wound inducible and auxin inducible. *Plant Cell* **2,** 225–233.

15. Benfey, P. N. and Chua, N.-H. (1990) The cauliflower mosaic virus promoter: combinational regulation of transcription in plants. *Science* **250,** 959–966.

16. Yenofsky, R., Fine, M., and Pellow, J. W. (1990) A mutant neomycin phosphotransferase II gene reduces the resistance of transformants to antibiotic selection pressure. *Proc. Natl. Acad. Sci. USA* **87,** 3435–3439.

17. Feldmann, K. A., Marks, M. D., Christianson, M. L., Quatrano, R. S. (1989) A dwarf mutant of *Arabidopsis* generated by T-DNA insertion mutagenesis. *Science* **243,** 1351–1354.

18. Herrera-Estrella, L., Depicker, A., Van Montagu, M., and Schell, J. (1983) Expression of chimeric genes transferred into plant cells using a Ti-plasmid-derived vector. *Nature* **303,** 209–213.

19. Ow, D. W., Wood, K. V., DeLuca, M., de Wet, J. R., Helinski, D. R., and Howell, S. H. (1986) Transient and stable expression of the firefly luciferase gene in plant cells and transgenic plants. *Science* **234,** 856–859.

20. Jefferson, R. A., Kavanagh, T. A., and Bevan, M. W. (1987) GUS fusion: β-glucuronidase as a sensitive and versatile gene fusion marker in higher plants. *EMBO J.* **6,** 3901–3907.

21. Martin, T., Schmidt, R., Altmann, T., and Frommer, W. B. (1992) Non-destructive assay systems for detection of β-glucuronidase activity in higher plants. *Plant Mol. Biol. Rep.* **10,** 37–46.

22. Mariani, C., De Beuckeller, M., Truettner, J., Leemans, J., and Goldberg, R. B. (1990) Introduction of male sterility in plants by a chimeric ribonuclease gene. *Nature* **347,** 737–741.

23. Green, P. J., Pines, O., and Inouye, M. (1986) The role of antisense RNA in gene regulation. *Annu. Rev. Biochem.* **55,** 569–597.

24. Oeller, P. W., Min-Wong, L., Taylor, L. P., Pike, D. A., and Theologis, A. (1991) Reversible inhibition of tomato fruit senescence by antisense RNA. *Science* **254,** 437–439.

25. Teeri, T. H., Herrera-Estrella, L., Depicker, A., Van Montagu, M., and Palva, E. T. (1986) Identification of plant promoters *in situ* by T-DNA-mediated transcriptional fusion to the *npt*-II gene. *EMBO J.* **5,** 1755–1760.

26. Fobert, P. R., Miki, B. L., and Iyer, V. N. (1991) Detection of gene regulatory signals in plants revealed by T-DNA-mediated fusions. *Plant Mol. Biol.* **17,** 837–852.

27. Gierl, A. and Saedler, H. (1992) Plant-transposable elements and gene tagging. *Plant Mol. Biol.* **19,** 39–49.

28. Koncz, C., Nemeth, K., Redri, G. P., and Schell, J. (1992) T-DNA insertional mutagenesis in *Arabidopsis*. *Plant Mol. Biol.* **20,** 963–976.

29. Feldmann, K. A. (1991) T-DNA insertion mutagenesis in *Arabidopsis:* mutational spectrum. *Plant J.* **1,** 71–82.

30. van Wordragen, M. F. and Dons, H. J. M. (1992) *Agrobacterium tumefaciens*-mediated transformation of recalcitrant crops. *Plant Mol. Biol. Rep.* **10,** 12–36.

CHAPTER 7

Leaf Disk Transformation

Ian S. Curtis, Michael R. Davey, and J. Brian Power

1. Introduction

Reliable and efficient methods of transferring cloned genes into plants are essential for engineering crops with desired traits. The Gram-negative soil bacteria, *Agrobacterium tumefaciens* and *A. rhizogenes*, are natural genetic engineers, capable of transforming a range of dicotyledonous plants by transferring plasmid-encoded genes into recipient plant genomes. Stable integration of these bacterial genes into the plant genome interferes with normal plant development, which is manifest as crown gall and hairy root diseases, respectively. Co-integration of chimeric genes into the T-DNA between the border sequences of wild-type Ti (tumor-inducing) or Ri (root-inducing) plasmids enabled such genes to be introduced into plant genomes. However, such crown gall tumors usually failed to regenerate plants capable of developing functional root systems, whereas plants regenerated from hairy roots exhibited the characteristic phenotype of shortened internodes, wrinkled leaves, and reduced fertility, together with negatively geotropic and/or plagiotropic roots. Replacement of the oncogenes located between the T-DNA borders of Ti plasmids or the *rol* genes of Ri plasmids with DNA sequences from *E. coli* cloning vectors, resulted in "disarmed" Ti or Ri plasmids. Introduction, by homologous recombination, of chimeric genes for antibiotic resistance into such disarmed vectors enabled transformed plant cells or roots to be selected on the basis of their antibiotic resistance. Plants regenerated from such tissues and transformed roots were pheno-

From: *Methods in Molecular Biology, Vol. 44:* Agrobacterium *Protocols*
Edited by: K. M. A. Gartland and M. R. Davey Humana Press Inc., Totowa, NJ

Table 1
Examples of Transgenic Plants Produced Using
A. *tumefaciens*-Mediated Transformation of Leaf Disks

Genus, species, cultivar	Ref.
Arabidopsis thaliana (Thale Cress) C24	20
Brassica napus (Oilseed Rape) cv. Westar	7
Dendranthema (Syn. *Chrysanthemum*) *indicum* (Chrysanthemum) Des Moul	21
Glycine max (Soybean) cv. Peking	22
Kalanchoë laciniata	11
Lactuca sativa (Lettuce) cv. Cobham Green	10
Lycopersicon esculentum (Tomato) cvs. Beefmaster, Betterboy, Roma	6
Morus alba (Mulberry) cv. Ohyutaka	23
Nicotiana glauca (Tobacco)	24
Nicotiana tabacum (Tobacco) cvs. Samson, Havana 425	4
Petunia hybrida (Petunia) cv. Mitchell	25
Phaseolus vulgaris (Green bean) cv. Dark Red Kidney	26
Solanum melongena (Aubergine) cv. Picentia	27
Solanum tuberosum (Potato) line NDD-277-2	28

typically normal and able to grow on their own roots *(1,2)*. The development of small binary vectors, in which the T-DNA is separated on an independent replicon from the remainder of the Ti or Ri plasmid, facilitated foreign gene insertion between the T-DNA borders *(1)*.

Protoplast-derived cells, callus and a variety of explants taken from leaves, cotyledons, hypocotyls, and roots have been used for transformation. In the case of explants, the procedure relies on wounded cells at the periphery of the explant being competent to receive genes delivered by *Agrobacterium* and to develop shoot buds with the minimum of callus formation. Since the first successful leaf disk transformations using tobacco, petunia, and tomato *(3; see* Section 3.1.), the method has been applied successfully to other members of the *Solanaceae* and, with various modifications, to species from other plant families (Table 1).

Studies have been performed to optimize leaf disk transformation for transgenic plant production. Seedling cotyledons provide a convenient,

readily available source of material; mature leaves can be difficult to sterilize unless the plants are grown under axenic conditions. There are conflict-
ing views on the transformation efficiency of different leaf types. For
example, in tomato, cotyledons and older true leaves were more readily
transformed than younger leaves *(4)*. In *Lotus corniculatus (5)*, young leaves
were more amenable to *Agrobacterium*-mediated gene delivery, although
the morphogenic potential of cotyledons was greater than mature leaves.
When disks are prepared from leaves excized from glasshouse-grown
plants of tomato, transformation and subsequent shoot regeneration from
such material may be influenced by the season *(6)*. In *Brassica napus*, the
presence of an intact petiole greatly increases the efficiency of transgenic
shoot production, which may be related to the regenerative potential of
petiole tissues and their susceptibility to *Agrobacterium* infection *(7)*. In
relation to explant size, there appears to be a requirement for a minimal
surface area per explant in relation to the number of shoots produced *(8)*.

Some investigators consider co-integrative (cis) vectors to be more
efficient in transforming plants compared to binary (trans) vectors *(6,9)*,
although such responses may result from the testing of limited numbers
of vectors, plant genotype-specificity and the effects of culture and physiological conditions. Bacterial concentration is considered to be a critical
variable in the transformation of lettuce *(10)*. In the presence of high
concentrations of bacteria, plant cells become stressed; in the presence
of a restricted number of bacteria, transgenic shoot production is low
because of an insufficient number of plant cells being transformed. Prolonging the time that Agrobacteria are in contact with the leaf tissues in
the absence of antibiotics (coculture), can have different effects depending on plant species. Generally, the coculture period is 2–4 d, after which
bacterial overgrowth can become a problem. However, in *Kalanchoë
laciniata*, prolonging the coculture period up to 7 d results in an increase
in both the number of transformed explants and the percentage area of
transformed tissue *(11)*. The use of nurse cultures during coculture can
increase gene transfer and improve the transformation efficiency by minimizing stress experienced by cultured leaf explants. Nurse cultures may
also stimulate the *vir* genes located on the *Agrobacterium* Ti or Ri plasmids and so increase bacterial virulence. Increasing the virulence of
Agrobacterium toward the target plant has been of benefit through the use of
the hypervirulent plasmid pTOK47 *(12)* and the *vir*-inducing phenolic
compound, acetosyringone *(13)*. Acetosyringone appears to improve

transformation efficiency only in those plants, such as *Arabidopsis*, that naturally produce insufficient amounts of *vir*-inducing compounds *(14)*. The pH of the cocultivation medium is also an important factor in optimizing the transformation response *(15)*. The potential for modifications to existing techniques to further improve transformation efficiencies in a range of dicotyledons is promising, although one procedure will probably not be universal for all species.

This chapter outlines the method of leaf disk transformation first reported using *A. tumefaciens* strain GV3 Ti11SE *(3)* and illustrates the way in which modifications have been made to the procedure for transforming all groups of lettuce.

2. Materials

2.1. General Procedure

2.1.1. A. tumefaciens *Strain*

A. tumefaciens strain GV3 Ti11SE containing a modified octopine Ti plasmid (pTiB6S3SE)::pMON 200. The cointegrative vector pMON 200 has the gene for neomycin phosphotransferase II (*npt*II), which can be used as a selectable marker because it encodes, in transformed plant cells, resistance to kanamycin sulfate and Geneticin (G418). The construct also carries the nopaline synthase (*nos*) coding sequence. Both *npt*II and the *nos* genes are under the control of the *nos* promoter and terminator.

2.1.2. Culture Media for A. tumefaciens

1. Luria broth (LB) medium: 10 g/L Oxoid Bacto tryptone, 5 g/L Oxoid Bacto-yeast extract, 10 g/L NaCl, 18 g/L agar (Sigma, St. Louis, MO), pH 7.2.
2. Kanamycin sulfate (Sigma): 10 mg/mL stock in water. Sterilize by filtration through a 0.2-μm membrane (Minisart, Epsom, UK). Dispense 20-mL aliquots into screw-capped bottles and store at –20°C.
3. Streptomycin sulfate (Sigma): 25 mg/mL stock in water. Sterilize by filtration and store at –20°C.
4. Chloramphenicol (Sigma): 10 mg/mL stock in absolute ethanol. Sterilize by filtration and store at –20°C.
5. Liquid culture medium: LB medium containing 50 mg/L kanamycin sulfate, 25 mg/L streptomycin sulfate, and 25 mg/L chloramphenicol.

2.1.3. Tissue Culture

1. Seeds of *Nicotiana tabacum* cvs. Samson and Havana 425, *Petunia hybrida* cv. Mitchell, and *Lycopersicon esculentum* cv. L2.

2. B5 vitamin stock *(16)*: 100 mg/mL *myo*-inositol, 10 mg/mL thiamine-HCl, 1 mg/mL nicotinic acid, 1 mg/mL pyridoxine-HCl.
3. MSB medium: 4.3 g/L Murashige and Skoog (MS) salts *(17)* (Flow Laboratories, Irvine, Scotland), 1 mL/L B5 vitamin stock solution, 30 g/L sucrose, 8 g/L Difco-Bacto agar, pH 5.7.
4. Half-strength MSB medium: 2.15 g/L MS salts, 1 mL/L B5 vitamins, 30 g/L sucrose, 8 g/L agar, pH 5.7.
5. α-Naphthalene acetic acid (NAA, Sigma): 1 mg/mL stock solution. Prepare by dissolving the powder in 70% vol of absolute alcohol and add double distilled water to make up to vol. Store at 4°C.
6. 6-Benzylaminopurine (BAP, Sigma): 50 mg/100 mL stock solution. Prepare by dissolving in 1 mL of 1N HCl, before making to 100 mL with water.
7. MS104 medium: MSB with 1 mg/L BAP, 0.1 mg/L NAA.
8. MS selection medium: MS 104 with 500 mg/L carbenicillin, 300 mg/L kanamycin sulfate.
9. MS rooting medium: MSB with 6 g/L agar (Sigma), 500 mg/L carbenicillin, 100 mg/L kanamycin sulfate.
10. Carbenicillin (Pyopen; Beechams Research Laboratories, Brentford, UK), 100 mg/mL aqueous stock solution. Sterilize by filtration through a 0.2-μm membrane and store at –20°C.
11. Kanamycin sulfate (Sigma): *see* Section 2.1.2., item 2.
12. MS suspension medium: 4.3 g/L MS salts, 1 mL/L B5 vitamins, 30 g/L sucrose, 4 mg/L *p*-chlorophenoxyacetic acid, 0.5 mg/L kinetin (Sigma), pH 5.7. Dispense 50-mL aliquots into 250-mL flasks.
13. Surface sterilant: 10% (v/v) "Domestos" bleach (Lever Bros., Kingston-upon-Thames, UK).

2.2. Lettuce
2.2.1. A. tumefaciens *Strain*

A. tumefaciens strain LBA4404 containing the binary vector pMOG23 *(18)* and hypervirulent pTOK47 *(12)*. The binary vector has both the *npt*II and β-glucuronidase *(gus)*-intron *(19)* coding sequences, driven by the *nos* and CAMV-35S promoters, respectively. This resulting strain is described as 1065.

2.2.2. Culture Medium for A. tumefaciens

1. LB medium: *see* Section 2.1.2., item 1, but adjusted to pH 7.0.
2. Rifampicin (Sigma): 4 mg/mL stock in methanol. Sterilize by filtration through a 0.2-μm membrane and store at –20°C.
3. Kanamycin sulfate (Sigma): *see* Section 2.1.2., item 2.

4. Tetracycline-HCl (Sigma): 5 mg/mL stock in water. Sterilize by filtration and store at 4°C.
5. Liquid culture medium: LB medium (*see* Section 2.1.2., item 1), with 50 mg/L kanamycin sulfate, 40 mg/L rifampicin, and 2 mg/L tetracycline-HCl.
6. Agar culture medium: LB medium (*see* Section 2.1.2., item 1), with 50 mg/L kanamycin sulfate, 100 mg/L rifampicin, 5 mg/L tetracycline-HCl.

2.2.3. Tissue Culture

1. Seeds of *Lactuca sativa* L. cvs. Lake Nyah, Cortina, Luxor.
2. Germination medium: MS salts and vitamins (Flow Laboratories), 10 g/L sucrose, 8 g/L agar (Sigma), pH 5.9.
3. UM medium: 4.71 g/L MS salts and vitamins, 30 g/L sucrose, 2 g/L casein hydrolysate, 2 mg/L 2,4-dichlorophenoxyacetic acid (2,4-D; Sigma), 0.25 mg/L kinetin, 9.9 mg/L thiamine-HCl, 9.5 mg/L pyridoxine-HCl, 4.5 mg/L nicotinic acid, 8 g/L agar (Sigma), pH 5.8. Omit the agar from liquid medium.
4. NAA: 1 mg/mL stock (*see* Section 2.1.3., item 5).
5. BAP (Sigma): 1 mg/mL stock. Prepare by dissolving the powder in a few drops of concentrated HCl and adding double distilled water to make to vol. Store at 4°C.
6. Cefotaxime (Claforan, Roussel Laboratories, Wembley, UK): 10 mg/mL aqueous stock. Sterilize by filtration and store at –20°C.
7. Carbenicillin (*see* Section 2.1.3., item 10).
8. Shoot-inducing (SI) medium: 4.71 g/L MS salts and vitamins, 30 g/L sucrose, 0.04 mg/L NAA, 0.5 mg/L BAP, 500 mg/L carbenicillin, 100 mg/L cefotaxime, 50 or 100 mg/L kanamycin sulfate, 8 g/L agar (Sigma), pH 5.8.
9. Rooting medium: 4.71 g/L MS salts and vitamins, 30 g/L sucrose, 50 or 100 mg/L kanamycin sulfate, 2.5 g/L Phytagel (Sigma), pH 5.8.
10. Surface sterilant: 10% (v/v) "Domestos" bleach (Lever Bros.).

All media are sterilized by autoclaving at 121°C for 20 min. Antibiotics are added to the medium after autoclaving (40°C for agar media).

3. Methods
3.1. General Procedure

1. Soak seeds in 10% (v/v) "Domestos" bleach solution for 20–30 min, followed by three rinses in sterile water. Retaining seeds on a 64-μm nylon mesh facilitates handling.
2. Sow seeds onto the surface of 25-mL aliquots of MSB agar medium contained in 9-cm diameter Petri dishes (Sterilin) and grow under moderate light intensity (48 μmol/s/m^2, 16 h photoperiod, daylight fluorescent tubes) at 24–28°C until cotyledons have fully expanded. Sow tomato seeds on half-strength MSB agar medium.

3. Excise cotyledons with their petioles intact from 7-d-old seedlings and cut the cotyledons in half to increase the wounded surface. For tomato, wound the leaf using fine forceps or a dissecting needle on the abaxial surface (*see* Note 1).

4. Place explants wounded-surface down on MS104 medium. Incubate for 1–2 d at a low light intensity (24 µmol/s/m^2, 16 h photoperiod, daylight fluorescent tubes) at 24–26°C, to encourage cell division at wound sites (*see* Note 2).

5. Take an overnight liquid culture of *A. tumefaciens* and dilute 1:10 (v:v) with MSB liquid medium. For tomato, a 1:20 (v:v) dilution giving 5 × 10^8 bacterial cells/mL is required.

6. Immerse the explants in the *A. tumefaciens* culture until all the wounded surfaces have been soaked; blot dry on sterile filter paper. Tomato leaves need to be soaked for 5 min to ensure inoculation (*see* Note 3).

7. Prepare nurse culture plates by distributing 1–1.5 mL vol of 5-d-old, log phase *Nicotiana tabacum* suspension cells over 25-mL aliquots of MS104 agar medium contained in 9-cm diameter Petri dishes (*see* Note 4). For the transformation of tomato, a 90% reduction in MS salts contained in MS104 is required. Swirl the suspension to cover the surface of the agar medium. Place a sterile 8.5-cm diameter Whatman (Maidstone, UK) filter paper (guard disc) over the cells; overlay the disc with another sterile 7.0-cm Whatman filter paper (transfer disc).

8. Place explants upside down on the transfer disc in the nurse culture plates and incubate (24–26°C) at a low light intensity (24–48 µmol/s/m^2, 16 h photoperiod, daylight fluorescent tubes) for 2–3 d. Check plates regularly for bacterial growth (*see* Note 5).

9. Place the transfer disc carrying the explants onto MS selection medium and incubate as before (*see* Note 6).

10. Shoots should begin to develop about 18 d after inoculation. Excise the shoots and transfer to MS rooting medium with selection.

11. When roots first appear, remove the plants, wash agar from their roots, and pot the plants into 7-cm diameter pots containing Levington M3 compost (Fisons Ltd., Ipswich, UK). Place pots in propagating trays, cover with plastic domes, and close the vents to retain humidity.

12. After 7 d, progressively open the vents until the plants have acclimated to the ambient humidity. Transfer the plants to the glasshouse and grow to maturity.

3.2. Lettuce

1. Immerse lettuce seeds in 10% (v/v) "Domestos" for 30 min and rinse twice with sterile distilled water.

2. Sow seeds onto 20-mL aliquots of germination medium contained in 9-cm diameter Petri dishes (40 seeds/dish). Incubate the dishes at 23 ± 2°C, 16 h

photoperiod, with a light intensity of 18 μmol/s/m^2 (daylight fluorescent tubes) for 7d.

3. Grow *A. tumefaciens* strain 1065 on LB agar culture medium (with antibiotics) at 28°C for 2–3 d. Take one loopful of bacteria and transfer to 10 mL of LB liquid culture medium (with antibiotics) and grow overnight at 28°C, in the dark, on an orbital shaker (180 rpm).

4. Pour 20-mL aliquots of UM agar medium into 9-cm diameter Petri dishes and allow to solidify.

5. Soak one sterile 7.0-cm diameter Whatman filter paper in liquid UM and place onto the surface of the UM agar medium.

6. Excise the cotyledons from 7-d-old seedlings, leaving the petiole intact, but removing the apices of the cotyledons (*see* Note 7). Score the abaxial surface, using a scapel, making shallow cuts (1-mm apart) transversely across the surface of the leaf. Float the cotyledons (10 min) with their wound-surfaces in contact with the surface of an overnight liquid culture of *Agrobacterium*, diluted 1:10 (v:v) with liquid UM medium. Control cotyledons should be scored and floated on liquid UM medium without *Agrobacterium*.

7. Remove the cotyledons, blot dry with sterile filter paper, and transfer to the prepared UM dishes (eight cotyledons/dish). Place one, sterile, dry, filter paper over the cotyledons to keep the explants flat. Incubate the explants for 2 d under the same conditions as for germinating seeds.

8. Set up test plates as follows:
   a. Control explants without *Agrobacterium* inoculation on:
      i. SI medium;
      ii. SI medium + 50 mg/L kanamycin sulfate; and
      iii. SI medium + 100 mg/L kanamycin sulfate.
   b. Explants inoculated with *Agrobacterium* on:
      i. SI medium + 50 mg/L kanamycin sulfate; and
      ii. SI medium + 100 mg/L kanamycin sulfate.

9. Transfer the explants to SI agar medium, submerging the petiolar ends of the cotyledons into the medium to a depth of about 2 mm. Incubate as for germinating seeds and subculture to fresh SI agar medium every 14 d (*see* Note 8). Shoots should appear in the treatments given in step 8 (Fig 1; *see* Note 8).

10. After 49 d, transfer those explants that have produced callus and shoots (Fig. 2) to 175-mL capacity glass jars (Beatson-Clarke, Rotherham, UK), each containing 40 mL of SI agar medium with 0.11% (w/v) 2[*N*-morpholino]ethanesulfonic acid (MES; Sigma) without carbenicillin (one explant/jar) (*see* Note 9).

11. Transfer shoots when approx 1 cm high to 175-mL jars each containing 40 mL of rooting medium (one to three shoots/jar), or to Magenta boxes

Figs. 1–3. (**Fig. 1**) Shoot regeneration from a wounded lettuce cotyledon cultured on SI medium with 50 mg/L kanamycin sulfate, 35 d post-inoculation. (**Fig. 2**) Regeneration of shoots from a wounded lettuce cotyledon inoculated with *A. tumefaciens* strain 1065, 49 d post-inoculation. (**Fig. 3**) A cotyledon-derived lettuce plant cultured on rooting medium.

(Sigma) each with 90 mL of medium. Incubate shoots at $20 \pm 2°C$, 12 h photoperiod, with a light intensity of 18 μmol/s/m^2 (daylight fluorescence tubes) (*see* Notes 10–14).

12. When rooted (Fig. 3), carefully remove the plants from the containers, wash the agar from the roots, and transfer individual plants to 9-cm diameter pots, containing a mixture of Levington M3 compost (Fisons), John Innes No. 3 compost (J. Bentley Ltd., Barrow-on-Humber, UK) and Perlite (Silvaperl, Gainsborough, UK) (3:3:2 [v:v:v]). Enclose plants in clear polythene bags and incubate at a light intensity of (29 μmol/s/m^2, 16 h photoperiod, daylight fluorescent tubes, $23 \pm 2°C$, 7 d). Remove one corner of each bag and a second corner 7 d later. After 21 d, transfer the plants to the glasshouse and grow to maturity. Assay regenerated plants for GUS and NPTII activities (*see* Note 15 and Chapters 17–19).

4. Notes

1. Instead of cotyledons as explants, newly expanded, healthy, unblemished leaves from glasshouse-grown plants may be used. Young leaves from 3-mo-old tobacco and petunia plants are ideal, although for tomato, the two subterminal leaflets from the 4th, 5th, and 6th nodes of 6-wk-old plants are preferable. Submerge the leaves in 10% (v/v) "Domestos" bleach contained in a sterile casserole dish, removing air bubbles from the leaf surfaces by agitation with a sterile spatula; leave for 15 min. Rinse leaves four times with sterile water (400 mL). Remove any white, bleach-damaged areas and the midrib with a scapel. Use a sterile paper punch or cork borer to produce leaf disks, each with a wounded edge.

2. During preculture of the explants on MS104 medium, eliminate any that show signs of having been damaged during sterilization or handling and those that fail to swell at the wounded edge.

3. Excessive soaking of the explants with the bacterial culture should be avoided to prevent extensive cell damage.

4. The use of nurse cultures is not essential for transformation, but they may increase the transformation efficiency.

5. A longer coculture period may improve transformation efficiency, but bacterial overgrowth may become a problem.

6. Explants often become twisted during culture, resulting in the cut edges losing contact with the medium. These disks should be sliced and the cut edges pressed about 2 mm into the agar medium to reduce the development of nontransformed shoots.

7. Select for cotyledons of even size for transformation to ensure uniformity both between and within treatments.

8. Shoots should appear from the explants about 35 d post-inoculation.

9. Kanamycin sulfate at a concentration of 50 mg/L is the lowest concentration required to inhibit growth of control cotyledons. Control cotyledons cultured without antibiotics produce many adventitious shoots (20–30 shoots/explant), although there are differences in the shoot regeneration response between cultivars.

10. Strains of *A. tumefaciens* carrying the hypervirulent plasmid pTOK47 induce the production of more shoots compared to strains lacking pTOK.

11. Kanamycin sulfate at 100 mg/L is more efficient for selection of transformed shoots compared to a concentration of 50 mg/L.

12. The number of adventitious shoots and the proportion of transgenic plants varies with different cultivars. The number of transgenic shoots from the total shoot population ranges from 14–62%.

13. Nurse cultures are of benefit to the transformation of some lettuce cultivars.

14. A 1:10 (v:v) dilution of an overnight culture of *A. tumefaciens* results in more transformants compared to a 1:1 dilution, especially when the strain carries pTOK47.

15. GUS positive plants should also exhibit NPTII activity following transformation by *A. tumefaciens* strain 1065. Plants regenerated from cotyledons not infected with *Agrobacterium* should not show GUS or NPTII activities.

References

1. Zambryski, P., Joos, H., Genetello, C., Leemans, J., Van Montagu, M., and Schell, J. (1983) Ti plasmid vector for the introduction of DNA into plant cells without alteration of their normal regeneration capacity. *EMBO J.* **2**, 2143–2150.

2. McInnes, E., Davey, M. R., Mulligan, B. J., Davies, K., Sargent, A. W., and Morgan, A. J. (1989) Use of a disarmed Ri plasmid vector in analysis of transformed root induction. *J. Exp. Bot.* **40**, 1135–1144.

3. Horsch, R. B., Fry, J. E., Hoffmann, N. L., Eichholtz, D., Rogers, S. G., and Fraley, R. T. (1985) A simple and general method for transferring genes into plants. *Science* **227**, 1229–1231.

4. Davis, M. E., Lineberger, R. D., and Miller, A. R. (1991) Effects of tomato cultivar, leaf age, and bacterial strain on transformation by *Agrobacterium tumefaciens*. *Plant Cell Tissue Org. Cult.* **24**, 115–121.

5. Armstead, I. P. and Webb, K. J. (1987) Effect of age and type of tissue on genetic transformation of *Lotus corniculatus* by *Agrobacterium tumefaciens*. *Plant Cell Tissue Org. Cult.* **9**, 95–101.

6. McCormick, S., Niedermeyer, J., Fry, J., Barnason, A., Horsch, R., and Fraley, R. (1986) Leaf disc transformation of cultivated tomato (*L. esculentum*) using *Agrobacterium tumefaciens*. *Plant Cell Rep.* **5**, 81–84.

7. Moloney, M. M., Walker, J. M., and Sharma, K. K. (1989) High eficiency transformation of *Brassica napus* using *Agrobacterium* vectors. *Plant Cell Rep.* **8**, 238–242.

8. Beck, M. J. and Camper, N. D. (1991) Shoot regeneration from petunia leaf disks as a function of explant size, configuration and benzyladenine exposure. *Plant Cell Tissue Org. Cult.* **26**, 101–106.

9. Guri, A. and Sink, K. C. (1988) *Agrobacterium* transformation of eggplant. *J. Plant Physiol.* **133**, 52–55.

10. Michelmore, R., Marsh, E., Seely, S., and Landry, B. (1987) Transformation of lettuce (*Lactuca sativa*) mediated by *Agrobacterium tumefaciens*. *Plant Cell Rep.* **6**, 439–442.

11. Jia, S. R., Yang, M. Z., Ott, R., and Chua, N.-H. (1989) High frequency transformation of *Kalanchoë laciniata*. *Plant Cell Rep.* **8**, 336–340.

12. Jin, S., Komari, T., Gordon, M. P., and Nester, E. W. (1987) Genes responsible for the supervirulence phenotype of *Agrobacterium tumefaciens* A281. *J. Bact.* **169**, 4417–4425.

13. Holford, P., Hernandez, N., and Newbury, H. J. (1992) Factors influencing the efficiency of T-DNA transfer during co–cultivation of *Antirrhinum majus* with *Agrobacterium tumefaciens*. *Plant Cell Rep.* **11**, 196–199.

14. Sheikholeslam, S. N. and Weeks, D. P. (1987) Acetosyringone promotes high efficiency transformation of *Arabidopsis thaliana* explants by *Agrobacterium tumefaciens*. *Plant Mol. Biol.* **8**, 291–298.
15. Godwin, I., Todd, G., Ford-Lloyd, B., and Newbury, H. J. (1991) The effects of acetosyringone and pH on *Agrobacterium*-mediated transformation vary according to plant species. *Plant Cell Rep.* **9**, 671–675.
16. Gamborg, O. L., Miller, R. A., and Ojima, K. (1968) Nutrient requirements of a suspension culture of soybean root cells. *Exp. Cell Res.* **50**, 151–158.
17. Murashige, T. and Skoog, F. (1962) A revised medium for rapid growth and bioassays with tobacco tissue cultures. *Physiol. Plant.* **15**, 473–497.
18. Sijmons, P. C., Dekker, B. M. M., Schrammeijer, B., Verwoerd, T. C., Van den Elzen, P. J. M., and Hoekema, A. (1990) Production of correctly processed human serum albumin in transgenic plants. *Bio/Technology* **8**, 217–221.
19. Vancanneyt, G., Schmidt, R., O'Connor-Sanchez, A., Willmitzer, L., and Rocha-Sosa, M. (1990) Construction of an intron-containing marker gene: splicing of the intron in transgenic plants and its use in monitoring early events in *Agrobacterium*–mediated plant transformation. *Mol. Gen. Genet.* **220**, 245–250.
20. Van Lijsebettens, M., Vanderhaeghen, R., and Van Montagu, M. (1991) Insertional mutagenesis in *Arabidopsis thaliana*: isolation of a T-DNA-linked mutation that alters leaf morphology. *Theor. Appl. Genet.* **81**, 277–284.
21. Ledger, S. E., Deroles, S. C., and Givens, N. K. (1991) Regeneration and *Agrobacterium*-mediated transformation of chrysanthemum. *Plant Cell Rep.* **8**, 336–340.
22. Owens, L. D. and Smigocki, A. N. (1988) Transformation of soybean cells using mixed strains of *Agrobacterium tumefaciens* and phenolic compounds. *Plant Physiol.* **88**, 570–573.
23. Machii, H. (1990) Leaf disc transformation of mulberry plant (*Morus alba* L.) by *Agrobacterium* Ti plasmid. *J. Sericult Sci. Jpn.* **59**, 105–110.
24. Mozo, T. and Hooykaas, P. J. J. (1992) Factors affecting the rate of T-DNA transfer from *Agrobacterium tumefaciens* to *Nicotiana glauca* plant cells. *Plant Mol. Biol.* **19**, 1019–1030.
25. Janssen, B. J. and Gardner, R. C. (1989) Localised transient expression of GUS in leaf discs following cocultivation with *Agrobacterium*. *Plant Mol. Biol.* **14**, 61–72.
26. Franklin, C. I., Trieu, T. N., Cassidy, B. G., Dixon, R. A., and Nelson, R. S. (1993) Genetic transformation of green bean callus via *Agrobacterium* mediated DNA transfer. *Plant Cell Rep.* **12**, 74–79.
27. Rotino, G. L. and Gleddie, S. (1990) Transformation of eggplant (*Solanum melongena* L.) using a binary *Agrobacterium tumefaciens* vector. *Plant Cell Rep.* **9**, 26–29.
28. Shahin, E. A. and Simpson, R. B. (1986) Gene transfer system for potato. *Hort. Sci.* **21**, 1199–1201.

CHAPTER 8

Agrobacterium-Mediated Transformation of Protoplasts from Oilseed Rape (*Brassica napus* L.)

Jürgen E. Thomzik

1. Introduction

Although a number of protocols are available for the regeneration of oilseed rape protoplasts *(1–4)*, there are almost no reports of successful *Agrobacterium tumefaciens*-mediated transformation of protoplasts resulting in transgenic plants. Plant regeneration from protoplast-derived calli of *Brassica napus* generally occurs at a low frequency and depends on the genotype used *(4–5)*. In addition, shoot regeneration is significantly reduced in protoplast-derived calli after cocultivation and kanamycin selection *(6)*. The methods for the transformation and regeneration of *B. napus* hypocotyl protoplasts described in this chapter are suitable for obtaining transgenic plants by an *A. tumefaciens*-mediated gene transfer. However, a prerequisite is the use of oilseed rape genotypes possessing a sufficient shoot regeneration capability, more than 20%, and the application of a three-step kanamycin selection. In this protocol a marker gene coding for kanamycin resistance is introduced into hypocotyl protoplasts of winter-type oilseed rape. Genes, possessing relevant qualities for improving the agronomic character of oilseed rape varieties, such as stilbene synthase and T4 phage lysozyme, have also been cointroduced together with a chimeric kanamycin resistance gene using this protocol *(6)*.

From: *Methods in Molecular Biology, Vol. 44:* Agrobacterium *Protocols*
Edited by: K. M. A. Gartland and M. R. Davey Humana Press Inc., Totowa, NJ

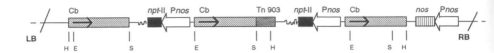

Fig. 1. Integration of *E. coli* plasmid pLGV1103neo(dim) into the Ti plasmid pGV3850: (/) left and right limit (LB and RB) of the Ti-plasmid specific sequences; (▨) pBR322-derived sequences, carbenicillin-resistance gene (Cb); (▨) represents a 1.2-kb-long Tn903-specific fragment; (■) the neomycin phosphotransferase-II coding region (*npt*II) from Tn5; (‑‑‑) polyadenylation signal isolated from octopine synthase gene; (⇐) nopaline synthase promoter region (Pnos); (▥) the coding region of the nopalin synthase gene *(nos)*; (|) restriction sites H, E, and S of *Hin*dIII, *Eco*RI, *Sal*I.

2. Materials

2.1. *Plants and* Agrobacterium *Strain*

1. Plant material: Winter type oilseed rape of double-low quality *(7)* line R117 (Kartoffelzucht Boehm KG, Langquaid, Germany) and cv. Ceres (Norddeutsche Pflanzenzucht Lembke KG, Holtsee, Germany).
2. *A. tumefaciens*: Strain C58CI harboring pVG3850:: 1103 neo dimer *(8)* is used (*see* Note 1). This disarmed Ti-plasmid contains the nopaline synthase gene, two tandem chimeric *npt*II genes flanked at the 5' end by the nopaline synthase promoter, and at the 3' end by the polyadenylation site of the octopine synthase gene conferring resistance to kanamycin in plants (Fig. 1).

2.2. *Solutions*

1. 70% (v/v) Ethanol.
2. Sodium hypochlorite solution (NaOCl, 13% active chlorine): Solution of 6% is prepared by dilution with water and the addition of 3 drops/L of Tween-80 as wetting agent.
3. ENZ 1/3/5: Enzyme solution consisting of 1% (w/v) Cellulase R-10, 0.3% (w/v) Macerozyme R-10, and 0.05% (w/v) Pectolyase Y-23 dissolved in W 13M, pH 5.5.
4. W 13M salt solution: 1500 mg/L $CaCl_2 \cdot 2H_2O$, 250 mg/L $MgSO_4 \cdot 7H_2O$, 100 mg/L KNO_3, and 30 mg/L KH_2PO_4, supplemented with 13% (w/v) mannitol, pH 5.5.
5. W 22S: W 13M salt solution but supplemented with 22% (w/v) sucrose instead of mannitol.
6. W5 salt solution *(9)*: 125 m*M* $CaCl_2$, 155 m*M* NaCl, 5 m*M* KCl, 5 m*M* glucose, pH 5.6.

2.3. Nutrient Media and Supplements

1. YEB medium (per L): 1 g yeast extract, 5 g beef extract, 5 g peptone, 0.5 g $MgSO_4 \cdot 7H_2O$, 5 g sucrose, pH 7.2.
2. Antibiotics: 50 mg/mL stocks of kanamycin acid sulfate and cefotaxime in distilled water, filter sterilized; 50 mg/mL stock of rifampicin in ethanol. Stocks for cefotaxime and rifampicin should be freshly prepared for each use.
3. Growth regulators: All growth regulators are prepared as stocks of 0.2 mg/ mL, filter sterilized, and stored in the refrigerator. 6-Benzylaminopurine (BAP) and zeatin have to be dissolved in some drops of $0.5M$ HCl before the volume is made up by adding distilled water. 2,4-Dichlorphenoxyacetic acid (2,4D), naphthaleneacetic acid (NAA), and 3-indole acetic acid (IAA) have to be dissolved in some drops of ethanol before the final volume is made up by adding distilled water.
4. Half-strength Murashige and Skoog (MS) medium: half-strength MS salts and vitamins *(10)* supplemented with 15 g/L sucrose and 0.25% (w/v) Gelrite, pH 5.8.
5. Protoplast culture (PC) medium: A modified 8p medium *(11)* containing, per liter, 1900 mg KNO_3, 600 mg NH_4NO_3, 600 mg $CaCl_2 \cdot 2H_2O$, 300 mg $MgSO_4 \cdot 7H_2O$, 300 mg KCl, 170 mg KH_2PO_4, 37 mg Na_2EDTA, 28 mg $FeSO_4 \cdot 7H_2O$, 10 mg $MnSO_4 \cdot H_2O$, 3 mg H_3BO_3, 2 mg $ZnSO_4 \cdot 7H_2O$, 0.75 mg KI, 0.25 mg $Na_2MoO_4 \cdot 2H_2O$, 0.025 mg $CuSO_4 \cdot 5H_2O$, 0.025 mg $CoCl_2 \cdot 6H_2O$, 100 mg m-inositol, 10 mg thiamine $\cdot$ HCl, 2 mg ascorbic acid, 1 mg nicotinic acid, 1 mg pyridoxine $\cdot$ HCl, 40 mg malic acid, 40 mg citric acid, 40 mg fumaric acid, 20 mg Na-pyruvate, 0.25 mg each rhamnose, sucrose, cellobiose, mannose, sorbitol, fructose, mannitol, ribose, xylose, 68.4 g glucose, 20 mL/L coconut water, 1 mg 2,4D, 0.1 mg NAA, 0.1 mg BAP, pH 5.8, filter sterilized.
6. 2X PC medium: double-concentrated PC-medium.
7. PC agarose medium: Autoclave 2.4% (w/v) agarose (Sea Plaque or Sigma [St. Louis, MO] Type VII). Immediately before use melt the agarose, cool down to 45°C, and mix with an equal volume of prewarmed 2X PC medium containing 1000 µg/mL cefotaxime. Keep the PC agarose medium at 40°C in a water bath until mixed with protoplasts.
8. Callus culture (CC) medium: MS salts and vitamins supplemented with 0.1 mg/L 2.4D, 0.1 mg/L NAA, 0.5 mg/L BAP, 20 g/L sucrose, and 500 mg/L cefotaxime, pH 5.8 (for solidification use 0.6% [w/v] agarose, Sigma Type I).
9. Shoot regeneration (SR) medium: MS salts and vitamins supplemented with 0.1 mg/L IAA, 1.0 mg/L BAP, 2.0 mg/L zeatin, 15 g/L sucrose, 250 mg/L cefotaxime, and 0.3% (w/v) Gelrite or 0.8% (w/v) agar (Difco, Detroit, MI), pH 5.8.

10. Shoot multiplication I (SMI) medium: MS salts and vitamins supplemented with 4 mg/L BAP, 0.2 mg/L NAA, 20 g/L sucrose, 250 mg/L cefotaxime, 15 mg/L kanamycin acid sulfate, and 0.8% (w/v) agar, pH 5.8.

11. Rooting (R) medium: Half-strength MS salts and vitamins containing 0.25 mg/L NAA, 15 g/L sucrose, 250 mg/L cefotaxime, 10 mg/L kanamycin acid sulfate, and 0.3% (w/v) Gelrite, pH 5.8.

3. Methods

3.1. Isolation of Protoplasts

1. Surface-sterilize seeds by immersion in 70% (w/v) ethanol for 1 min followed by an incubation in a 6% solution of NaOCl on a rotary shaker (60 rpm) at room temperature for 25 min. Wash four times with autoclaved deionized water.

2. Place seeds on half-strength MS medium to germinate in the dark at 18°C for 3–4 d.

3. Cut 80 etiolated hypocotyls from 3-d-old seedlings into 0.5–1.0-mm segments and preplasmolyse in a 9-cm diameter Petri dish containing 20 mL of W 13M for 1 h.

4. Replace W 13M by 20 mL of an enzyme solution (ENZ 1/3/5) and agitate in the dark on a rotary shaker (40 rpm) at 26°C for about 15 h.

5. Sieve the enzyme–protoplast mixture through a 50-μm mesh nylon sieve and centrifuge the filtrate in 15-mL centrifuge tubes at 100g for 8 min. Suspend the protoplast pellet in 12 mL W 22S and centrifuge at 100g for 8 min. The protoplast band floating on the surface is transferred to 12 mL W5 and again centrifuged.

6. Suspend the pellet in PC medium at a density of 2.5×10^5 protoplasts/mL and place the freshly isolated protoplasts in 3-cm diameter Petri dishes containing 1.5 mL PC medium. Culture in diffuse light at 26°C.

3.2. Cocultivation

1. Transfer a single colony of the *A. tumefaciens* strain into 3 mL YEB medium containing 50 μg/mL rifampicin and place the culture in a rotary wheel (100 rpm). Incubate in the dark at 28°C overnight.

2. Three days after protoplast isolation (*see* Note 2), add 5 μL of an overnight culture of *A. tumefaciens* for cocultivation to 1 mL PC medium containing 2.5×10^5 protoplasts (100–200 bacteria/protoplast). Incubate at 20°C in the dark or dim light for 48–72 h.

3. Remove free bacteria by washing protoplasts in W5 and centrifuge at 100g for 8 min.

3.3. Kanamycin Selection

1. Resuspend the protoplast pellet in PC medium supplemented with 500 μg/ mL cefotaxime and mix with an equal volume of PC agarose medium.

2. Culture 1 mL of the mixture as droplets of approx 80–100 μL/5.5-cm Petri dish. Let the agarose droplets solidify and add 1.0 mL of PC medium with 500 μg/mL cefotaxime to each Petri dish. Keep the embedded protoplasts at 26°C in a 16-h photoperiod of 1500 lx intensity.
3. Nine days after protoplast isolation (*see* Note 3) replace the liquid medium surrounding the embedded cells by a volume of liquid CC medium containing 50 μg/mL kanamycin. The final concentration of kanamycin in the agarose droplets amounts to 25 μg/mL.
4. Increase the final kanamycin concentration in intervals of 5 days via 35 μg/mL to a concentration of 50 μg/mL (*see* Note 4).
5. Transfer about 150 microcalli of 0.5–1 mm in diameter onto solidified CC medium containing 50 μg/mL kanamycin (*see* Note 5), 28–30 days after protoplast isolation.

3.4. Regeneration of Transformants

1. Place 10 calli of 1–3 mm in diameter per 9-cm Petri dish containing SR medium and illuminate with 3000 lx at 26°C (*see* Note 6). Place the calli onto fresh medium at intervals of 2 wk four times.
2. Cut off shoots (Fig. 2) with a small piece of callus at the basal end and transfer to 100-mL powder jars containing SMI medium for further growth and multiplication.
3. Shoots are rooted on R medium and then "hardened off" in 1-L preserving jars containing a soil mixture consisting of clay (28%), peat (66%), and sand (6%).

3.5. Confirmation of Transformation

Confirmation of transformation requires the identification of the transferred gene coding for neomycin phosphotransferase in the genome of transformed plants using standard protocols for detecting NPTII enzyme activity *(12)*, DNA isolation *(13)*, Southern blot hybridization, and Northern blot hybridization *(14)*. Starting with 2.5×10^5 protoplasts per experiment, up to six transgenic shoots can be obtained per experiment using line R117. Among tranformants analyzed some plants can show a reduced fertility and a poor seed set. After back-crossing these plants, a normal seed set with no abnormalities in the progeny is observed. Some regenerants, however, which possess altered floral structures, are sterile and no seed set is produced after selfing and back-crossing.

4. Notes

1. Vectors derived from nopaline plasmids are especially useful for the transformation of *B. napus (6)*.

Fig. 2. Shoot regeneration on protoplast derived transgenic callus.

2. A proportion of at least 30% dividing protoplasts is essential before starting cocultivation between the second and fourth day of culture. During the
coculture period, 30–40% of cells die.
3. Optimal starting time for the three-step selection with kanamycin is
between 8 and 10 d after isolation of protoplasts just before colonies of
6–8 cells start to form elongated cells (Fig. 3). A too early addition of
kanamycin reduces the surviving population dramatically. A one-step
selection with 50 µg/mL kanamycin leads to almost total inhibition of
growth when added earlier than 9 d after protoplast isolation, whereas the
same concentration has almost no effect when added after 18 d.
4. By increasing the kanamycin concentration, nontransformed colonies
quickly turn brown and gradually stop growing, whereas transformed calli
continue their growth. Transformed calli possess a yellow brownish color
or become green (Fig. 4).
5. Most of the microcalli transferred onto solidified CC medium continue to
grow and some turn green.

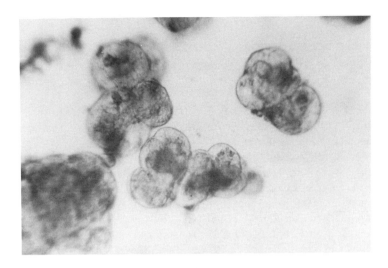

Fig. 3. Microcolonies with cells before starting to elongate 9 d after isolation.

Fig. 4. Agarose droplets with actively growing transgenic calli (dark spots) in the presence of kanamycin.

6. When transferred to SR medium most calli turn green. Kanamycin has to be omitted in SR medium because it has a negative effect on shoot regeneration *(15)*.

Acknowledgments

The author wishes to thank G. Werner, B. Nelke, and C. Ebke for their excellent technical assistance, and P. Loesel for critically reading the manuscript.

References

1. Pelletier, G., Primard, C., Vedel, F., Chetrit, P., Remy, R., Rousselle, R., and Renard, M. (1983) Intergeneric cytoplasmic hybridization in Cruciferae by protoplast fusion. *Mol. Gen. Genet.* **191**, 244–250.
2. Glimelius, K. (1984) High growth rate and regeneration capacity of hypocotyl protoplasts in some *Brassicaceae. Physiol. Plant.* **61**, 38–44.
3. Barsby, T. L., Yarrow, S. A., and Shepard, J. F. (1986) A rapid and efficient alternative procedure for the regeneration of plants from hypocotyl protoplasts of *Brassica napus. Plant Cell Rep.* **5**, 101–103.
4. Thomzik, J. E. and Hain, R. (1988) Transfer and segregation of triazine tolerant chloroplasts in *Brassica napus* L. *Theor. Appl. Genet.* **76**, 165–171.
5. Jourdan, P. S. and Earle, E. D. (1986) Influence of genotype on the regeneration of plants from seedling mesophyll protoplasts of three *Brassica* species. Crucifer genetics workshop III (Proceedings:58), University of Guelph, Canada.
6. Thomzik, J. E. (1993) Transformation in oilseed rape (*Brassica napus* L.), in *Biotechnology in Agriculture and Forestry, Vol 23, Plant protoplasts and Genetic Engineering IV* (Bajaj, Y. P. S., ed.), Springer Verlag, Heidelberg, Germany.
7. Thomzik, J. E. and Hain, R. (1990) Introduction of metribuzin resistance into German winter oilseed rape of double-low quality. *Pflanzenschutz-Nachrichten Bayer* **43(1,2)**, 61–87.
8. Czernilofsky, A. P., Hain, R., Herrera-Estrella, L., Lörz, H., Goyvaerts, E., Baker, B. J., and Schell, J. (1986) Fate of selectable marker DNA integrated into the genome of *Nicotiana tabacum. DNA* **5**, 101–113.
9. Menczel, L. and Wolfe, K. (1984) High frequency of fusion induced in freely suspended protoplast mixtures by polyethylene glycol and dimethylsulfoxide at high pH. *Plant Cell Rep.* **3**, 196–198.
10. Murashige, T. and Skoog, F. (1962) A revised medium for rapid growth and bioassays with tobacco tissue culture. *Physiol. Plant.* **15**, 473–497.
11. Glimelius, K., Djupsjöbacka, M., and Fellner-Feldegg, H. (1986) Selection and enrichment of plant protoplast heterokaryons of *Brassicacea* by flow sorting. *Plant Sci.* **45**, 133–141.
12. McDonnell, R. E., Clark, R. D., Smith, W. A., and Hinchee, M. A. (1987) A simplified method for the detection of neomycin phosphotransferase II activity in transformed plant tissues. *Plant Mol. Biol. Rep.* **4**, 380–386.
13. Dellaporta, S. L., Wood, J., and Hicks, J. B. (1983) A plant DNA minipreparation: version II. *Plant Mol. Biol. Rep.* **1**, 19–21.
14. Sambrook, J., Fritsch, E. F., and Maniatis, T. (1989) *Molecular Cloning, A Laboratory Manual.* Cold Spring Harbor Laboratory Press, Cold Spring Harbor, NY.
15. Thomzik, J. E. and Hain, R. (1990) Transgenic *Brassica napus* plants obtained by cocultivation of protoplasts with *Agrobacterium tumefaciens. Plant Cell Rep.* **9**, 233–236.

CHAPTER 9

Agrobacterium-Mediated Transformation of Stem Disks from Oilseed Rape (*Brassica napus* L.)

Jürgen E. Thomzik

1. Introduction

The commercial importance of oilseed rape (*Brassica napus* L.) makes it desirable to have an efficient transformation protocol for practical applications. With the exception of a few reports, transformation experiments with *B. napus* using *Agrobacterium tumefaciens* yielding transgenic plants have been performed principally with the spring variety *B. napus* L. ssp. oleifera cv. Westar *(1–5)*. Transformation protocols are relatively specific for cv. Westar and cannot be extended to most of the other spring and none of the winter varieties tested *(6)*. There are only isolated reports on transgenic winter oilseed rape plants *(6–8)*.

Internodal stem disks *(9)* appeared to be most suitable for an *Agrobacterium*-mediated transformation of winter oilseed rape. Slight modifications of the transformation protocol described for cv. Westar *(1)* led to a simple transformation system, which also works with high efficiency for winter oilseed rape. These modifications minimize tissue necrosis after cocultivation with *A. tumefaciens.*

Using this modified protocol, two genes coding for stilbene synthase have been cointroduced with a chimeric kanamycin resistance gene. Stilbene synthase is an enzyme present in a few plant species that synthesize the phytoalexin trans-resveratrol *(10,11)*. Transgenic plants and their seed progeny express the introduced genes. Starting with 50-stem disks

From: *Methods in Molecular Biology, Vol. 44:* Agrobacterium *Protocols*
Edited by: K. M. A. Gartland and M. R. Davey Humana Press Inc., Totowa, NJ

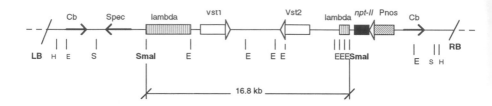

Fig. 1. Integration of stilbene synthase genes via pS-Vit into the Ti plasmid pGV3850: (LB, RB) left and right border of the Ti-plasmid specific sequences; (Cb) carbenicillin-resistance gene; (Spec) spectinomycin resistance gene; λ phage derived DNA sequence; (*Vst1, Vst2*) two stilbene synthase genes derived from *Vitis vinifera* var. Optima; (*npt*II) the neomycin phosphotransferase-II coding region; (P*nos*) L nopaline synthase promoter region; (H,E,S,*Sma*I) restriction sites of *Hin*dIII, *Eco*RI, *Sal*I, and *Sma*I.

derived from 2 plants, approx 20–30 transgenic plants potted in soil can be obtained within 9–12 wk. Transgenic plants derived from stem disks are fertile, show a normal seed set, and are phenotypically normal.

2. Materials

1. Plant material: Winter oilseed rape line SEM1, line R117 (Kartoffelzucht Boehm KG, Langquaid, Germany), line R109 (KWS Kleinwanzlebener Saatzucht AG, Einbeck, Germany), cv. Libravo (DSV Deutsche Saatvcredelung GmbH, Lippstadt, Germany), and cv. Ceres (NPZ Norddeutsche Pflanzenzucht Lemke KG, Holtsee, Germany) possessing double-low (low in erucic acid and glucosinolate content) quality *(12)* are grown in the greenhouse in 18-cm pots containing a soil mixture of peat (66%), clay (28%), and sand (6%) fertilized with an N:P:K ratio of 450:200:250 mg/L. Greenhouse conditions are a day/night temperature of 20/15°C and a daylight period of 14 h at 12,000 lx. Plants with 4–6 leaves are vernalized in a climate room for 9 wk at 4°C.
2. Feeder plates: Mix 1 g (fw) of a 4-d-old tobacco or potato cell suspension culture with 15 mL MSF medium (*see* item 10) prior to solidification in a 9-cm diameter Petri dish and incubate for 2 d at 26°C.
3. *A. tumefaciens*: Strain C58CI harboring the disarmed plasmid pGV3850. This plasmid contains two linked full-length stilbene synthase genes (*Vst*1 and *Vst*2) under control of its original promoter that were isolated from grapevine *(10,11)* and a chimeric *npt*II gene conferring resistance to kanamycin in plants (Fig. 1).
4. YEB medium (per liter): 1 g yeast extract, 5 g beef extract, 5 g peptone, 0.5 g MgSO$_4$ · 7H$_2$O, 5 g sucrose, pH 7.2.

5. Antibiotics: 50 mg/mL stocks of kanamycin acid sulfate and cefotaxime in distilled water, filter sterilized; 50 mg/mL stock of rifampicin in ethanol. Stocks for cefotaxime and rifampicin should be freshly prepared for each use *(12)*.
6. 70% (v/v) Ethanol.
7. Sodium hypochlorite solution (13% active chlorine): Solution of 3% is prepared by dilution with water and the addition of 3 drops/L of Tween-80 as a wetting agent.
8. Growth regulators: 0.2 mg/mL stocks of 6-benzylaminopurine (BAP) and α-naphthalene acetic acid (NAA) in distilled water, filter sterilized.
9. MS0 medium: MS salts and vitamins *(13)*, 3% (w/v) sucrose, pH 5.8.
10. MSF medium: MS0 medium supplemented with 1 mg/L BAP and 0.3% (w/v) Gelrite (*see* Note 2). Use for feeder plates.
11. MSS medium: MSF medium containing 500 µg/mL cefotaxime and 50 µg/mL kanamycin acid sulfate.
12. Shoot multiplication (SMI) medium: MS salts and vitamins, 20 g/L sucrose, 4 mg/L BAP, 0.2 mg/L NAA, 0.8% (w/v) agar, 15 mg/L kanamycin acid sulfate, 250 mg/L cefotaxime, pH 5.8.
13. Rooting (R) medium: Half-strength MS0 supplemented with 0.25 mg/L NAA, 0.3% (w/v) Gelrite, 250 mg/L cefotaxime, 10 mg/L kanamycin, pH 5.8.
14. MS/Km medium: MS0 medium containing 2 mg/L NAA, 2 mg/L BAP, 0.8% (w/v) agar, and 100 mg/L kanamycin acid sulfate.
15. MS/SKm medium: MS salts and vitamins, 10 g/L sucrose, 125 mg/L kanamycin acid sulfate, 0.8% (v/w) agar, pH 5.8.
16. Autoclaved distilled water.
17. Sterile filter papers.
18. Dissecting instruments: scalpels, forceps, and so on.

3. Methods

3.1. Transformation of Stem Disks

1. Transfer a single colony of the *A. tumefaciens* strain into 3 mL YEB medium containing 50 µg/mL rifampicin and incubate the culture on a rotary wheel (100 rpm) in the dark at 28°C overnight.
2. Remove the four upper internodes from oilseed rape plants, which are in the process of bolting, but which have not yet flowered *(1)* and surface sterilize them in 70% (v/v) ethanol for 30 s and in 3% NaOCl for 15 min followed by three washes in autoclaved distilled water.
3. Cut stem disks of 5–8 mm of thickness from the internodes and place these disks with the basal cut downwards in a 9-cm diameter Petri dish (*see* Note 3).
4. Dilute an overnight culture of *A. tumefaciens* (to $OD_{600} = 0.5–0.8$) with MS0 medium (*see* Note 4). Add a thin film of the bacterial suspension to

Fig. 2. Shoot regeneration on stem disks after cocultivation with *A. tumefaciens* in the presence of 50 µg/mL kanamycin.

the Petri dish but cover only the bottom around the stem disks (*see* Note 5). Incubate at room temperature for 5 min.
5. Blot the stem disks dry with autoclaved filter paper and place them so that the inoculated side is in contact with the MSF medium (feeder plate). Cocultivate at 26°C in dim light for 48 h.
6. After cocultivation wash the inoculated side of stem disks in cefotaxime (500 µg/mL) and thoroughly blot them dry with autoclaved filter paper.

3.2. Selection and Regeneration of Transformants

1. Transfer the disks to the MSS medium (*see* Note 6) and incubate in the light (3000 lx) at 26°C. Transfer the stem disks to fresh MSS medium (*see* Note 7) at two weekly intervals (*see* Note 8).
2. Excise green shoots of 2-cm from stem disks (*see* Fig. 2) and transfer them onto SMI medium for further growth or place them directly onto R medium for rooting.
3. Transfer rooted plants to small soil-filled pots (*see* Section 2.) and place them in preserving jars for hardening-off.

3.3. Confirmation of Transformation

Resistance to kanamycin can be assayed by placing small stem sections or leaf pieces of the regenerants on a callus-inducing medium containing kanamycin. This allows the elimination of regenerants that escaped the selection on MSS medium.

1. Cut off small stem sections of 5 mm or leaf pieces of 5 mm^2 and place them onto MS/Km medium for callus formation. Culture in the light (3000 lx) at 26°C and score for kanamycin resistance after 14 d (*see* Note 9).
2. The transgenic character of regenerants selected on kanamycin has to be confirmed by using standard protocols for detecting *npt*II enzyme activity (*14*), DNA isolation (*15*), RNA isolation (*16*), Southern blot hybridization, and Northern blot hybridization (*17*) (*see* Note 10).

3.4. Inheritance of Kanamycin Resistance

The inheritance of kanamycin resistance by the seed progeny from self-pollinated transformants can be assayed by germinating seeds in the presence of kanamycin.

1. Seeds from self-pollinated transformants and nontransformed plants are surface-sterilized in 70% (v/v) ethanol (1 min) followed by continuous shaking in a 6% solution of NaOCl for 20 min.
2. Germinate seeds in the light at 26°C on MS/SKm medium that contains 125 mg/L kanamycin acid sulfate and score for resistance or sensitivity to kanamycin 10–14 d after germination (*see* Note 11).

4. Notes

1. Most stem disks of *B. napus* loose the capability to form shoots or become necrotic after cocultivation with *A. tumefaciens*. This is owing to a hypersensitive response to *A. tumefaciens* in connection with an inadequate concentration of antibiotics. Particularly after transfer of stem disks to medium containing 500 µg/mL cefotaxime, bacteria start to form a slimy dark-brown film covering the whole cut end of the stem disks. Beneath this film a small cell layer of the plant tissue becomes necrotic and turns black. Unfortunately, tissue necrosis is remarkably intensive in the shoot producing area around the cambium layer.
2. Avoid using agar for medium solidification. The use of 0.3% (v/w) Gelrite or 0.7% (w/v) agarose minimizes tissue necrosis and improves the shoot regeneration.
3. Compact stem disks with no tears or gaps in the tissue should be used.
4. Bacterial growth can not be inhibited by 500 µg/mL cefotaxime when the density is $>OD_{600} = 0.8$.
5. Avoid inoculating more of the tissue than the basal cut of the stem disks from which the shoots will appear.
6. Kanamycin acid sulfate concentrations <50 µg/mL in the MSS medium leads to a drastic increase in escapes. Transgenic shoots remain green in the presence of 50 µg/mL kanamycin whereas nontransformed shoots bleach or turn violet. Almost no shoots are formed when a concentration of

75 µg/mL is applied. For some genotypes, for instance cv. Libravo, a kanamycin acid sulfate concentration of 60 µg/mL is optimal.

7. Keep the inoculated side of the disks in constant contact with the medium after all transfer steps to fresh medium.
8. White nodules are formed around the cambium layer of the inoculated side of stem disks on MSS- medium. After 3–4 wk some green, but predominantly white and violet shoots emerge from these nodules when the correct kanamycin concentration is applied. However, most green shoots appear after another transfer to fresh MSS medium and most of the transgenic shoots are found among these (*see* Fig. 2).
9. Explants of transgenic plants remain green and produce a uniform dark green and compact callus. Explants from nontransformed plants form almost no callus and turn brown or bleach.
10. Stilbene mRNA synthesis can be induced in callus and leaves of transformants by inoculation with *Phytophthora infestans*, *Botrytis cinerea* (2×10^5 spores/mL), wounding, a water-soluble fungal elicitor preparation of *Phytophtora megasperma* or UV illumination.
11. Seedlings possessing kanamycin resistance develop normally and form green primary leaves, whereas nontransformed seedlings show reduced growth, form violet primary leaves, and finally bleach.

Acknowledgments

The author wishes to thank B. Nelke, C. Ebke, and P. C. Reinhard for excellent technical assistance; P. H. Schreier for the bacterial strain that was used; and P. Loesel for critically reading the manuscript.

References

1. Fry, J., Barnason, A., and Horsch, R. B. (1987) Transformation of *Brassica napus Agrobacterium tumefaciens* based vectors. *Plant Cell Rep.* **6,** 321–325.
2. Pua, E. C., Mehra-Plata, A., Nagy, F., and Chua, N. H. (1987) Transgenic plants of *Brassica napus* L. *Bio/Technology* **5,** 815–817.
3. Radke, E. S. E., Andrews, B. M., Moloney, M. M., Crouch, M. L., Kridl, J. C., and Knauf, V. C. (1988) Transformation of *Brassica napus* using *Agrobacterium tumefaciens*: developmentally regulated expression of a reintroduced napin gene. *Theor. Appl. Genet.* **75,** 685–694.
4. Charest, P. J., Holbrook, L. A., Gabard, J., Iyer, V. N., and Miki, B. L. (1988) *Agrobacterium*-mediated transformation of thin cell layer explants from *Brassica napus* L. *Theor. Appl. Genet.* **75,** 438–445.
5. Moloney, M. M., Walker, J. M., and Sharma, K. K. (1989) High efficiency transformation of *Brassica napus* using *Agrobacterium* vectors. *Plant Cell Rep.* **8,** 238–242.
6. De Block, M., Brouwer, D. D., Tenning, P. (1989) Transformation of *Brassica napus* and *Brassica oleracea* using *Agrobacterium tumefaciens* and the expression of the bar and neo genes in transgenic plants. *Plant Physiol.* **91,** 694–701.

7. Thomzik, J. E. and Hain, R. (1990) Transgenic *Brassica napus* plants obtained by cocultivation of protoplasts with *Agrobacterium tumefaciens*. *Plant Cell Rep.* **9,** 233–236.

8. Boulter, M. E., Croy, E., Simpson, P., Shields, R., Croy, R. R. D., and Shirsat A. H. (1990) Transformation of *Brassica napus* L. (oilseed rape) using *Agrobacterium tumefaciens* and *Agrobacterium rhizogenes*—a comparison. *Plant Sci.* **70,** 91–99.

9. Stringham, G. R. (1977) Regeneration in stem explants of haploid rapeseed (*Brassica napus* L.). *Plant Sci. Lett.* **9,** 115–119.

10. Hain, R., Reif, H. J., Langebartels, R., Schreier, P. H., Stöcker, R. H., Thomzik, J. E., and Stenzel, K. (1992) Foreign phytoalexin expression in plants results in increased disease resistance. Brighton crop protection conference. *Pest Dis.* **7B-5,** 757–766.

11. Hain, R., Reif, H. J., Krause, E., Langebartels, R., Kindl, H., Vornam, B., Wiese, W., Schmelzer, E., Schreier, P. H., and Stenzel, K. (1993) Disease resistance results from foreign phytoalexin expression in a novel plant. *Nature* **361,** 153–156.

12. Thomzik, J. E. and Hain, R. (1990) Introduction of metribuzin resistance into German winter oilseed rape of double-low quality. *Pflanzenschutz-Nachrichten Bayer* **43(1,2),** 61–87.

13. Murashige, T. and Skoog, F. (1962) A revised medium for rapid growth and bioassays with tobacco tissue culture. *Physiol. Plant* **15,** 473–497.

14. McDonnell, R. E., Clark, R. D., Smith, W. A., and Hinchee, M. A. (1987) A simplified method for the detection of neomycin phosphotransferase II activity in transformed plant tissues. *Plant Mol. Biol. Rep.* **4,** 380–386.

15. Dellaporta, S. L., Wood, J., and Hicks, J. B. (1983) A plant DNA minipreparation: version II. *Plant Mol. Biol. Rep.* **1,** 19–21.

16. Schröder, G., Brown, J. W. S., and Schröder, J. (1988) Molecular analysis of resveratrol synthase: cDNA, genomic clones and relationship with chalconsynthase. *Eur. J. Biochem.* **172,** 161–169.

17. Sambrook, J., Fritsch, E. F., and Maniatis, T. (1989) *Molecular Cloning, A Laboratory Manual.* Cold Spring Harbor Laboratory, Cold Spring Harbor, NY.

CHAPTER 10

Peanut Transformation

Elisabeth Mansur, Cristiano Lacorte, and William R. Krul

1. Introduction

Peanut or groundnut (*Arachis hypogaea* L.) is a member of the legume family, originated in South America, and today is largely cultivated in many tropical and subtropical areas worldwide. Peanut seeds are an important source of protein, carbohydrates, and oil for humans and animals. In addition to its use as food, the oil can be used directly to fuel diesel motors and phenolic resins obtained from the shells are excellent binding agents. The remainder of the plant is also useful as animal fodder and as a green cover crop providing fixed nitrogen to the soil (*1*).

The world production of peanut in 1991 was 23,366,000 tons, with the five largest producers being India, China, United States, Nigeria, and Indonesia (*2*). Although the maximum recorded productivity is about 6700 kg/ha, yields of half this are routinely obtained in the United States and much lower values are obtained in lesser developed countries of the world (*3*).

There is a need to improve both productivity and quality of peanut. Among the more important factors are development of varieties tolerant to water stress, and those with uniform maturity of the fruits. In addition, there is a need for resistance to nematodes, viruses, insects, leaf spot diseases, and to fungi that produce carcinogenic aflatoxins. The quality of the seeds could be improved by enhancing protein quality (methionine, tryptophan, and isoleucine contents) and by reduction of the linolenic acid content (*1*).

From: *Methods in Molecular Biology, Vol. 44:* Agrobacterium *Protocols*
Edited by: K. M. A. Gartland and M. R. Davey Humana Press Inc., Totowa, NJ

Many of these improvements can be accomplished by conventional breeding methods. However, objectives such as improvement of protein quality of the seeds and introduction of insect resistance might be done easier via genetic engineering methods. For example, the methionine rich 2S albumin gene from Brazil nut (4) may provide a useful gene for the enhancement of peanut seed quality and insects might be controlled by introducing genes of *Bacillus thuringiensis (5)*. However, to accomplish this and other genetic engineering objectives, the dual problems of efficient gene delivery to isolated cells and the subsequent regeneration of transformed plants must still be solved for commercially important, generally recalcitrant, peanut cultivars.

High in vitro regeneration rates are a basic prerequisite for the production of transgenic plants. Several reports on tissue culture of peanut have been published and in vitro regeneration has been achieved through morphogenetic calli, direct organogenesis, and somatic embryogenesis, usually from seed or seedling explants. Morphogenesis with an intervenient callus phase has been described from immature leaves (6–8). Direct organogenesis was described from cotyledons (9), leaves (10), intact mature seeds (11), and several seedling explants (12). Different somatic embryogenesis systems have also been reported from immature embryo explants (13–16), mature zygotic embryo axis (17), seedling explants (18,19), and intact seedlings (20). In these reports, improvement of regeneration efficiency has been investigated through examination of variations of the composition of the culture medium and the selection of suitable explants. However, factors such as temperature, pH, and medium rigidity strongly influence the number of plantlets produced from both leaf and cotyledon explants (Freitas et al., unpublished data).

Introduction of foreign genes into plant tissues via *Agrobacterium tumefaciens*-based vectors requires specific knowledge of host–pathogen compatibility. Although *Agrobacterium* has a broad host range in dicotyledoneous plant genera, susceptibility to a given strain varies among species and even varieties of the same species. Conversely, *Agrobacterium* strains show different degrees of virulence for a given species. Therefore an initial approach to establishing a transformation protocol is basically the determination of vector strain-plant genotype compatibility. Susceptibility to and gene transfer by *Agrobacterium* have been demonstrated in peanut whole plants (21,22). Strain A281 was found to be the most virulent to peanut cultivars representing Valencia, Virginia, and Spanish types (22).

Successful transformation using *Agrobacterium* is determined by several plant- and bacterial-related factors *(23)*. Susceptibility at explant level as well as the optimization of parameters that regulate transformation efficiency is especially important for peanut and other recalcitrant species in order to establish suitable transformation strategies.

In peanut, factors that regulate transformation have been studied for leaf and cotyledon explants, using a tumor induction system based on strain A281 harboring a binary vector. Among those factors tested, the explant type, the length of the cocultivation period, and the cocultivation medium were the most critical in affecting explant transformation efficiency *(23)*.

Factors that influence explant transformation efficiency can be evaluated by the frequencies of transgenic plant production or tumor induction in different conditions. Tumor induction systems also provide a rapid method to access optimal transformation conditions and determine the resistance level conferred by selectable marker genes. However, tumor formation frequencies may not reflect the actual transformation rates at the cellular level, as initially transformed cells may not respond to the phytormones encoded in the T-DNA. Differences between the processes of regeneration and tumor formation must also be considered. Tumor formation is mainly the product of cell dedifferentiation and division, whereas the production of transformed regenerants depends both on the competence of transformed cells for regeneration and on the selection schedule adopted.

Transformation efficiency can also be determined at early stages by using modified β-glucuronidase (GUS) genes, as, for example, the chimeric GUS-intron gene. This gene contains a portable intron derived from the second intron of the ST-LS1 gene from potato and it is not expressed in the bacterium. Due to this, it allows an efficient discrimination between agrobacterial and plant gene expression during the early events of transformation *(24)*. The use of GUS INT gene, however, does not accurately determine the transformation frequency as evaluation of early GUS expression includes transient expression. Besides, transformation efficiency measured by modified GUS genes expression may not correlate with transgenic plant production as the selective schedule used can affect the recovery of transformed cells. Alternatively, untransformed cells that drive the pattern of regeneration may be inhibited by the selection agent with a consequent loss of the regeneration capacity of the explant

(25). Nevertheless, modified GUS genes, which allow the identification of single transformed cells, are a useful tool for the development of efficient transformation strategies especially for recalcitrant species.

In peanut, histological identification of cells that are both susceptible to *Agrobacterium* and competent for in vitro organogenesis using the GUS INT gene *(24)* indicates that regeneration from seed explants occurs from cells that are not susceptible to *Agrobacterium*. On the other hand, transformation and morphogenesis processes in leaf explants seem to involve the same cell types (Freitas et al., in preparation). Further studies on factors that influence transformation rates and regeneration efficiency may provide conditions for the regeneration of transformable cells or conversely transformation of cells involved in regeneration.

In this chapter we describe general protocols for studying peanut transformation and evaluating transformation efficiency by *A. tumefaciens*, which can be taken as a model for other recalcitrant species.

2. Materials
2.1. Tissue Culture

1. Peanut seeds cv. Tatu (Valencia type).
2. Growth regulators: 6-benzylaminopurine (BAP) (stock solution 10 mg/ mL) and α-naphthalene acetic acid (NAA) (stock solution 1 mg/mL). Dissolve each in $0.1M$ NaOH and store in aliquots at –20°C.
3. Cefotaxime (Cx): 100 mg/mL in water. Filter sterilize and store in aliquots at –20°C.
4. MS0 medium: MS salts and vitamins *(26)*, 30 g/L sucrose, 7 g/L agar, pH 5.8.
5. MC medium: MS salts and B5 vitamins *(27)*, 30 g/L sucrose, 25 mg/L BAP, 7 g/L agar, pH 5.5.
6. MF medium: MS salts and B5 vitamins, 30 g/L sucrose, 5 mg/L BAP, 7 g/L agar, pH 5.5.
7. 0.1% $HgCl_2$, 5 mM EDTA, sterile distilled water.

Media pH is adjusted before adding the agar and autoclaving. Filter-sterilized antibiotics are added to the media after autoclaving.

2.2. Agrobacterium *Strains*

A. tumefaciens tumorigenic strain A281 and its derivative strain EHA101, both carrying plasmid pGUS INT as binary vector *(24)*. The avirulent strain A136 containing the same plasmid is used as control (Table 1). Plasmid pGUS INT harbors the GUS gene containing a plant derived intron, under the transcriptional control of the CaMV 35S pro-

Table 1
Commonly Used Wild-Type and Disarmed Derivative Strains
of *Agrobacterium tumefaciens*

Wild-type strains[a]	Plasmids	Chromosomal background	Main opines[b]	Disarmed derivatives	Virulence helper plasmid	Refs.
C58	pTiC58	C58	NOP	GV3101[c]	pMP90[d]	28,31
A281	pTiBo542	C58	AGR	EHA101	pEHA101[e]	28–30
A208	pTiT37	C58	NOP	A208-SE	pTiT37-SE[e,f]	28,31
Ach5	pTiAch5	Ach5	OCT	LBA4404[g]	pAL4404	32
A136	—	C58	—	—	—	33

[a]Rif 100 μg/mL (chromosomal resistance).
[b]NOP, nopaline; AGR, agropine; OCT, octopine.
[c] Corresponds to C58C1 Rif^R.
[d]Gm 50 μg/mL.
[e]Km 50 μg/mL.
[f]Co-integrative host *vir* plasmid.
[g]Sm 300 μg/mL, Sp 100 μg/mL.

moter and the neomycin phosphotransferase gene (*npt*II) under the control of the nopaline synthase (*nos*) promoter. *Agrobacterium* wild-type strains commonly used to evaluate strain-genotype compatibility, and their disarmed derivatives are listed in Table 1.

2.3. Plasmid Mobilization into Agrobacterium by Triparental Mating

1. YEB medium: 5 g/L meat extract, 1 g/L yeast extract, 5 g/L peptone, 5 g/L sucrose, 493 mg/L Mg SO$_4$ · 7H$_2$O, pH 7.2. For solid YEB medium add 15 g/L agar before autoclaving.
2. MinA medium: 10.5 g/L K$_2$HPO$_4$, 4.5 g/L KH$_2$PO$_4$, 1 g/L (NH$_4$)$_2$SO$_4$, 0.5 g/L sodium citrate · 2H$_2$O. Add these salts as 5X stock to 15 g/L agar solution. For 1 L, add to the autoclaved medium 1 mL of 20% MgSO$_4$ · 7H$_2$O and 10 mL of 20% glucose.
3. Antibiotics: 50 mg/mL kanamycin (Km), 100 mg/mL streptomycin (Sm), 100 mg/mL spectinomycin (Sp), and 100 mg/L rifampicin (Rif). Dissolve in water, filter sterilize, and store in aliquots at –20°C.

2.4. Foreign Gene Expression Assays

2.4.1. Nopaline and Octopine Detection

1. Electrophoresis buffer: 15% (v/v) acetic acid, 5% (v/v) formic acid, pH 1.8.
2. Staining solutions:
   a. Solution A: 0.02% (w/v) phenanthrenoquinone in absolute ethanol.

b. Solution B: 10% (w/v) NaOH in 60% (v/v) ethanol.

Solutions A and B can be stored at $-20°C$. Mix A and B (1:1) just prior use.

3. Marker solutions: orange G, xylene cyanol, and methyl green (10 μg/ mL each).

2.4.2. Agropine Detection

1. Solution A: Dissolve 1 g $AgNO_3$ in 1 mL H_2O. Add 200 μL of $AgNO_3$ solution to 100 mL acetone.
2. Solution B: Add 10 mL 20% (w/v) NaOH to 90 mL absolute ethanol. Prepare fresh.
3. Solution C: 10% (w/v) sodium thiosulfate and 0.5% (w/v) sodium metabisulfite or any photographic fixer.

2.4.3. Neomycin Phosphotransferase (NPTII) Assay

1. Extraction buffer: 125 mM Tris-HCl, pH 6.8, 10% (v/v) β-mercaptoethanol, 0.2% (w/v) sodium dodecyl sulfate (SDS), 20% (v/v) glycerol.
2. Reaction buffer: (5X stock) 335 mM Tris-HCl, 210 mM $MgCl_2$, 2M NH_4Cl, pH 7.1.
3. 10 mM Adenosine triphosphate (ATP).
4. γ^{32}P-ATP (10 Ci/μL).
5. 22 mM Kanamycin sulfate.
6. 1M NaF or KF.
7. ATP/pyrophosphate solution: 20 mM ATP, 100 mM pyrophosphate.
8. 10 mM Sodium phosphate buffer pH 7.5.

2.4.3.1. PREPARATION OF *E. COLI* EXTRACT

1. *E. coli* strains HB101::Tn5 and HB101.
2. LB medium: 10 g/L meat extract, 5 g/L yeast extract, 10 g/L NaCl, pH 7.2.
3. Extraction buffer: 25 mM Tris-HCl, pH 8.0, 10 mM EDTA, 50 mM glucose, 1 mg/mL lysozyme.

3. Methods

3.1. Regeneration from Leaf and Cotyledon Explants

Here we describe an optimized protocol in which maximum regeneration efficiency is obtained when explants are incubated at 35°C and media pH is adjusted to 5.5 (Freitas et al., unpublished data).

1. Surface sterilize peanut seeds in 0.1% $HgCl_2$ for 30 min with constant agitation.
2. Wash three times with sterile distilled water, soak in 5 mM EDTA for 5 min and rinse three more times with water.
3. For leaf culture, incubate seeds on cotton wool wetted with 1/2 strength liquid MS medium for 7–10 d.

4. Excise dark green, 10–13-mm long leaflets from the seedlings, keeping the petiolules, and inoculate on ML medium.
5. For cotyledon culture, separate cotyledons from ungerminated seeds into embryonated and deembryonated halves, using a scalpel. Transfer explants to MC medium.
6. Incubate explants at 35°C, 80 μE/m^2/s and a photoperiod of 16 h/d.

3.2. Conjugation of Plasmids into Agrobacterium by Triparental Mating

Triparental mating is a simple and convenient method for transferring plasmids to *Agrobacterium (34)*. This method is based on the use of plasmid pRK2013, which provides the transfer and mobilization functions to nonconjugative plasmids. Transconjugants are selected by the resistance to rifampicin, which is encoded by a chromosomal gene of *Agrobacterium* and resistance conferred by the donor plasmid.

1. Inoculate one colony of the following bacteria into 3-mL culture medium and shake overnight:
   a. *A. tumefaciens* strains A281, EHA101, and A136—liquid YEB medium with the correspondent antibiotics (*see* Table 1).
   b. *E. coli* HB101(pRK 2013)—LB medium, 50 μg/mL Km, 37°C.
   c. *E. coli* MC1061(pGUS INT)—LB medium, 20 μg/mL Sm and 50 μg/mL Sp, 37°C.
2. Plate 25 μL of each *E. coli* culture and 25 μL of the *Agrobacterium* culture together onto a YEB plate. Mix well by spreading and incubate overnight at 28°C.
3. Prepare one plate with the *E. coli* strains and one with the *Agrobacterium* strain as controls.
4. Collect mating cells with a loop and streak on MinA agar plates containing 100 μg/mL Rif, 50 μg/mL Km, 300 μg/mL Sm, and 100 μg/mL Sp. Incubate at 28°C for 48 h.
5. Collect isolated colonies of putative transconjugants, inoculate in liquid MinA medium containing the antibiotics in step 4, and incubate on a gyratory shaker at 28°C for 48 h.
6. Streak the cultures on YEB agar medium plus selective antibiotics and incubate at 28°C for 48 h. Store the plates at 10°C for up to 1 mo (*see* Note 1).

3.3. Tumor Induction on Explants

The following protocol is an optimized procedure for leaf and cotyledon explants transformation, taking into account plant and bacterial factors that regulate transformation efficiency *(23)*.

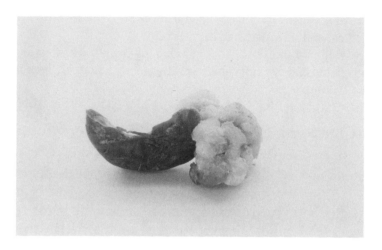

Fig. 1. Detection of nopaline in tumors induced by strain A208. A, arginine; N, nopaline; 1, cv. Tatu; 2, cv. Tupã; 3, cv. Tatu branco; 4, cv. Penápolis; C, nontransformed callus.

1. Grow an overnight *Agrobacterium* culture under selection in liquid YEB medium. Pellet cells and resuspend in liquid MS medium (OD = 1.0–1.2).
2. Soak the explants in the *Agrobacterium* culture for 10 min.
3. Blot-dry the explants on sterile filter paper to remove excess bacteria.
4. Transfer the explants to MS0 medium, when using tumorigenic strains, or to morphogenesis induction medium, for disarmed strains. Incubate in dim light at 28°C.
5. After 48 h, transfer explants to the same media used for cocultivation plus 500 mg/L cefotaxime.
6. Score tumor formation frequency after 4–6 wk. A typical tumor, induced on a deembryonated cotyledon explant is shown in Fig. 1.

3.4. Characterization of Transformed Tissues

Cells transformed by wild strains of *Agrobacterium* are characterized by the production of opines, which are amino acids and sugar derivatives. Detection of opine production as well as of the activity of screenable (GUS) and selectable (*npt*II) genes present in binary or cointegrate vectors are used as evidence of transformation. T-DNA integration is confirmed by Southern or dot blot hybridization using T-DNA specific probes *(35)*.

3.4.1. Opine Detection

Wild-type *Agrobacterium* strains produce different opines (Table 1). This protocol describes specific methods for the detection of nopaline, octopine, and agropine.

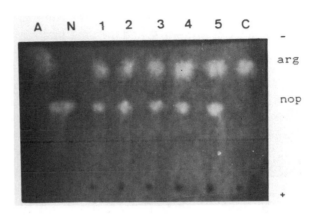

Fig. 2. Tumor induced on deembryonated cotyledon explant (cv. Tatu) by strain A281.

3.4.1.1. NOPALINE AND OCTOPINE DETECTION *(36)*

1. Test the sterility of putative tumors incubating segments on liquid YEB medium for 48 h at 28°C.
2. For octopine detection, incubate 50–100 mg of tissue from axenic tumors for 16–20 h in 5 mL of MS0 medium supplemented with 10 mM arginine and 10 mM sodium pyruvate.
3. Grind the tissue in an Eppendorf tube, centrifuge at 12,000 rpm for 2 min, and transfer the supernatant to another tube.
4. Apply 5 µL of the sample with a capillary tube on 3MM Whatman chromatography paper, 5 cm above the paper's lower end. Apply 2 µL of arginine, nopaline, or octopine standard solutions (500 µg/mL) and color marker solutions. The distance between samples should be about 2 cm.
5. Carefully wet the paper with the extraction buffer using a Pasteur pipette and submit it to electrophoresis at 400 V (20 V/cm) for 2 h.
6. Dry the paper with hot air. Spray the staining solution (A + B) inside a fume hood. **(Note: Phenanthrenoquinone is toxic.)**
7. Dry the paper with cold air, observe with a UV transilluminator, and photograph. The result of a typical nopaline detection assay is shown in Fig. 2.

3.4.1.2. AGROPINE DETECTION *(37)*

1. Follow Section 3.4.1.1., steps 1 and 3–5. Use an agropine standard solution at step 4.

2. Dry the paper with hot air. Mark the reference marker positions.
3. Dip in solution A for 1 min.
4. Transfer to solution B until the bands are visible (5–10 min).
5. Wash the paper in solution C and then in running water to remove excess silver (3–4 h).
6. Dry at room temperature.

3.4.2. GUS Gene Expression Assays (38)

The GUS reporter gene is the most frequently used method for studies in plant gene expression owing to its simplicity and versatility. An additional benefit is that most plants lack endogenous GUS activity. Histochemical assays can be used to investigate tissue-specific expression of GUS gene fusion and sensitive quantitative assays may be performed using the fluorimetric method *(38)*. A simple qualitative assay using the same fluorigenic substrate is convenient for screening large numbers of samples. These assay methods are described in Chapters 17 and 18 (*see* Notes 2 and 3).

3.4.3. NPTII Assay (39)

The dot-blot assay described in the next section is simpler and faster than the *in situ* detection method *(40)*. It allows the screening of large numbers of transformed tissues and requires the use of smaller amounts of radioactive label. It relies on antibiotic phosphorylation in conditions where hydrolysis of neomycin phosphate and ATP is prevented (high levels of cold ATP in combination with fluoride, a phosphatase inhibitor), suppressing nonspecific adsorption to phosphocellulose (*39, see also* Chapter 19).

3.4.3.1. Dot-Blot NPTII Assay

1. Freeze 200 mg tissue in liquid nitrogen. Smash to a fine powder with mortar and pestle. Add 200 µL extraction buffer and mix well. Keep the samples in an ice bath.
2. Controls are extracts of nontransformed plant tissue, *E. coli* HB101::Tn5 (positive control) and HB101 (negative control).
3. Transfer the mixture to an Eppendorf tube, vortex vigorously for 1 min.
4. Spin at 12,000 rpm during 5 min and transfer the supernatant to another tube.
5. Determine protein concentration of plant and bacterial samples using the Bradford method (*see* Note 4).
6. Prepare the reaction mixture:

Reaction buffer	4.936 mL
ATP (stock solution 10 mM)	5 µL
γ^{32}P-ATP (stock solution 10 µCi/µL)	1.5 µL

Km sulfate (stock solution 22 m*M*)	7 μL
NaF or KF (stock solution 1*M*)	50 μL
Total volume	5 mL

7. Pipet 15 μL of each sample in an Eppendorf tube (20 μg of plant and 1 μg of bacterial protein) and add 15 μL of the reaction mixture. Incubate at 37°C for 30 min.
8. Cut a piece of Whatman paper P81 in an appropriate size. Soak the paper on ATP/pyrophosphate solution. Dry at room temperature.
9. Centrifuge the reaction mixture at 12,000 rpm. Apply 20 μL of the supernatant onto the dry Whatman P81 paper, leaving a distance of approx 2 cm between spots.
10. Dry at room temperature and wash for 2 min in 10 m*M* sodium phosphate buffer pH 7.5 at 80°C.
11. Wash the paper 3–5 times in 10 m*M* sodium phosphate buffer at room temperature, monitoring background counts on the negative controls with a Geiger counter.
12. Dry the paper and expose to X-ray film for 16 h.

3.4.3.2. PREPARATION OF *E. COLI* EXTRACT

1. Spin 1 mL of an *E. coli* overnight culture in LB medium in a microcentrifuge.
2. Resuspend the pelleted cells in 200 μL extraction buffer.
3. Incubate on ice for 10–15 min. Add 40 μL Triton X-100.
4. Spin in a microcentrifuge for 10 min. Transfer the supernatant to another tube.
5. Determine protein concentration using the Bradford method (*see* Note 4). *E. coli* extracts can be maintained at –20°C without loosing NPTII activity for up to 3 mo.

4. Notes

1. To confirm the presence of the conjugated plasmid into *Agrobacterium*, prepare plasmid DNA using the boiling method *(35)*. Plasmid yield can be improved by using exponentially growing cultures and washing the cells two to three times with 50 m*M* Tris, 50 m*M* EDTA before the lysis step.
2. When analyzing seed tissues histochemically, after the incubation period with X-Gluc, transfer slices to water and then to 5% sodium hypochlorite to remove excess oil and starch.
3. When screening tumors it should be noted that GUS negative samples are not necessarily escapes, as tumors are not homogeneous. Transformed cells can either contain the Ti plasmid alone or associated with the T-DNA of pGUS INT.
4. Use Bio-Rad Laboratories (Richmond, CA) reagent according to the manufacturer's instructions.

References

1. Woodroof, J. G. (1983) Summary and future outlook, in *Peanut: Production, Processing, Products* (Woodroof, J. G., ed.), Avi Publishing, CT, pp. 369–379.
2. Food and Agriculture Organization of the United Nations (1991) *FAO Yearbook*, vol. 45, Statistics Series No. 91.
3. Pattee, H. E. and Young, C. T. (eds.) (1982) A look to the future, in *Peanut Science and Technology,* American Peanut Research and Education Society, Yoakum, TX, pp. 754–764.
4. Altenbach, S. B. and Simpson, R. B. (1990) Manipulation of methionine-rich protein genes in plant seeds. *Tibtech* **8,** 156–160.
5. Gasser, C. S. and Fraley, R. T. (1989) Genetically engineering plants for crop improvement. *Science* **244,** 1293–1299.
6. Mroginski, L. A., Kartha, K. K., and Shyluk, J. P. (1981) Regeneration of peanut (*Arachis hypogaea*) plantlets by in vitro culture of immature leaves. *Can. J. Bot.* **59,** 826–830.
7. Pittmann, R., Banks, D. J., Kirby, J. S., and Richardson, P. E. (1983) *In vitro* culture of immature peanut (*Arachis* spp.) leaves: morphogenesis and plantlet regeneration. *Peanut Sci.* **10,** 21–25.
8. Narasimhulu, S. B. and Reddy, G. M. (1983) Plantlet regeneration from different callus cultures of *Arachis hypogaea* L. *Plant Sci. Lett.* **31,** 157–163.
9. McKently, A. H., Moore, G. A., and Gardner, F. P. (1990) *In vitro* plant regeneration of peanut from seed explants. *Crop Sci.* **30,** 192–196.
10. McKently, A. H., Moore, G. A., and Gardner, F. P. (1991) Regeneration of peanut and perennial peanut from cultured leaf tissue. *Crop Sci.* **32,** 833–837.
11. Daimon, H. and Mii, M. (1991) Multiple shoot formation and plantlet regeneration from cotyledonary node in peanut (*Arachis hypogaea* L). *Jpn. J. Breed.* **41,** 461–466.
12. Cheng, M., Hsi, D. C. H., and Phillips, G. C. (1992) *In vitro* regeneration of valencia-type peanut (*Arachis hypogaea* L.) from cultured petiolules, epicotyl sections and other seedling explants. *Peanut Sci.* **19,** 82–87.
13. Ozias-Akins, P. (1989) Plant regeneration from immature embryos of peanut. *Plant Cell Rep.* **8,** 217,218.
14. Hazra, S., Sathaye, S. S., and Mascarenhas, A. F. (1989) Direct somatic embryogenesis in peanut (*Arachis hypogaea* L.). *Bio/Technology* **7,** 949–951.
15. Sellars, R. M., Southward, G. M., and Phillips, G. C. (1991) Adventitious somatic embryogenesis from cultured immature zygotic embryos of peanut and soybean. *Crop Sci.* **30,** 408–414.
16. Durham, R. E. and Parrott, W. (1992) Repetitive somatic embryogenesis from peanut cultures in liquid medium. *Plant Cell Rep.* **11,** 122–125.
17. McKently, A. (1991) Direct somatic embryogenesis from axes of mature peanut embryos. *In vitro Cell Dev. Biol.* **27P,** 197–200.
18. Baker, C. M. and Wetzstein, H. Y. (1992) Somatic embryogenesis and plant regeneration from leaflets of peanut, *Arachis hypogaea. Plant Cell Rep.* **11,** 71–75.
19. Gill, R. and Saxena, P. K. (1992) Direct somatic embryogenesis and regeneration of plants from seedling explants of peanut (*Arachis hypogaea*): promotive role of thidiazuron. *Can. J. Bot.* **70,** 1186–1192.

20. Saxena, P. K., Malik, K. A., and Gill, R. (1992) Induction by thidiazuron of somatic embryogenesis in intact seedlings of peanut. *Planta* **187,** 421–424.
21. Dong, J. D., Bi, Y. P., Xia, L. S., Sun, S. M., Song, Z. H., Guo, B., Jiang, X. C., and Shao, Q. Q. (1990) Teratoma induction and the nopaline synthase gene transfer in peanut. *Acta Genet. Sin.* **17,** 13–16.
22. Lacorte, C., Mansur, E., Timmerman, B., and Cordeiro, A. R. (1991) Gene transfer into peanut (*Arachis hypogaea* L.) by *Agrobacterium tumefaciens*. *Plant Cell Rep.* **10,** 354–357.
23. Mansur, E., Lacorte, C., Freitas, V. G., Oliveira, D. E., Timmerman, B., and Cordeiro, A. R. (1993) Regulation of transformation efficiency of peanut (*Arachis hypogaea* L.) explants by *Agrobacterium tumefaciens*. *Plant Sci.* **89,** 93–99.
24. Vancanneyt, G., Schmidt, R., O'Connor-Sanchez, A., Willmitzer, L., and Rocha-Sosa, M. (1990) Construction of an intron-containing marker gene: splicing of the intron in transgenic plants and its use in monitoring early events in *Agrobacterium*-mediated plant transformation. *Mol. Gen. Genet.* **220,** 245–250.
25. Colby, S. M., Juncosa, A. M., and Meredith, C. P. (1991) Cellular differences in *Agrobacterium* susceptibility and regenerative capacity restrict the development of transgenic grapevines. *J. Am. Soc. Hort. Sci.* **116,** 356–361.
26. Murashige, T. and Skoog, F. (1962) A revised medium for rapid growth and bioassays with tobacco tissue cultures. *Physiol. Plantarum* **15,** 473–497.
27. Gamborg, O. L., Miller, R. A., and Ojima, K. (1968) Nutrient requirements of suspension cultures of soybean root cells. *Exp. Cell Res.* **50,** 151–158.
28. Sciaky, K., Montoya, A. L., and Chilton, M.-D. (1978) Fingerprints of *Agrobacterium* Ti plasmids. *Plasmid* **1,** 238–253.
29. Koncz, C. and Schell, J. (1986) The promoter of Ti-DNA gene 5 controls the tissue-specific expression of chimeric genes carried by a novel type of *Agrobacterium* binary vector. *Mol. Gen. Genet.* **204,** 383–396.
30. Hood, E. E., Helmer, G. L., Fraley, R. T., and Chilton, M.-D. (1986) The hypervirulence of *Agrobacterium tumefaciens* A281 is encoded in a region of pTiBo542 outside of T-DNA. *J. Bacteriol.* **168,** 1291–1301.
31. Hinchee, M. W. M., Connor-Ward, D. V., Newell, C. A., McDonnel, R. E., Sato, J. S., Gasser, C. S., Fischhoff, D. A., Re, D. B., Fraley, R. T., and Horsh, R. B. (1988) Production of transgenic soybean plants using *Agrobacterium*-mediated DNA transfer. *Bio/Technology* **6,** 915–922.
32. Ooms, G., Hooykaas, P. J. J., Moolenaar, G., and Schilperoort, R. A. (1981) Crown-gall plant tumours of abnormal morphology, induced by *Agrobacterium tumefaciens* carrying mutated octopine Ti-plasmids: analysis of T-DNA functions. *Gene* **14,** 33–50.
33. Watson, B., Currier, T. C., Gordon, M. P., Chilton, M.-D., and Nester, E. W. (1975) Plasmid required for virulence of *Agrobacterium tumefaciens*. *J. Bacteriol.* **123,** 255–264.
34. Ditta, G., Stanfield, S., Corbin, D., and Helinski, D. R. (1980) Broad host range DNA cloning system for Gram-negative bacteria—construction of a gene bank of *Rhizobium meliloti*. *Proc. Natl. Acad. Sci. USA* **77,** 7347–7351.
35. Scott, R. (1988) DNA restriction and analysis by Southern hybridization, in *Plant Genetic Transformation and Gene Expression* (Draper, J., Scott, R., and Armitage, P., eds.), Alden, Oxford, UK, pp. 237–261.

36. Otten, L. A. B. M. and Schilperoort, R. A. (1978) A rapid micro scale method for the detection of lysopine and nopaline dehydrogenase activities. *Biochim. Biophys.* **527,** 497–500.

37. Reynaerts, A., De Block, M., Hernalsteens, J. P., and Van Montagu, M. (1988) Selectable and screenable markers, in *Plant Molecular Biology Manual* (Gelvin, S. B., Shilperoort, R. A., and Verma, D. P. S., eds.), Kluwer, Academic, Dordrecht, pp. A9, 1–16.

38. Jefferson, R. A. (1987) Assaying chimeric genes in plants: the GUS gene fusion system. *Plant Mol. Biol. Rep.* **5,** 387–405.

39. McDonnell, R. W., Clark, R. D., Smith, W. A., and Hinchee, M. A. (1987) A simplified method for the detection of neomycin phosphotransferase II activity in transformed plant tissues. *Plant Mol. Biol. Rep.* **5,** 380–386.

40. Reiss, B., Sprengel, R., Will, H., and Schaller, H. (1984) A new sensitive method for qualitative and quantitative assay of neomycin phosphotransferase in crude cell extracts. *Gene* **30,** 211–218.

CHAPTER 11

Transformation of Soybean (*Glycine max*) via *Agrobacterium tumefaciens* and Analysis of Transformed Plants

Paula P. Chee and Jerry L. Slightom

1. Introduction
1.1. General Aspects of Soybean Transformation

The transfer of genetic material into soybean (*Glycine max* L. Merr.) plant tissues has been accomplished by several methods: electroporation *(1)*, microprojectiles *(2)*, and by the more widely used *Agrobacterium*-mediated T-DNA transfer *(3,4)*. The transformation of soybean by electroporation-mediated gene transfer appears to be efficient for obtaining stable transformants in soybean cells *(1)*. However, the disadvantage of this method is that the efficiency is affected by many variables, such as voltage, duration, and spacing of electrical pulses, size and number of protoplasts, buffer composition, temperature, concentration, and form of DNA. Thus, optimum electroporation must be determined for each protoplast type. The transformation of soybean by particle bombardment *(2)*, in which the microprojectiles are used to transfer the genetic material into meristematic cells of immature seeds, appears to be simple and very effective. However, it requires a knowledge of the design of the acceleration instruments for those who wish to build their own, or at least access to an instrument that can control the acceleration of microprojectiles. These limitations presently restrict genetic engineering of soybean to some university and corporate laboratories.

From: *Methods in Molecular Biology, Vol. 44:* Agrobacterium *Protocols*
Edited by: K. M. A. Gartland and M. R. Davey Humana Press Inc., Totowa, NJ

The use of the natural gene transfer mechanism of *Agrobacterium tumefaciens* to transform soybean would appear to have some advantages because soybean is a dicotyledonous plant species and thus, it is susceptible to *A. tumefaciens* infection *(3,4)*. The *Agrobacterium* strains that have been found to be the most effective for the infection of soybean cultivars are those of the nopaline-type that encode for the synthesis of the novel amino acid nopaline, in particular strains such as C58 and A208. The first report to describe the use of *Agrobacterium*-mediated gene transfer to obtain a transformed soybean plant line was by Hinchee et al. *(3)*. This report followed (by 5 yr) the first report describing *Agrobacterium*-mediated gene transfers of engineered bacterial *(5)* or plant-derived genes *(6)* into a dicotyledons plant species (tobacco). The major reason for this delay is attributed in part to the difficulty in developing a method in soybean tissues where the transformation and regeneration events occurred in the same cells *(3)*.

During this time, an alternative method of *A. tumefaciens*-mediated transformation of soybean was developed, which is independent of the problems inherent to the tissue-culture procedure. This method involves the infection of germinating soybean seeds with *A. tumefaciens (4)*. This alternative method does not require the use of any tissue-culture steps and is not limited by genotype specificities because the *Agrobacterium* strains C58 and A208 infect most soybean cultivars *(3,4)*. This method has been used for the transformation of many crop species, which include corn *(7)*, *Arabidopsis thaliana (8,9)*, petunia *(10)*, *Phaseolus vulgaris* L., *P. coccineus* L. *(11)*, and sunflower *(12)*. The purpose of this chapter is to describe this procedure as it has been used for the transformation of soybean, and also to describe the molecular biology techniques (genomic DNA isolation, PCR amplification, and genomic blot hybridization) used for the analyses of transgenic soybean plants.

1.2. Transformation of Germinating Soybean Seeds

The ability to obtain transformed plants by infecting germinating seeds with *A. tumefaciens* indicates that some undifferentiated cells are present within the vicinity of the plumule and cotyledonary node regions of germinating seeds. Some of these cells can be transformed by *A. tumefaciens* and can differentiate, eventually yielding transformed ovules or pollen cells. Cellular events of this nature would have to occur to obtain transformed plant tissue and seed progeny as a result of infecting germinating

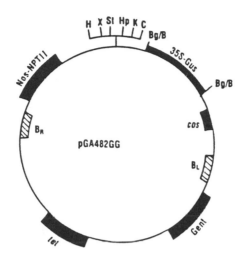

Fig 1. Structure of the *Agrobacterium* binary plasmid pGA482. It contains the T-DNA border fragments of pTi37 (labeled B_R and B_L), the *cos* site of bacterial phage λ, a restriction enzyme polylinker, and the *nos-npt*II fusion gene, *npt*II, driven by the *nos* promoter *(13)*. Restriction enzyme sites are: C – *Clu*I; H = *Hind*III; Hp = *Hpa*; K = *Kpn*; St = *Stu*. Not all site locations are shown.

seeds with *A. tumefaciens*. A general method for the transformation of germinating soybean seeds with *A. tumefaciens* is described in Section 3.1.

1.3. Identification of Transformed Soybean Plants

1.3.1. Neomycin Phosphotransferase II (NPTII) Activity Assay

The methods used for the identification of a transformed plant are dependent on the genetic composition of the T-DNA region transferred into the plant genome. In most cases, binary-type micro T-DNA vectors are utilized for this transfer and these vector contain various chimeric genes that can be used for selection or screening. The most widely used gene is the bacterial transposon Tn5-derived gene that encodes NPTII, which confers resistance to the antibiotic kanamycin. For the binary T-DNA vector pGA482 (Fig. 1), the *npt*II gene has been engineered for plant expression using the nopaline gene promoter *(13)* and this chimeric gene is referred to as the *nos-npt*II gene. However, in the case of transforming germinating soybean seeds, selective growth on kanamycin-containing medium is not effective. In this case, NPTII enzyme activity can be used for the identification of transformed soybean plants *(14)*. The NPTII enzyme activity assay procedure is described in Section 3.2. and in Chapter 19.

1.3.2. Polymerase Chain Reaction (PCR) Analysis
of the nptII Gene

Within the last few years a revolutionary technique has been developed that allows the in vitro synthesis and amplification of specific segments of DNA. This has become feasible by using the technique known as PCR *(15,16)*. This technique can enzymatically amplify a specific region of DNA that is flanked by two oligonucleotide primers that share identity to the opposite DNA strands. Amplification is achieved by a repetitive series of cycles involving template denaturation, primer annealing, and DNA synthesis by DNA polymerase from the annealed primers. By using repetitive cycles, where the primer extension products of the previous cycle serve as new templates for the following cycle, the number of target DNA copies has the potential to double each cycle. Thus if 20 cycles are preformed, theoretically the target DNA could be amplified by a millionfold (2^{20}).

Having the means to amplify a specific region of target DNA by a millionfold is indeed a very powerful method to determine the identity and structural integrity of a transferred target gene. The uniqueness of the PCR technique is that only small amounts of target DNA (usually from 100 ng–1 µg) are needed (*see* Section 3.4.). The DNA need not be pure for amplification (provided the samples do not contain components that inhibit *Taq* polymerase, *see* Section 3.3. for DNA isolation) and the amplified DNA product can generally be detected by gel electrophoresis followed by staining with ethidium bromide. Since the sequence of the target gene and surrounding DNA is generally known (Fig. 1), specific oligomer primers can be synthesized to answer specific structural questions about the integrated target gene. Section 3.4. describes the use of PCR for the identification of the *npt*II gene in putative transformed soybean plants.

1.3.3. Genomic Blot Analysis of Integrated nos-nptII Gene

The identification of which plants are transgenic by using PCR amplification of the target gene is an important step because it eliminates the need to carry forward a large number of putative transformed plants, thus saving space and labor. This allows one to focus the investigation on plants that are indeed transformed by the establishment of transformed plant lines, in which the nature of the transfer event resulting in a particular plant can be investigated in more detail. The primary question concerning a newly transferred target gene is whether or not it is functional. The ability to answer this question depends on the stage of plant development that the target gene is expressed; if the gene is expressed in root, stem, or leaf tissues, the answer can be obtained early. However, if the target gene is expressed later in plant development (flower, seed pod, or seed), the answer to this question may be delayed for some time. In the meantime, structural information can be obtained that describes the target gene in its new environment, and this information could be important in predicting whether the gene will be functional. Although the PCR analysis just described indicates that the target gene is intact, a knowledge of its flanking DNA could be used to determine if its transfer occurred as expected or if it rearranged during the transfer process; T-DNA transfers would be expected to proceed from the right border (T_R) to the left border (T_L) *(17)*. In addition, a knowledge of the number of copies of the target gene in a particular transgenic plant line is important so that the level of expression obtained from each transferred target gene can be determined. Multiple gene transfers have been found using the *Agrobacterium*-mediated gene transfer mechanism *(18)*, and DNA transfers made by electroporation *(19)* and microprojectile bombardment *(20)* quite frequently yield long tandem arrays of interdispersed target gene and plasmid vector DNAs.

2. Materials
2.1. NPTII Enzyme Assay

1. Extraction buffer: 40 mM EDTA, 150 mM NaCl, 100 mM NH$_4$Cl, 10 mM Tris-HCl, pH 7.5, 2.31 mg/mL DTT, 0.12 mg/mL leupeptin (L-2884, Sigma, St. Louis, MO), 0.21 mg/mL trypsin-chymotrypsin inhibitor (Sigma T-9777).
2. Sample buffer: 30 mM NaCl, 15 mM NH$_4$Cl, 5 mM EDTA, 3 mM Tris-HCl, pH 7.5, 0.031 mg/mL DTT, 36 mg/mL sucrose.
3. Reaction buffer: 67 mM Tris-maleate, pH 7.2, 42 mM MgCl$_2$, 400 mM NH$_4$Cl.
4. Others: Kanamycin sulfate (25 mg/mL), [γ-^{32}P]ATP (3000 Ci/mmole), 0.1% bromophenol blue, phosphocellulose paper (Whatman [Maidstone, UK] P81), filter paper (Whatman 3MM), X-ray film (Kodak [Rochester, NY], X-OMAT AR), cellophane paper (Saran Wrap), intensifying screen (Du Pont [Wilmington, DE] Quanta III).

2.2. DNA Isolation

1. CTAB extraction buffer (stock 1 L): 750 mL of sterile, double-distilled water, 100 mL of 1M Tris-HCl, pH 7.5, 140 mL of 5M NaCl, 20 mL 0.5M

EDTA (pH 8.0), 10 g of hexadltrimethylammonium bromide (CTAB, Sigma H-5882), 10 mL of 140 mM β-mercaptoethanol. Add CTAB to solution and stir at warm temperature until completely dissolved. Add β-mercaptoethanol to buffer just prior to use; 140 mM = 1% by volume.

2. TE buffer (stock 100 mL): 1 mL of 1M Tris-HCl, pH 8.0, 0.2 mL of 0.5M EDTA (pH 8.0), 98.9 mL of sterile, double-distilled water.
3. DNA precipitation solution (stock 500 mL): 380 mL of absolute ethanol, 40 mL of 2.5M sodium acetate (pH 6.3), 80 mL of sterile, double-distilled water.
4. DNA washing solution (stock 100 mL): 76 mL of absolute ethanol, 1 mL of 1M ammonium acetate, 23 mL of sterile, double-distilled water.
5. Others: Chloroform/octanol (24:1 [v:v]), cold isopropanol, Thomas-Wiley mill (Baxter, Scientific Products Division, McGaw Park, IL), TKO 100 Mini-Fluorometer (Hoefer, San Francisco, CA).
6. Thomas-Wiley mill or pestle and mortar.
7. Ground glass.
8. Fluorometer (e.g., TKO 100 Hoefer).

2.3. PCR

1. PCR buffer (10X): 500 mM KCl, 100 mM Tris-HCl, pH 8.3, 15 mM MgCl$_2$, 0.1% gelatin (w:v) (Perkin Elmer, Norwalk, CT).
2. dNTP mix: 125 µL of 10 mM dATP, dGTP, dCTP, and dTTP, 500 µL of sterile, double-distilled water (Perkin Elmer).
3. Others: *Taq* polymerase (Perkin Elmer), mineral oil, DNA Thermal Cycler 480 or 9600 (Perkin Elmer), or equivalent.

2.4. Genomic Blot Transfer and Hybridization

1. Denaturation solution: 0.5M NaOH, 1.5M NaCl.
2. Neutralization solution: 1.5M NaCl, 0.5M Tris-HCl, pH 7.0.
3. 10X TBE buffer (stock 4 L): 484.4 g of Tris, 205.4 g of Boric acid, 14.9 g EDTA, dissolve in a final vol of 4 L water, expected pH 8.3. Final concentrations: 4M Tris, 3.3M boric acid, 5 mM EDTA.
4. 6X SSC: 0.9M NaCl, 0.09M Na citrate, pH 7.2.
5. 1X Denhardt's solution: 0.02% (w:v) BSA (bovine serum albumin, Sigma A-4503), 0.02% (w:v) Ficoll (Sigma, F-4375), 0.02% (w:v) PVP (polyvinylpyrrolidone, Sigma, PVP-40).
6. Others: Hoefer Model TE 52 or 42 Transphor (Hoefer), nylon filters (Nytran [Schleicher and Schuell, Keene, NH], Hybond-N [Amersham, Arlington Heights, IL], Genescreen [Du Pont], or others), polyadenylic acid (Sigma P-9403), filter paper (Whatman 3MM), X-ray film (Kodak X-OMAT AR), intensifying screens (Du Pont Quanta III).
7. Electroblotter (e.g., TE 53 or 42 Transphor, Hoefer).

Fig. 2. The *Agrobacterium*-inoculation target area of a germinated soybean seed is shown, the plumule, cotyledonary node, and adjacent regions. The arrows point to the region of the germinating seed where *Agrobacteria* was applied using three separate inoculations with the aid of a syringe fitted with a 30 $^1/_2$ gage needle.

3. Methods

3.1. Soybean Seed Transformation

The following procedure was developed for the transformation of germinating soybean seeds with *A. tumefaciens*. This procedure was previously published by Chee et al. *(4)*.

1. Sterilize the seeds (*Glycine max* L. Merr) by soaking in a 15% Clorox bleach solution for 15 min, follow by several rinses with sterile-distilled water.
2. Germinate the seeds on sterile moistened paper towels in Petri dishes for 18–24 h, at 26°C, in the dark.
3. Remove the seed coats, and one of the two cotyledons of each germinated seed.
4. Inoculate the half seed, with the embryonic axis attached, using overnight liquid cultures ($OD_{600} = 0.5$) of an *Agrobacterium* strain such as C58Z707/pGA482G (Fig. 1) *(4)*. Inoculations are done at three different points (between the arrows shown in Fig. 2) by forcing a $30^1/_2$-gage needle into the plumule, cotyledonary node, and adjacent regions, and injecting about 30 µL of *Agrobacterium* cell solution at each injection point.
5. Place the *Agrobacterium*-infected seeds on a sterile (H_2O) moistened paper towel and incubate at 26°C, in the dark, for about 4 d and then transfer to soil for full development.

3.2. Identification of Transformed R₀ Soybean Plants by Determining the NPTII Enzyme Activity

Identification of soybean plants that are putatively transformed is accomplished by assaying each plant for NPTII enzyme activity *(4,14)*. The enzyme activity of NPTII is determined as follows.

1. Collect 100 mg (2–3 young trifoliate leaves) of material from each of the infected plants. Remove leaves from plants at the position approx 10 or more inches above the inoculation sites.
2. Macerate the tissue (*see* Section 4.1.) with 20 µL of extraction buffer in a 1.5-mL Eppendorf tube and centrifuge for 15 min at 4°C. Remove the supernatant and add to a clean Eppendorf tube.
3. To a 35 µL of the supernatant add 3.5 µL of sample buffer and 1 µL of 0.1% bromophenol blue.
4. Load into the pockets of 10% nondenaturing polyacrylamide gel (ndPAGE) (*see* Notes 1–3). Run ndPAGE at room temperature and do not run the dye front off the bottom as the enzyme runs near the dye.
5. After electrophoresis, rinse the gel twice with water and equilibrate the gel for 30 min in 100 mL of reaction buffer.
6. Transfer the gel onto a glass plate and cover with a layer of 1% agarose containing kanamycin sulfate (10 mM) and [γ-^{32}P]ATP (*see* Notes 1–3). Incubate for 30 min at room temperature.
7. Place one sheet of phosphocellulose paper (Whatman P81) on top of gel, followed by two sheets of Whatman 3MM paper, a stack of blotting paper, and a weight (1 kg). Transfer for 3 h.
8. Remove the P81 paper from the sandwich, wash four or five times with hot water (80°C), followed by several times with cold water. Wash each for 5 min.
9. Wrap the paper in cellophane paper or Saran Wrap and expose to X-ray film along with an intensifying screen overnight or up to 1 wk at –70°C.

3.3. DNA Isolation

This section describes a procedure for the isolation of total plant DNA that is relatively simple, yet yields DNA pure enough to be used for PCR amplifications (*see* Section 3.4.) and restriction enzyme digestions. This procedure also yields sufficient quantity of DNA to allow for the analysis of the target gene by the more laborious blot hybridization technique (*see* Section 3.5.). The preparation described in the next section generally yields from 100–200 µg of DNA, much more DNA than is necessary for PCR detection of the target gene. If target gene detection is all that is required, this DNA isolation procedure can be scaled down by a factor of 10 or more. We recommend that at least 2 µg of DNA be isolated in case several PCR amplifications are needed. Once a particular transformed plant is identified as containing the target gene, the determination as to whether it is integrated within the plant genome by genomic blot hybridizations can proceed rapidly (*see* Section 3.5.).

1. Collect leaves and wrap in cheese cloth, mark with appropriate identification numbers, then freeze in liquid nitrogen.
2. Transfer frozen tissue bundles into a freeze-dryer container and then connect to the freeze-dryer before leaves begin to thaw. Freeze-dry until the container reaches ambient room temperature (if cold to the touch, liquid is still present). This freeze-drying step can usually be done overnight depending on the strength of the vacuum source.
3. Grind dried tissues to a fine powder by using a Thomas-Wiley mill (*see* Note 4). After each sample is ground, clean the Thomas-Wiley mill so that the next sample will not be contaminated by residues from the previous samples (nontransformed and other special control tissues should be ground first).
4. Weigh out 300–400 mg of fine powder in a 15-mL polypropylene centrifuge tube. Distribute tissue powder along sides of tube to avoid a clump of dry tissue in the bottom, and add 9.0 mL CTAB extraction buffer. Mix several times by inversion and vortex briefly. (If the amount of tissue is less, use appropriately reduced volumes of CTAB extraction buffer and other components listed throughout the procedure.)
5. Incubate for 90 min in a 65°C water bath; mix samples every 30 min by inverting several times.
6. Remove tubes from bath, wait 4–5 min to allow the tube to cool, and then add 4.5 mL chloroform/octanol (24:1 [v:v]). Mix gently by inverting several times.
7. Spin at room temperature in tabletop centrifuge for 10 min at full speed (3000–5000 rpm).
8. Pour off the top aqueous layer into new 15 mL polypropylene centrifuge tubes, add 4.5 mL chloroform/octanol (24:1), mix gently, and repeat step 7.
9. Pipet the top aqueous layer into 15-mL polypropylene tubes containing 6.0 mL cold isopropanol (2-propanol). Mix *very* gently by inverting.
10. After 10–15 min, remove precipitated DNA with a glass hook and transfer to 5-mL polypropylene tubes containing 3 mL of 76% EtOH, 0.2M NaOAc. Leave DNA on the hook in tubes for about 20 min.
11. Rinse DNA on hook briefly in 1–2 mL of 76% EtOH, 10 mM NH$_4$OAc and immediately transfer DNA to 5-mL polypropylene tubes containing 1.0 mL of TE buffer.
12. After the DNA is resuspended, determine the DNA concentration by the use of a TKO 100 Mini-Fluorometer. This size of DNA preparation generally yields from 100–200 µg of DNA.

3.4. Identification of a Target Gene
by PCR Amplification

This section describes the use of PCR amplifications to identify the presence of a specific region of the chimeric plant expressible *nos-npt*II gene in *npt*II positive plants identified in Section 3.2. This chimeric *npt*II gene uses the *Agrobacterium* T-DNA-derived nopaline gene *(nos)* regulatory regions (promoter and polyadenylation signal) to regulate the expression of the *E. coli* transposon Tn5-derived *npt*II gene *(13,21,22)* (Fig. 1). The oligomer primers used (21 and 24 nucleotides in length) were selected to PCR amplify a DNA product of a size that would indicate that the transferred *nos-npt*II gene was intact (containing the *nos* promoter and complete *npt*II gene coding region). The 5'-primer, referred to as primer A (5'-CCCCTCGGTATCCAATTAGAG-3'), shares identity with the *nos* promoter region 33 bp 5' of the *nos* gene translation initiation codon (ATG) *(21)*. The 3'-primer, referred to as primer B (5'-CGGGGGGTGGGCGAAGAACTCCAG-3'), shares identity with the opposite strand of 3'-flanking region of the *npt*II gene, 150 bp 3' of the *npt*II translation termination codon (TAA) *(22)*. The DNA product amplified by these primers is about 1000 bp in length; this was confirmed by PCR amplification using 10 ng of pGA482 DNA as a positive control template.

1. From the DNA concentration of each sample, determine the volume needed to obtain 200 ng; also from the concentration of each PCR primer (concentrations are determined by measuring OD_{260}, and 1 OD = 20 µg/mL), determine the volume needed to obtain a final concentration of 0.5–1 µM in a total reaction volume of 100 µL (*see* Notes 5 and 6).
2. For each PCR sample, use a sterile 0.5-mL centrifuge tube; label each tube accordingly. The order in which the reaction components is added is important; generally, add the component with the largest volume first (usually water), followed by the buffer and dNTP mix, then add the DNAs (genomic and oligomer primers), and add the *Taq* polymerase last.
3. Assuming a genomic DNA concentration of 200 µg/mL and the concentration of each oligomer primer is 200 µg/mL, the following components are added to each PCR sample tube (*see* Table 1).
4. Mix gently, then briefly spin in a centrifuge. If necessary, overlay each PCR sample with 100 µL of mineral oil to prevent evaporation.
5. Ensure that the temperature controls are in place, i.e., if water baths are used, they should be at the appropriate temperatures; if an automated temperature cycler is used, the appropriate program should already be

Table 1
PCR Components

Component	Order of addition	Volume (µL)
Double-distilled, sterile water	1	66.5
PCR Buffer, 10X	2	10.0
dNTPs Mix, 1.25 m*M*, each dNTP	3	16.0
Primer A, 200 µg/mL	4	3.0
Primer B, 200 µg/mL	5	3.0
Genomic DNA, 200 µg/mL	6	1.0
Taq polymerase, 2.5 U/µL	7	0.5
Total		100.0

selected. Recommended temperatures and times used for PCR amplification of the *nos-npt*II gene given in Chee et al. (4) are as follows: duplex denaturation, 94°C, 2 min; annealing, 55°C, 2 min; and DNA synthesis, 72°C, 3 min.

6. Repeat the temperature cycles for the appropriate number of times; in the case of PCR amplification of the *nos-npt*II gene, 30 cycles were used. On the last cycle, omit the heat denaturation step and extend the DNA synthesis step to 10 min to aid strand completion of the accumulated DNAs. At this point the samples can be analyzed, or, if desired, the samples can be stored at refrigerator temperatures or frozen at –20°C. If a refrigerated, automated, thermal cycler is used, it can be programmed to maintain the samples at 4°C overnight for analysis the following day.

7. Analysis of the PCR reaction products is very straightforward depending on the size of the expected DNA product. About half (50 µL) of each sample is electrophoresed on either a 0.7% agarose gel (if the DNA product is larger than 700 bp) or 7% acrylamide 15% glycerol gel (if the DNA product is smaller than 700 bp). After electrophoresis, the gels are stained with ethidium bromide (a solution containing about 0.5 ng/mL) and observed by placing on a UV light box and photographed (Fig. 3). If an amplified DNA fragment is not visible, *see* Notes 5 and 6.

3.5. Analysis of Transgenic R_0 Soybean Plants for Integration of nos-nptII Gene

Following the identification of transformed soybean plants, the structural integrity of a transferred gene can best be determined using the genomic blot hybridization technique (23). For this analysis, total genomic DNA is isolated (*see* Section 3.3.), digested with one or more restriction enzyme(s), followed by electrophoresis through a gel matrix,

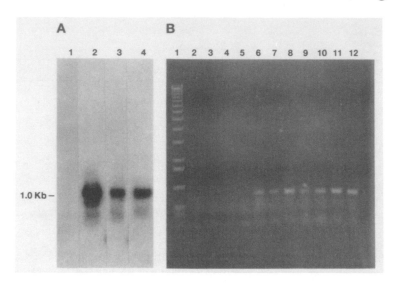

Fig. 3. PCR amplification of the *nos-npt*II gene region from DNA isolated from transformed soybean plants. (**A**) DNA were isolated from two putative transformed R_0 (numbered 5 and 6) soybean plants. DNAs (200 ng) were subjected to PCR amplification using two oligomer primers specific for amplification of a fragment of the *nos-npt*II gene of approx 1000 bp in length. One primer annealed to the 5'-*nos* promoter region while the other annealed to the 3'-flanking region of the *npt*II gene, thus the entire *npt*II coding could be amplified. After PCR amplification, one-third of each sample was electrophoresed in a 0.7% agarose gel followed by electroblotting onto a nylon filter. The filter was hybridized against the ^{32}P-labeled 600 bp, *Bgl*II to *Nco*I, *npt*II gene fragment *(4,22)* and hybridizing bands detected by autoradiography. Lane 1 contains the PCR amplification of 200 ng of soybean DNA isolated from a nontransformed DNA (negative control) and lane 3 contains DNA from the amplification of 10 ng of pGA482 (positive control). Lanes 3 and 4 contain PCR amplification results obtained from putative transformed soybean plants numbers 5 and 6. (**B**) The ethidium bromide staining pattern for the agarose gel analysis of PCR amplified samples. DNA were isolated from three R_1 plants derived from putative transformed soybean plant number 5. DNA size standard (BRL 1 kb ladder) is shown in the left lane. Lanes labeled 1–3 are the result of PCR amplifying soybean DNA isolated from three independent nontransformed soybean plants. Lanes 4 and 5 contain the PCR amplification results using DNAs isolated from three progeny plants derived from plant number 5, 5-1, 5-2, and 5-3, respectively. Lane 7 contains the PCR amplification of 10 ng of pGA482, the positive control.

and transferred onto a filter membrane (nitrocellulose or nylon). After fixing the DNA onto the filter membrane, it is hybridized against a specific ^{32}P-labeled nucleotide probe (the complete or partial fragment of the gene or its cDNA component) followed by exposure to film and an analysis of this autoradiograph to determine if the targeted gene is contained within the expected size restriction enzyme fragments. Information gained from the genomic blot hybridization technique can be used to show integration of a transferred gene within a plant genome.

This section first describes the electroblotting of restriction enzyme digested genomic DNA onto nylon filters followed by the hybridization of these electroblotted nylon filters against ^{32}P-labeled DNA probes.

3.5.1. Electroblotting Genomic DNA onto Nylon Filters

1. After electrophoresis is completed (*see* Notes 7–10), stain the DNA with ethidium bromide and photograph the gel. Place a ruler alongside the gel so that the distance that any given band of DNA has migrated can be read directly from the photographic image.
2. Transfer the gel to a glass Pyrex dish or suitable plastic box and trim away any unused areas of the gel with a razor blade.
3. Denature the DNA by immersing the gel in several volumes of 0.25N NaOH for 20 min at room temperature with constant agitation. If the DNA fragments to be transferred are large (>10 kb), the DNA should be fragmented by first treating with 0.25M HCl at 22°C for 15 min to break the DNA strands by depurination.
4. Discard the solution and rinse gel briefly in deionized water, then soak in 0.5X TBE buffer for 20 min at room temperature with constant agitation.
5. Fill the electroblot tank (Hoefer Model Transphor unit TE 52 or 42) with about 4 L of 0.5X TBE.
6. Open the cassette and place one-half in a big tray such that the hook faces up.
7. Place one wet porous pad (e.g., Scotch-brite) on the cassette half. Press on the pad to force out any trapped bubbles.
8. Lay two pieces of blot paper prewet with 0.5X TBE on the sponge.
9. Lay the gel on the blot paper. Make sure there are no air bubbles between the blot paper and the gel.
10. Cut a piece of nylon filter to the size of the gel. Wear gloves when handling the nylon filter. Use a ballpoint pen (permanent ink) to draw a line on the filter to indicate the position of the gel wells, also write down your name, date, and experiment number.
11. Float filter in deionized water until completely wet, then soak in 0.5X TBE until ready to use.

12. Place the wet nylon filter on top of gel so that the line on the filter is on the line of slots at the top of the gel. Be careful to remove all air bubbles that are trapped between the gel and the filter.

13. Place one or two additional pieces of blot paper on top of the nylon filter. Again, remove all air bubbles.

14. Place another wet porous pad on top of the blot paper without trapping air bubbles.

15. Place the second half of the cassette on top of the stack such that the hook is down and faces the hole near the edge of the bottom half.

16. Press the two halves together and slide them toward each other, making sure both hooks engage with the opposite half. Assembled cassette can be secured tightly with pins, clamps, or string.

17. Insert the assembled cassette into the TE unit with the membrane on the side of the gel facing the anode.

18. Add enough 0.5X TBE to cover the blotting assembly.

19. Place the TE Power Lid on top of the chamber for TE 52 unit.

20. If using the TE 42 unit with a separate power supply, place the lid on the chamber and connect the leads to the power supply.

21. Plug in the power supply and turn on the power switch. Electroblot the DNA sample onto the nylon filter by using 20 V for 12 h at 4°C. Do not permit buffer in chamber to heat. If DNA fragments larger than 20 kb are to be electroblotted, *see* Section 4.4.

22. After electroblotting, pull the assembled cassette out, dismantle blotting stack down to filter, remove filter, and gel together. Place filter side down on a clean glass Pyrex dish or suitable plastic container.

23. Peel off gel and restain in ethidium bromide to determine if most of the DNA has been transferred. Soak filter in denaturation solution ($1.5M$ NaCl, $0.5M$ NaOH) for 20 min with slow agitation.

24. Discard solution and replace with neutralization solution ($1.5M$ NaCl, $0.5M$ Tris-HCl, pH 8.0). Leave filter in neutralization solution for 20 min.

25. Place wet filter with DNA side up under a UV source such as a transilluminator for 3 min or an ultraviolet crosslinking device to crosslink the DNA to the filter.

26. The filter can be used immediately for hybridization. If not used immediately the filter can be wrapped with Saran Wrap and stored at 4°C.

3.5.2. Genomic Blot Hybridization

1. Place filter in flat-bottomed plastic container containing prehydridization solution (6X SSC, 1X Denhardt's, 0.5% SDS), and agitate at 68°C for at least 2 h.

2. Denature the probe by adding 0.1 vol $5N$ NaOH and incubate at 65°C for 15 min.

3. Add 0.1 vol of 5N HCl followed by 0.1 vol 2M Tris-HCl (pH 7.5).
4. Remove prehybridization solution from container, add hybridization solution (6X SSC, 1X Denhardt's reagent, 2.0% SDS, 50 µg/mL polyadenylic acid), then add the probe at a concentration of 1×10^6 cpm/mL of hybridization solution for the detection of single copy genes. More abundant DNA sequences can be detected using 5×10^5 cpm/mL.
5. Agitate gently at 68°C for 12 h.
6. Remove hybridization solution, rinse with 3X SSC, and wash filter in 300 mL of 3X SSC for 1 h at 65°C with constant gentle agitation.
7. Discard solution and rewash filter in 300 mL of 1X SSC for 1 h at 65°C with constant gentle agitation.
8. Dry the filter at room temperature on a sheet of Whatman 3MM paper.
9. Wrap the filter in Saran Wrap and apply to X-ray film to obtain an autoradiographic image. Store film exposure at –80°C (*see* Section 4.4.). The film exposure resulting from the genomic blot hybridization of the ^{32}P-labeled *npt*II gene against DNAs isolated from soybean plants that were putatively transformed with the T-DNA of the binary vector pGA482G is shown in Fig. 4.
10. After exposure, save the filters as they can be used many times, either for hybridizations against the same probe (if a higher specific activity probe is needed) or against a different gene probe (*see* Notes 7–10).

4. Notes

4.1. npt*II Enzyme Activity Assay*

1. Include controls of uninfected plant and an extract of the *npt*II enzyme prepared from an *E. coli.* strain that contains the bacterial expressed *npt*II gene, for example *E. coli.* HB101 strain containing the plasmid pKS4 *(6)*. The *E. coli* is prepared as follows: Grow the cells in 20 mL of LB (Luria-Bertani) medium (10 g Bacto-tryptone, 5 g bacto-yeast extract, 10 NaCl, adjust to pH 7.5 with NaOH) to a density of 4×10^8 cells/mL. Centrifuge and resuspend in 50 µL of extraction buffer. Sonicate the cells to obtain a crude extract and centrifuge to remove cellular debris. The supernatant can be added at various dilutions in extraction buffer.
2. The 10% ndPAGE is carried out as follows: *Separating gel*: 10 mL of 29% acrylamide, 1% bisacrylamide, 12.2 mL of H_2O, 7.5 mL of 1.5M Tris-HCl pH 8.8, 10 µL of TEMED, 200 µL of 10% ammonium persulfate. *Stacking gel*: 1.5 mL of 29% acrylamide, 1% bisacrylamide, 1.25 mL of 1M Tris-HCl pH 6.8, 7.0 mL of H_2O, 10 µL of TEMED, 30 µL of 10% ammonium persulfate. Electrophoresis buffer is made by dissolving 3.0 g of glycine and 0.605 g of Tris base in 1 L of water.
3. The agarose gel overlay is prepared by mixing 15 mL of reaction buffer, 15 mL of melted 2% agarose in water, 40 µL of kanamycin sulfate (25 mg/

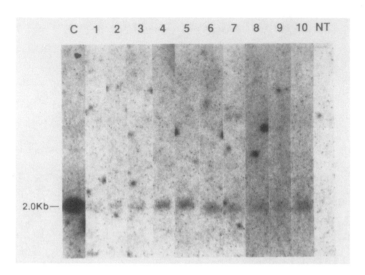

Fig. 4. Blot hybridization analysis of genomic DNA isolated from R_0 soybean plants. Total genomic DNAs were isolated from putative transformed soybean plants and digested with *Hind*III and *Bam*HI. Digested DNAs were electrophoresed in a 0.7% agarose gel and then blotted onto a nylon filter. The blotted filter was hybridized against the ^{32}P-labeled *npt*II gene 600 bp *Bgl*II to *Nco*I fragment *(4,22)*. After washing, the filter was autoradiographed and the resulting film revealed hybridizing fragments of approx 2 kb in length. The size of these hybridizing fragments agrees with that expected from a *Hind*III and *Bam*HI digest of the *nos-npt*II gene contained within the T-DNA region of pGA482 *(13)*. Lane C contains the positive control, this control represents a one gene copy per soybean haploid genome reconstruction using pGA482 DNA digested with *Hind*III and *Bam*HI. Lanes labeled 1–10 contain DNAs from soybean plants which showed *npt*II enzyme activity and were found to hybridize to the ^{32}P-labeled *npt*II gene fragment just described. The lane labeled NT contains DNA isolated from a nontransformed soybean plant.

mL), and 100 µCi of [γ-^{32}P]ATP (3000 Ci/mmole). This will be enough to cover a 200 cm^2 gel and should be scaled accordingly for larger gels. The use of more [γ-^{32}P]ATP can increase the sensitivity.

4.2. DNA Isolation

4. If a Thomas-Wiley mill is not available, the plant tissues can be ground using a pestle and mortar and ground glass. However, for a large number of samples this routine is quite tedious and thus a smaller sample size should be used (20–100 mg of tissue).

4.3. PCR Amplification

5. For oligomer primers in the range of 20 nucleotides, use between 300 and 600 ng per reaction. If the final concentration of the oligomer is to be about 1 μM, then the number of moles of the oligomer is calculated as follows: $(1 \times 10^{-6} \text{ mol/L}) \times (1 \times 10^{-4} \text{ L}) = 1 \times 10^{-10}$ mol. Assuming the average molecular weight of a nucleotide of 300 g/mol, the molecular weight of a oligomer of 20 nucleotides is about 6000 g/mol. Thus, the mass of the oligomer to be added is $(1 \times 10^{-10} \text{ mol}) \times (6000 \text{ g/mol}) = 600$ ng.

6. PCR products from plants containing the target gene should yield a visible DNA band at the expected size, as shown in Fig. 3 for the amplification of the *nos-npt*II gene transferred into the soybean genome. The lack of a visible band should not be concluded that the sample does not contain the target gene, as it could be an indication of a poor PCR amplification owing to having impure genomic DNA, or having made an error in measuring its concentration. Before discarding the sample (and putative transgenic plant) the gel containing the sample should be challenged by hybridization against ^{32}P-labeled target DNA, in this case, a DNA fragment from the *npt*II gene. Hybridization results should determine if there is any small amount of PCR product present and also confirm the identity of the band in samples with visible DNA bands. We have found the 600 bp fragment generated by a *Bgl*II and *Nco*I digest is an excellent radioactive probe fragment because of its low background when hybridized against PCR-generated products or total plant DNAs. Before discarding the sample, PCR amplifications should be repeated in which the amount of genomic DNA is increased (after rechecking that its concentration was correctly measured).

4.4. Genomic Blot Transfer and Hybridization

7. About 10 μg of total genomic DNA (haploid genome = 3×10^9 bp) must be applied to a single gel-slot in order to be able to detect sequences that occur only at the single-copy level.

8. The electroblot settings described are sufficient to transfer DNAs in the range of 20 kb. If larger DNA fragments are to be transferred, the voltage or the electroblotting time should be increased. However, if the voltage is increased, be sure that the apparatus is kept cool.

9. Film exposures are generally done using two Quanta III intensifying screens (Du Pont) and exposure time range between 24 and 72 h.

10. Hybridized DNA probes can be removed from the filters by washing in distilled water at 68°C for about 30 min, or by a brief treatment with 0.2N NaOH at room temperature.

5. Concluding Comments

Transformation of soybean can be accomplished by infecting the cells of the plumule, cotyledonary node, and adjacent cotyledonary tissues of germinating seeds with *Agrobacteria*. The frequency of transformation is about 0.7% for obtaining transformed tissues in the R_0 generation and approx 1 in 10 of these R_0 plants will yield transformed R_1 progeny. The overall frequency of this method is low, about 0.07%. Despite the low frequency of this method for obtaining transformed soybean plants, it offers certain advantages, such as avoiding tissue-culture techniques and in being technically simple. Disadvantages inherent to this transformation technique are directly related to the low frequency of obtaining transgenic plants, which increases the amount of labor, glasshouse space, and time needed to obtain a reasonably large number of stable transgenic plants.

References

1. Dhir, S. K., Dhir, S., Savka, M. A., Belanger, F., Kritz, A. L., Farrand, S. K., and Widholm, J. M. (1992) Regeneration of transgenic soybean (*Glycine max* L.) plants from electroporated protoplasts. *Plant Physiol.* **99,** 81–88.
2. McCabe, D. E., Swain, W. F., Martinell, B. J., and Christou, P. (1988) Stable transformation of soybean (*Glycine max*) by particle acceleration. *Bio/Technology* **6,** 923–926.
3. Hinchee, M. A. W., Cannor-Ward, D. V., Newell, C. A., McDonnell, R. E., Sato, S. J., Gasser, C. S., et al. (1988) Production of transgenic soybean plants using *Agrobacterium*-mediated DNA transfer. *Bio/Technology* **6,** 915–923.
4. Chee, P. P., Fober, K. A., and Slightom, J. L. (1989) Transformation of soybean (*Glycine max*) by *Agrobacterium tumefaciens*. *Plant Physiol.* **91,** 1212–1218.
5. Fraley, R. T., Rogers, S. G., Horsch, R. B., Sanders, P. R., Flick, J. S., Adams, S. P., et al. (1983) Expression of bacterial genes in plant cells. *Proc. Natl. Acad. Sci. USA* **80,** 4803–4807.
6. Mauri, N., Sutton, D. W., Murray, M. G., Slightom, J. L., Merlo, D. J., Reichert, N. A., et al. (1983) Phaseolin gene from bean is expressed after transfer to sunflower via tumor-inducing plasmid vectors. *Science* **222,** 476–482.
7. Graves, A. C. F. and Goldman, S. L. (1986) The transformation of *Zea mays* seedlings with *Agrobacterium tumefaciens*. *Plant Mol. Biol.* **7,** 43–50.
8. Felmann, K. A. and Marks, M. D. (1987) *Agrobacterium*-mediated transformation of germinating seeds of *Arabidopsis thaliana*: a non-tissue culture approach. *Mol. Gen. Genet.* **208,** 1–9.
9. Feldmann, K. A., Marks, M. D., Christianson, M. L., and Quatrano, R. S. (1989) A dwarf mutant of *Arabidopsis* generated by T-DNA insertion mutagenesis. *Science* **243,** 1351–1354.
10. Ulian, E. C., Smith, R. H., Gould, J. H., and McKnight, T. D. (1988) Transformation of plants via the shoot apex. *In Vitro Cellular Dev. Biol.* **24,** 951–954.

11. Mariotti, D., Fontana, G. S., and Santini, L. (1989) Genetic transformation of grain legumes: *Phaseolus vulgaris* and *P. coccineus* L. *J. Genet. Breed.* **43**, 77–82.

12. Schrammeijer, B., Sijmons, P. C., van den Elzen, P. J. M., and Hoekema A. (1990) Meristem transformation of sunflower via *Agrobacterium. Plant Cell Rep.* **9**, 55–60.

13. An, G. (1986) Development of plant promoter expression vectors for analysis of differential activity of nopaline synthase promoter in transformed tobacco cells. *Plant Physiol.* **81**, 86–91.

14. Reiss, B., Sprengel, R., Will, H., and Schaller, H. (1984) A new sensitive method for qualitative and quantitative assay of neomycin phosphotransferase in crude cell extracts. *Gene* **30**, 211–218.

15. Mullis, K. B., Faloona, F., Scharf, S. J., Saiki, R. K., Horn, G. T., and Erlich, H. A. (1986) Specific enzymatic amplification of DNA *in vitro*: the polymerase chain reaction. *Cold Spring Harbor Symp. Quant. Biol.* **51**, 263–273.

16. Saiki, R. K., Scharf, S., Faloona, F., Mullis, K. B., Horn, G. T., Erlich, H. A., and Arnheim, N. (1985) Enzymatic amplification of β-globin genomic sequences and restriction site analysis for diagnosis of sickle cell anemia. *Science* **230**, 1350–1354.

17. Wang, K., Herrera-Estrella, L., Van Montagu, M., and Zambryski, P. (1984) Right 25 bp terminus sequence of the nopaline T-DNA is essential for and determines direction of DNA transfer from *Agrobacterium* to the plant genome. *Cell* **38**, 455–462.

18. Jorgensen, R., Snyder, C., and Jones, J. D. G. (1987) T-DNA is organized predominantly in inverted repeat structures in plants transformed with *Agrobacterium tumefaciens* C58 derivatives. *Mol. Gen. Genet.* **207**, 471–477.

19. Riggs, C. D. and Bates, G. W. (1986) Stable transformation of tobacco by electroporation: evidence for plasmid concatenation. *Proc. Natl. Acad. Sci. USA* **83**, 5602–5606.

20. Christou, P., Swain, W. F., Yang, N.-S., and McCabe D. E. (1989) Inheritance and expression of foreign genes in transgenic soybean plants. *Proc. Natl. Acad. Sci. USA* **86**, 7500–7504.

21. Depicker, A., Stachel, S., Dhaese, S., Zambryski, P., and Goodman, H. M. (1982) Nopaline synthase: transcript mapping and DNA sequence. *J. Mol. Appl. Genet.* **1**, 561–573.

22. Mazodier, P., Cossart, P., Giraud, E., and Gasser, F. (1985) Completion of the nucleotide sequence of the central region of Tn*5* confirms the presence of three resistance genes. *Nucl. Acid Res.* **13**, 195–205.

23. Southern, E. M. (1975) Detection of specific sequences among DNA fragments separated by gel electrophoresis. *J. Mol. Biol.* **98**, 503–517.

CHAPTER 12

Agrobacterium-Mediated Transformation of Potato Genotypes

Amar Kumar

1. Introduction

The potato (*Solanum tuberosum*) is the world's fourth major food crop. The potato is ideally suited for improvement by modern genetic manipulation methods, as it is highly amenable to a wide range of various tissue-culture techniques *(1)*. In recent years, a combination of cellular and molecular approaches, including somatic hybridization and genetic transformation, have been employed to improve existing potato cultivars *(2)*. Genetic transformation, however, provides a more direct and controlled method for manipulating the genome of potato for both basic and applied research purposes *(3)*. This chapter describes the *Agrobacterium*-mediated transformation method for potato genotypes. *Agrobacterium tumefaciens* and its Ti-plasmid have been used extensively as vectors to introduce engineered genes into plants *(4)*. In order for genetic modification via *Agrobacterium* to be successful, both a reproducible regeneration system and an efficient DNA delivery system directed toward those cells capable of regeneration are mandatory. An ability to select the transformed cells from a large number of untransformed cells is also essential, because transformation frequency is usually between 5 and 50%.

An efficient and reliable transformation system of potato has been achieved mainly for tetraploid genotypes, especially among the European cultivars Desiree, Bintje, Pentland Squire, Maris Piper, Maris Bard, and the American cultivar Russet Burbank *(5–12)*. However, transgenic plants can also be produced from dihaploid potato lines and from several

From: *Methods in Molecular Biology, Vol. 44:* Agrobacterium *Protocols*
Edited by: K. M. A. Gartland and M. R. Davey Humana Press Inc., Totowa, NJ

Table 1
Flow Chart for Potato Transformation

Day	Events
1	Cut and precondition explants. Start *Agrobacterium* culture.
2	Subculture *Agrobacterium* culture and add acetosyringone.
3	Inoculate explants with *Agrobacterium* and culture explants on LSR-1 medium without antibiotics.
4	Transfer explants on to selection medium (LSR-1) with antibiotics.
15	Transfer explants to fresh LDR-1 selection medium.
30	If green calli appear on explants, transfer to LSR-2 selection medium. Otherwise, subculture explants to fresh LSR-1 medium.
45	Transfer explants with green calli to LSR-2 medium for shoot production. Transfer individual shoots to MS30 medium with antibiotic for rooting.
60	Subculture shoot tips fresh to MS30 with antibiotics.
75	Transfer well-rooted plants to soil.

wild *Solanum* species *(3)*. Several transformation methods have been used that rely on the cocultivation of *Agrobacterium* strains with various types of explants, including discs from mature tubers *(8)* or in vitro microtubers *(12)*, and leaf and stem segments *(5–7,9–11)*. Also, both binary and cointegrate vectors derived from *A. tumefaciens* strains such as LBA4404 and C58, which have been disarmed, are capable of efficiently transforming potato genotypes *(5,6)*.

In this chapter the use of leaf and stem explants from in vitro-grown plant materials for transforming potato cultivars Desiree and Pentland Squire via the *Agrobacterium*-mediated transformation method is described. This transformation protocol has produced transgenic plants within 6–8 wk (Table 1). The main advantage of using in vitro-grown plant material for transformation is that it is sterile and is already conditioned to grow in culture. As just mentioned, establishment of an efficient and reliable plant regeneration system from leaf and stem explants is a prerequisite for any transformation protocol. Unfortunately, plant regeneration from explants is dependent on the potato genotype used, together with the specific concentration and combination of growth regulators in the culture medium. It is therefore virtually impossible to give a regeneration and transformation protocol that works well with every genotype. However, the regeneration and transformation protocol provided here can be employed for a range of tetraploid and dihaploid potato genotypes.

2. Materials

2.1. Potato Shoot Cultures

1. Tubers of potato (*Solanum tuberosum L. ssp. tuberosum*) cultivars Desiree and Pentland Squire, from the germplasm collection of Scottish Crop Research Institute (Dundee, Scotland).
2. Sterilizing solution for plants: "Domestos" bleach (Lever Bros., Kingston-upon-Thames, UK).
3. Sterilizing media, solutions, and containers: All media, solutions, containers, and equipment should be sterilized by standard autoclaving methods. All plant growth substances, antibiotics and vitamins should be filter sterilized.
4. Nescofilm: Bando Chemical Co. (Kobe, Japan).
5. MS10 medium: 4.3 g/L MS salts (Sigma, St. Louis, MO), 1 mL/L B5 vitamins (Sigma), 20 g/L sucrose (BDH, Poole, UK), 8 g/L agar (Difco, Surrey, UK), pH 5.8.
6. MS30 medium: 4.3 g/L MS salts, 1 mL/L B5 vitamins, 30 g/L sucrose, 8 g/L agar, pH 5.8.

2.2. Transformation of Potato Cultivars

1. Disarmed *A. tumefaciens* strains LBA 4404 containing the binary vectors pSLJ 4D4. 81 or KIWI 105 both carrying the neomycin phosphotransferase (*npt*II) gene and the β-glucuronidase (GUS) gene *(3)*, or containing pSCR107 carrying the *npt*II gene and the coat protein gene of potato leafroll virus gene *(6)*. Also, disarmed *A. tumefaciens* strain C58 containing cointegrate vector pKU2 carrying the *npt*II and hygromycin phosphotransferase II (*hpt*II) genes *(6)*.
2. MS30 liquid medium: 4.3 g/L MS salts, 1 mL/L B5 vitamins, 30 g/L sucrose, pH 5.8.
3. High hormones (HH) medium: MS30 liquid with 10 mg/L α-naphthaleneacetic acid (NAA: Sigma), 10 mg/L zeatin riboside (ZR), or zeatin (Sigma).
4. Leaf and stem regeneration (LSR-1) medium: MS30 with 0.2 mg/L NAA, 0.02 mg/L gibberellic acid (GA$_3$; Sigma), 2.00 mg/L ZR, 0.8% agar (Difco).
5. LSR-2 medium: MS30 with 0.02 mg/L NAA, 0.02 mg/L GA$_3$, 2.00 mg/L ZR, 0.8% agar.
6. Transgenic callus selection medium: LSR-1 with 250 mg/L cefotaxime (Roussel, Uxbridge, UK) or augmentin (Beecham, Middlesex, UK), 50–100 mg/L kanamycin sulfate, or 10–30 mg/L hygromycin B hydrochloride (Sigma).
7. Transgenic shoot selection medium: LSR-2 with 250 mg/L cefotaxime or 200 mg/L augmentin, 50–100 mg/L kanamycin sulfate, or 20–30 mg/L hygromycin B hydrochloride.

8. Transgenic root selection medium: MS30 with 250 mg/L cefotaxime or augmentin, 100 mg/L kanamycin sulfate or 20–30 mg/L hygromycin B hydrochloride.
9. Bacterial culture medium: LB (Luria Broth) medium, 1% Bactopeptone, 0.5% Bacto-yeast extract, 1% (w/v) NaCl, pH 7.2.
10. Plant growth hormones: GA_3 dissolved in water at 1.0 mg/mL, NAA dissolved in dimethyl sulfoxide (DMSO, Sigma) at 1 mg/mL, ZR in DMSO at 1 mg/mL.
11. Antibiotics: Kanamycin sulfate is dissolved in water at 100 mg/mL, hygromycin B hydrochloride in water at 100 mg/mL, cefotaxime in water at 100 mg/mL, augmentin in water at 100 mg/mL, rifampicin in water at 500 mg/mL, tetracycline in DMSO at 10 mg/mL, and carbenicillin in DMSO at 100 mg/mL.
12. Acetosyringone: Acetosyringone (3,5-dimethoxy-4-hydroxyacetophenone, Aldrich Chemical, Dorset, UK) is dissolved in DMSO at 10 mM.
13. Chemicals dissolved in DMSO do not require sterilization. All other solutions must be filter sterilized using 0.2-µm filters and stored at 4°C for short periods (a few days) and at –20°C for longer periods.

3. Methods
3.1. Axenic Shoot Cultures

1. Sterilize axillary buds in 7% (v/v) "Domestos" for 30 min, followed by four to six washes in sterile water (*see* Note 1).
2. Culture the axillary buds on MS10 or MS30 medium in a temperature (20°C) and light (100 µE/m^2/s, White fluorescent tubes [Phillips, Eindhoven, Holland], 16 h photoperiod) controlled growth room (*see* Note 2).
3. Propagate in vitro plants by subculturing the internodal segments, with axillary buds, and apical buds from 3–4-wk-old plants on MS10 and MS30 media.
4. Excise well-grown leaves and stems from 3–4-wk-old in vitro plants and cut leaves into two to four pieces. Remove stem segments of about 1–2 cm in length devoid of axillary buds, with a sharp scalpel and immediately transfer to a 9-cm Petri dish containing 20 mL of MS30 liquid medium to prevent the cut ends from drying (*see* Note 3).
5. Transfer 50 leaf and 50 stem explants into a 9-cm Petri dish containing 20 mL of preconditioning liquid HH medium, seal with Nescofilm, and culture under low light (60 µE/m^2/s; White fluorescent tubes, Phillips) at 20°C for 1–3 d before coculturing with *Agrobacterium* (*see* Note 4).

3.2. Transformation

1. Grow *Agrobacterium* strains (a single colony from a stock plate) in 5 mL liquid LB, with appropriate concentration of antibiotics (e.g., Strain LBA

4404 containing binary vector pSLJ 4D4:81 should be grown in LB medium with 1 µg/mL tetracycline and the binary vector KIWI 105 in the presence of 50 µg/mL kanamycin sulfate) to maintain the vectors in the bacterial cells, for 24 h at 28°C, on a rotary shaker.

2. Subculture 50–100 µL of *Agrobacterium* into 5 mL of liquid LB medium with antibiotics and 20 µ*M* acetosyringone and grow for 18 h as in step 1 (*see* Notes 5 and 6).

3. Centrifuge to pellet *Agrobacterium* at 1000g for 5 min.

4. Resuspend the bacterial pellet in 2 mL of MS30 liquid medium. Remove 50–100 µL and mix with 20 mL of MS30 medium containing 20 µ*M* acetosyringone in a 9-cm Petri dish.

5. Inoculate the preconditioned explants by immersion in the diluted *Agrobacterium* culture and coculture for 30–60 min under low light (60 µE/m²/s). To ensure even inoculation of the explants, swirl the contents of the Petri dish every 5–10 min.

6. Transfer the explants on to LSR-1 medium containing 250 mg/L cefotaxime plus an appropriate antibiotic (e.g., 50–100 mg/mL kanamycin) for the selection of transgenic tissues and incubate under high light (100 µE/m²/s; white fluorescent tubes, Phillips) at 20°C. Ensure that explants are in contact with the medium (*see* Note 7).

7. Subculture explants every 10–14 d to fresh LSR-1 medium with the appropriate antibiotics. Observe every 1–2 d for any *Agrobacterium* growth around the explants. If *Agrobacterium* growth is observed, transfer the explants immediately to fresh medium. After 14–28 d of culture, small green calli should appear at the cut end of the explants (*see* Notes 8, 10, 15, and 16).

8. Remove the green calli from the explants and transfer to LSR-2 medium containing antibiotics and culture as before (*see* Note 9).

9. Subculture the green calli every 10–14 d to fresh LSR-2 medium with the antibiotics. Green shoots usually regenerate from the calli within 14–28 d of culture on this medium.

10. Excise the regenerated green shoots (2–4 cm) from the calli and transfer to MS30 medium containing 250 mg/mL cefotaxime, and double the amount of transgenic selection antibiotic (i.e., 100 µg/mL kanamycin sulfate or 20 µg/mL hygromycin B hydrochloride). This will ensure that only transgenic shoots will grow further and, more important, that they regenerate roots (*see* Notes 11 and 12).

11. Putative transgenics can be analyzed at this stage for the presence and expression of the transgenes by PCR and Northern blotting (*see* Note 13).

12. Subculture, at least two to three times, the apical region of shoots from in vitro-grown healthy plantlets in MS30 medium with the appropriate anti-

biotics to ensure that the transgenic plants are free from *Agrobacterium* and to micropropagate the shoots.

13. Transfer in vitro plantlets to compost (Universal Compost, SCRI, Dundee, Scotland). Wash the agar from the roots and transfer into sterile compost in 5-cm diameter pots. Place the potted plants in a humidity chamber for 3–7 d.

14. After 7 d, transfer the plants into 9-cm pots and maintain under containment in the glasshouse.

4. Notes

1. Always select axillary buds from young, healthy plants and avoid plants from contaminated environment.

2. Axillary buds and apical shoots can be cultured in 9-cm Petri dishes or 120 mL capacity screwcapped glass jars (6 explants per container). Culture containers should be covered with transparent plastic lids and sealed with nescofilm. It is essential that two to four small cuts are made in the Nescofilm to vent the containers and to avoid ethylene accumulation. Gaspermeable closures (Suncaps, Sigma) can be used instead of plastic lids.

3. Avoid excessive damage of explant tissues during cutting; explant size should not be less than 0.5 cm.

4. Preculture of explants in a high auxin and cytokinin medium for 1–3 d greatly improves the transformation efficiency.

5. *Agrobacterium* liquid cultures should be initiated 2 d before the day of explant inoculation. Subculture *Agrobacterium* and grow for 16 h in the presence of 20 μM acetosyringone before using the bacteria for inoculation.

6. Coculture of explants with *Agrobacterium* should be for 30–60 min on a shaker. Coculture of explants with *Agrobacterium* in the presence of 20 μM acetosyringone increases the efficiency of transformation.

7. Inoculated explants should be cultured for 1–2 d on LSR-1 medium lacking antibiotics. Longer coculture can boost transformation, but can also result in more bacterial overgrowth during later selection.

8. Cefotaxime at 250 mg/L should be used to suppress *Agrobacterium* growth in the selection medium. Cefotaxime should be increased to 500 mg/L in the event of excessive *Agrobacterium* growth. Other antibiotics such as carbenicillin (400 mg/L) or augmentin (200 mg/L) can be used instead of cefotaxime.

9. Kanamycin sulfate at 50–100 mg/L should be used to select transgenic cells and calli. Other antibiotics used in selecting transgenic potato tissues include hygromycin B hydrochloride (10–25 mg/L) and phosphinothricin (5–10 mg/L), depending on the selectable marker genes carried by the vectors.

10. The concentration of antibiotics for selecting potato transgenic tissues may vary from genotype to genotype and may depend on the ploidy of the genotypes.

11. Remove only one shoot from each individual callus. Do not transfer any callus at the base of the shoots.
12. Use a concentration of the antibiotic (e.g., 100–500 mg/L of kanamycin sulfate or 20–50 mg/L of hygromycin B hydrochloride), which is known to prevent rooting of nontransformed shoots.
13. Be aware that PCR analysis and GUS staining on in vitro plant material could give false positives owing to possible *Agrobacterium* contamination of the regenerated plants.
14. Addition of silver thiosulfate to shoot culture and shoot regeneration media can lead to a high efficiency of callus and plant regeneration from explants for certain potato genotypes *(6)*.
15. Develop a plant regeneration system before initiating transformation experiments with new potato genotypes. This can usually be achieved by changing the levels of auxin and cytokinin in LSR-1 and LSR-2 media.
16. Culture of explants in the presence of exponentially growing tobacco or potato cells as nurse can improve transformation efficiency in some potato genotypes.

References

1. Bajaj, Y. P. S. (1987) *Biotechnology in Agriculture and Forestry—3: Potato.* Springer-Verlag, Heidelberg, Germany.
2. Kumar, A., Cooper-Bland, S., and Powell, W. (1992) Transfer of disease resistance genes into crop plants: the role of biotechnology, in *Proceedings of the Symposium on Interactions Between Plants and Microorganisms* (Wolf, J. N., ed.), International Foundation for Science, Stockholm, Sweden, pp. 475–490.
3. Kumar, A., Graham, J., Whitty, P., and Lyon, J. (1992) Genetic transformation in plants, in *Annual Report of the Scottish Crop Research Institute, 1991* (Perry, D. A. and Heilbronn, T. D., eds.), SCRI, Dundee, UK, pp. 29–32.
4. Klee, J., Horsch, R., and Rogers, S. (1987) *Agrobacterium*-mediated plant transformation and its further application to plant biology. *Annu. Rev. Plant. Physiol.* **38,** 467–486.
5. De Block, M. (1988) Genotype-dependent leaf disc transformation of potato (*Solanum tuberosum*) using *Agrobacterium tumefaciens. Theor. Appl. Genet.* **76,** 767–774.
6. Barker, H., Reavy, B., Kumar, A., Webster, K. D., and Mayo, M. A. (1992) Restricted virus multiplication in potato transformed with the coat protein gene of potato leafroll luteovirus: similarity with a type of host gene-mediated resistance. *Ann. Appl. Biol.* **120,** 55–64.
7. Tavazza, R., Tavazza, M., Ordas, R. J., Ancora, G., and Benvenuto, E. (1988) Genetic transformation of potato *(Solanum tuberosum)*: an efficient method to obtain transgenic plants. *Plant Sci.* **59,** 175–181.
8. Sheerman, S. and Bevan, M. W. (1988) A rapid transformation method for *Solanum tuberosum* using binary *Agrobacterium tumefaciens* vectors. *Plant Cell Rep.* **7,** 13–16.

9. Newell, C. A., Rozman, R., Hinchee, M. A., Lawson, E. C., Haley, L., Sanders, P., et al. (1991) *Agrobacterium*-mediated transformation of *Solanum tuberosum* L. cv. Russet Burbank. *Plant Cell Rep.* **10,** 30–34.
10. Visser, R. G. F. (1991) Regeneration and transformation of potato by *Agrobacterium tumefaciens*, in *Plant Tissue Culture Manual: Fundamentals and Applications* (Lindsey, K., ed.), Kluwer, Dordrecht, The Netherlands, pp. 1–9.
11. Higgin, E. S., Hulme, J. S., and Shields, R. (1992) Early events in transformation of potato by *Agrobacterium tumefaciens*. *Plant Sci.* **82,** 109–118.
12. Ishida, B. K., Snyder, G. W., Jr., and Belknap, R. (1989) The use of *in vitro*-grown microtuber discs in *Agrobacterium*-mediated transformation of Russet Burbank and Lemhi Russet potatoes. *Plant Cell Rep.* **8,** 325–328.

CHAPTER 13

Agrobacterium-Mediated Transformation of Soft Fruit *Rubus, Ribes,* and *Fragaria*

Julie Graham, Ronnie J. McNicol, and Amar Kumar

1. Introduction

Gene transfer technology is particularly useful for fruit crops where breeders are faced with long generation times because of the relatively long periods of juvenility associated with perennial plants. Also, the highly heterozygous nature of the germplasm requires the evaluation of large seedling populations and maintenance of the genotype by vegetative propagation.

A transformation technique that permits improvement without alteration of overall genetic makeup is extremely valuable. Transformation of soft fruit (raspberries [*Rubus*], blackberries [*Rubus*], strawberries [*Fragaria*], and blackcurrants [*Ribes*]) has proven problematic because of difficulties in the development of efficient shoot regeneration systems; the latter are a prerequisite to successful transformation. The plant regeneration systems already developed tend to be very genotype-specific and thus not directly applicable to other cultivars. The toxicity of commonly used antibiotics, such as kanamycin sulfate, to select transgenic tissues and cefotaxime or carbenicillin to inhibit growth of *Agrobacterium* on the explant tissues following cocultivation, has also led to regeneration difficulties. Initially, these problems resulted in the development of transformation systems that did not rely on kanamycin sulfate selection, but that were based on the expression of the β-glucuronidase (GUS) gene as a visual marker for identifying transgenic tissues. The unwanted growth of *Agrobacterium* on inoculated explant tissues can be suppressed

From: *Methods in Molecular Biology, Vol. 44:* Agrobacterium *Protocols*
Edited by: K. M. A. Gartland and M. R. Davey Humana Press Inc., Totowa, NJ

by regular treatment of plant tissue in antibiotic solution every few weeks following the transformation procedure. This method is less toxic to the tissues than the use of antibiotic-containing medium and does not inhibit the growth of explants. The GUS-intron construct employed for transformation has proven extremely valuable in nonantibiotic selection systems, allowing the identification and selection of transformed cells from leaf discs on media containing either 5,bromo-4-chloro-3-indolyl-β-D-glucuronide (X-gluc) or 4-methylumbelliferyl-β-D-glucuronide (MUG). A drawback of this system is that the large number of plants that are produced must be analyzed to detect the few transgenics. Additionally, the possibility of generating transgenic chimeric plants exists with this transformation system. Improved transformation techniques have therefore been developed using more efficient regeneration systems and low concentrations of kanamycin sulfate (10–40 mg/L) to select transgenics, in conjunction with the use of ticaricillin to inhibit growth of *Agrobacterium* on the inoculated explants *(1,2)*.

In *Rubus*, the most effective transformation has been achieved for blackberry cv. Loch Ness and for the hybrid berries cvs. Tayberry and Sunberry. Red raspberries have proven more difficult because of inefficient regeneration systems. Transformation can be achieved both with and without antibiotic selection, although the use of antibiotic selection enables easier identification of transgenics and reduces the risk of chimeras *(3,4)*. Strawberry transformation (*Fragaria*) is easier to obtain than that of raspberry owing to more efficient shoot regeneration, especially from stem tissue explants. A range of techniques and cultivars have been adopted by many workers around the world *(5–7)*.

Blackcurrants (*Ribes* species) have proven the most difficult to transform because an efficient regeneration system is not available. There have been no reports of organogenesis or embryogenesis from leaf tissue. However, peeled internodal segments from in vitro-grown plantlets of the cultivars Ben More and Baldwin have been induced to regenerate *(8)*. The regeneration and transformation protocols described here can be used to produce transgenic plants for most of the soft fruit genotypes.

2. Materials

2.1. Axenic Shoot Cultures and Medium

1. Micropropagated plantlets of soft fruit cultivars were obtained from Scottish Crop Research Institute (Dundee, Scotland). *Rubus* cultivars: Tayberry,

Sunberry, Loch Ness Blackberry; *Fragaria* cultivars: Rhapsody, Melody, Symphony; *Ribes* cultivar: Ben Lomond.

2. "Domestos" bleach (Lever Bros., Kingston-upon-Thames, UK) and "Chloros" (BDH, Poole, UK).
3. MS0 medium: Murashige and Skoog (1962)-based medium (Flow Laboratories Ltd., Irvine, Scotland) with 20 g/L sucrose and 7.0 g/L agar (Merck, Darmstadt, Germany), pH 5.8.
4. Nescofilm: Bando Chemical Co. (Kobe, Japan).
5. All chemicals purchased from Sigma (Poole, Dorset, UK) unless otherwise stated previously. All growth regulators are dissolved in a minimum volume of ethanol and made up to 0.1 or 1 mg/mL and stored in the dark at 0–4°C for no more than 1 mo.
6. All media, solutions, and equipment should be autoclaved; all plant growth regulators, antibiotics, and vitamins should be filter sterilized using 0.2-μm filters. Distilled water is used throughout and the pH of the medium is adjusted prior to the addition of agar and before autoclaving.

2.2. Soft Fruit Multiplication Medium

MS medium: MS salt, with 100 mg/L inositol, 2 mg/L glycine, 1 mg/L nicotinic acid, 1 mg/L pyridoxine HCl, 1 mg/L thiamine HCl, 1 mg/L calcium pantothenate, 1 mg/L cysteine HCl, 0.01 mg/L biotin, 30 g/L sucrose, 7 g/L agar (Merck), 1 mg/L BAP (6-benzylamino purine; Sigma), pH 5.6.

2.3. Rooting Medium

Follow the multiplication medium, but without BAP.

2.4. Rubus *and* Ribes *Regeneration Medium*

MS medium (Sigma or ICN, Thames, Oxfordshire, UK) with 20 g/L sucrose, 7 g/L agar, 0.1 mg/L IBA (indole butyric acid; Sigma), 2 mg/L BAP, pH 5.6.

2.5. Fragaria *Regeneration Medium*

MS medium with 20 g/L sucrose, 7 g/L agar, 0.2 mg/L BAP, 0.2 mg/L 2,4-D (2,4-dichlorophenoxyacetic acid; Sigma), pH 5.6.

2.6. Transformation of Soft Fruit Cultivars

1. A disarmed *Agrobacterium tumefaciens* strain LBA 4404, containing the binary vectors pPBI 121.X or Bin 19, which has the *npt*II and the GUS genes *(3)*.
2. Bacterial medium: LB (Luria broth); 1% bactopeptone (Difco, Detroit, MI), 0.5% (w/v) yeast extract (Sigma), 1% (w/v) NaCl, pH 7.0.

3. Agrobacterium coculture medium contains MS salts and 20 g/L sucrose at pH 5.6.
4. Antibiotics: Kanamycin sulfate (Sigma) dissolved in water at 10 mg/mL, filter sterilized and stored at –20°C, Ticarcillin (Beecham Research, Brentford, Middlesex, UK) dissolved in water at 125 mg/L; filter sterilize and use immediately.
5. Whatman No. 1 filter papers; place in a screwcapped glass jar, autoclave.

3. Methods

3.1. Axenic Shoot Cultures

1. For all plant material, sterilize axillary buds in 15% (v/v) "Domestos" for 15 min, followed by three or four rinses in sterile distilled water.
2. Culture the axillary buds on MS medium at 20°C under warm white fluorescent tubes at 70 $\mu E/m^2/s$ with a 16-h photoperiod.
3. Propagate in vitro plants by subculturing the internodal segments with axillary buds to fresh multiplication medium every 4–6 wk.
4. Use 4-wk-old leaves of *Rubus* and strawberry, 6-wk-old internodal segments of blackcurrant, and 8-wk-old stem segments of strawberry for transformation. Excise leaf explants with a sterile cork borer, peel the stems of blackcurrant, and cut into 5-mm lengths; peel the stems of strawberry and cut into 2-mm lengths. Place all explants on the appropriate regeneration medium for 3–5 d before inoculation (*see* Notes 1 and 5).

3.2. Transformation

1. Grow *Agrobacterium tumefaciens* strain LBA4404 containing a binary vector PBI 121.X or Bin 19 in 5 mL liquid LB medium, containing the antibiotics 50 $\mu g/mL$ rifampicin and 50 $\mu g/mL$ kanamycin sulfate to maintain the binary vector, at 28°C overnight.
2. Wash overnight culture of *Agrobacterium* twice by centrifugation and resuspension in liquid plant regeneration medium without hormones, at pH 5.6.
3. Select explants showing signs of expansion on regeneration medium for inoculation.
4. Inoculate the explants by immersing in the *Agrobacterium* suspension in a Petri dish for 20 min with occasional agitation (*see* Note 3).
5. Culture inoculated explants on sterile filter paper moistened with regeneration medium for 2 d.
6. Transfer the explants onto the appropriate selection medium (regeneration medium containing 125 mg/L ticarcillin and 10–40 mg/L kanamycin sulfate, depending on the genotype) (*see* Notes 2 and 5).
7. Regeneration usually occurs within 12–16 wk and plantlets remaining green on antibiotic selection medium are grown on for further analysis (*see* Note 4).
8. Subculture shoot tips on to rooting medium with kanamycin sulfate selection.

9. Once regenerated plants that remain green have been subcultured, they can be analyzed further using Southern blotting, Northern blotting, or PCR.

4. Notes

1. Before initiating transformation experiments with a new genotype, determine that the shoot regeneration system functions efficiently.
2. Determine the level of toxicity of the antibiotics to be used for selection of transgenic shoots on in vitro nontransformed transgenic shoots.
3. If *Agrobacterium* contamination becomes a problem, ticarcillin at 250–500 mg/L can be employed, but only for a few days.
4. TDZ (Thidiazuron) has been shown to induce organogenesis at high frequency in *Rubus* genotypes *(4,9)*.
5. Inoculated explants should be cultured for not more than 2 d on moistened filter paper laid on agar containing regeneration medium. Longer coculture can boost transformation, but may also result in more bacterial overgrowth during later selection.
6. Ensure that only one shoot is removed from each independent transformed region.

References

1. Graham, J. (1990) *The Development and Application of Methods for Using* Agrobacterium *spp. as DNA Vectors in Soft Fruit Plants.* Unpublished doctoral dissertation, University of St. Andrews, Scotland, UK.
2. McNicol, R. J. and Graham, J. (1993) Temperate small fruits, in *Biotechnology of Perennial Fruit Crops* (Hammcroschlag, F. A. and Litz, R. E., eds.), CAB International, Wallingford, Oxon, UK, pp. 303–321.
3. Graham, J., McNicol, R. J., and Kumar, A. (1990) Use of GUS gene as a selectable marker for *Agrobacterium*-mediated transformation of *Rubus. Plant Cell Tissue Org. Cult.* **20,** 35–39.
4. Hassan, M. A., Swartz, H. J., Inamine, G., and Mullineau, P. (1993) *Agrobacterium tumefaciens*-mediated transformation of several *Rubus* genotypes and recovery of transformed plants. *Plant Cell Tissue Org. Cult.* **33,** 9–17.
5. Graham, J. and McNicol, R. J. (1994) Towards genetic based insect resistance in strawberry using the Cowpea trypsin inhibitor gene. *Ann. Appl. Biol.,* submitted.
6. James, D. J., Passey, A. J., and Barbara, P. J. (1990) *Agrobacterium*-mediated transformation of the cultivated strawberry (*Fragaria x ananassa* Duch.) using disarmed binary vectors. *Plant Sci.* **69,** 79–94.
7. Nehra, N. S., Chibbar, R. N., Kartha, K. K., Datla, R. S. S., Crosby, W. L., and Stushnoff, C. (1990) Genetic transformation of strawberry by *Agrobacterium tumefaciens* using a leaf disc regeneration system. *Plant Cell Rep.* **9,** 293–298.
8. Graham, J. and McNicol, R. J. (1991) Regeneration and transformation of *Ribes. Plant Cell Tissue Org. Cult.* **24,** 91–95.
9. Fiola, J. A., Hassan, M. A., Swartz, H. J., Bors, R. H., and McNicol, R. J. (1990) Effect of thidiazuron, light fluence rates and kanamycin on *in vitro* shoot organogenesis from excised *Rubus* cotyledons and leaves. *Plant Cell Tissue Org. Cult.* **20,** 223–228.

CHAPTER 14

High-Frequency and Efficient *Agrobacterium*-Mediated Transformation of *Arabidopsis thaliana* Ecotypes "C24" and "Landsberg *erecta*" Using *Agrobacterium tumefaciens*

Mehdi Barghchi

1. Introduction

Arabidopsis thaliana has been widely used in studies on basic plant physiology and biochemistry as well as in plant molecular genetic manipulations and developmental biology research because of its small genome, low chromosome number, short regeneration time (4–6 wk), availability of many mutants and genetic maps, sexual self-compatibility, and prolific seed production. More extensive use of *Arabidopsis* has been hampered because of difficulties in efficient and rapid regeneration and transformation procedures. Several methods for plant regeneration have been reported *(1–7)*. Transformed plants have been recovered from various explants, such as leaf *(8)*, stem *(9)*, callus tissue *(10)*, germinating seeds *(11)*, root *(12)*, and by using direct gene transfer to protoplasts *(13)*. Despite these reports, the frequency of regeneration of transgenic plants was still low and took at least a few months to produce transgenic plants. Also the long period of in vitro incubation during shoot regeneration in these methods may increase the possibility of somaclonal variation and increases in ploidy level. Many reports have indicated the high regeneration potential of cotyledon explants at various stages of development in maturing embryos or after seed germination *(14–17)*. This

From: *Methods in Molecular Biology, Vol. 44:* Agrobacterium *Protocols*
Edited by: K. M. A. Gartland and M. R. Davey Humana Press Inc., Totowa, NJ

chapter presents a new procedure for rapid and prolific regeneration of shoots from cotyledon explants of *Arabidopsis* in ecotypes "Landsberg *erecta*" and "C24." Furthermore, this regeneration procedure is developed establishing a method for rapid production of transgenic *Arabidopsis* shoots using disarmed *Agrobacterium tumefaciens* within 2–3 wk *(18–20)*. A transformation procedure using root explants is also presented as a separate method.

2. Materials
2.1. Abbreviations

1. IAA = Indole-3-acetic acid.
2. NAA = l-Naphthalene acetic acid.
3. 2,4-D = 2,4-dichlorophenoxyacetic acid.
4. BAP = 6-Benzylaminopurine.
5. 2iP = N^6-(2-isopentenyl)adenine.
6. Km = Kanamycin.
7. GUS = β-Glucuronidase.
8. NB = Nutrient broth.
9. *npt*-II = Neomycin phosphotransferase II.
10. X-Gluc = 5-Bromo-4-chloro-3-indolyl glucuronide.
11. CTAB = Cetyl triethylammonium bromide.
12. MS0 = GM = Germination medium: Murashige and Skoog medium *(21)* without hormonal supplements.
13. CSR = Shoot regeneration medium: MS medium supplemented with 0.1–0.4 mg/L NAA + 1.0 mg/L BAP.
14. CSS1 = Shoot regeneration selection medium: CSR + 50 mg/L Km + 500 mg/L vancomycin or 400 mg/L augmentin.
15. CSS2 = CSR + 50 mg/L Km.
16. RCI = Callus-inducing medium: Gamborg's B5 medium + 0.5 mg/L 2,4-D + 0.05 mg/L kinetin.
17. RSR = Shoot regeneration medium: Gamborg's B5 medium + 5.0 mg/L 2iP + 0.15 mg/L IAA.
18. RSS1 = Shoot regeneration selection medium: RSR + 750 mg/L vancomycin + 50 mg/L Km.
19. RSS2 = Shoot regeneration selection medium: RSR + 500 mg/L vancomycin + 50 mg/L Km.

2.2. Tissue-Culture Conditions

1. Culture room: Growth room facilities with lighting (16-h photoperiod, 25–100 μE/m²/s) and air conditioning to maintain steady temperature (±0.5°C) within the range of 23–28°C.

2. Media preparation: Murashige and Skoog and Gamborg's B5 were adjusted to pH 5.7 before addition of 0.7% agar (Technical Standard, Oxoid, Basingstoke, UK) and sterilized by autoclaving at 121°C for 20 min. Hormones were added to the medium prior to autoclaving. Antibiotics were dissolved in water, filter sterilized (0.22 μm), and added to the medium after autoclaving and cooling to 60°C.

3. Aseptic seed germination:
   a. Surface sterilize seeds of *Arabidopsis* for 20 min in a 20-mL sterile glass or plastic Universal bottle (Sterilin Ltd., Milton Keynes, UK) containing 20% (v/v) Domestos (commercial bleach) followed by several times of washing with sterilized tap water. Distribute seeds thinly over the surface of 9-cm Petri dishes containing agar solidified GM. Seal Petri dishes with cling film tapes (use a fine saw to divide an ordinary role of household cling film into 2-cm wide sections to provide 2 cm wide sealing tape). Use these tapes for sealing around the rim of various tissue-culture containers as required and incubate them for 3–4 d in the dark at 4°C (in vitro aseptic seed vernalization). Cold treatment (vernalization) will improve uniform germination.
   b. Return seed to culture room to allow germination in light. Use suitable parts of the aseptically germinated seeds as explants for transformation. Cotyledon explants are best 4–8 d after germination. Root explants can be used 8–10 d after germination.

3. Methods

A. thaliana seeds, ecotypes "Landsberg *erecta*" and "C24" can be used for these protocols. Aseptically germinated seeds are the source of explants (i.e., cotyledons, roots, etc.) for in vitro culture.

3.1. Preparation of Agrobacterium Inoculant

To establish bacterial inoculum, using a flamed bacterial loop and observing sterile procedures throughout, transfer a single colony of disarmed *A. tumefaciens* strain C58Cl (pGV3850) (nopaline-type) from a fresh bacterial plate containing appropriate selective antibiotics, to 5 mL of sterile nutrient broth (NB) medium containing the same selective antibiotics in a 20-mL plastic disposable Universal bottle or a glass bottle. This strain of *Agrobacterium* carries the binary vector pBI 121 that contains a neomycin phosphotransferase II (*npt*II) gene driven by the *nos* promoter conferring resistance to Km (kanamycin), and GUS (β-glucuronidase) gene driven by the CaMV 35S promoter *(22)*. Incubate Universal bottle on an orbital shaker (a throw of 5–10 cm and shake rate of

150–200 rpm) in the growth room overnight or until a good growth is achieved. Use this overnight *Agrobacterium* culture for inoculation of explants.

3.2. Transformation of Arabidopsis Cotyledon Explants

1. Explant preparation: Pull aseptically germinated seedlings out of the germination medium, put them in a sterile Petri dish, and cut off cotyledons from the rest of the seedling using a sharp scalpel without including any of the adjacent meristematic cotyledonary axillary bud. Use these cotyledon explants for culture and ensure that they have good contact with medium to induce growth at all stages of culture.
2. Explant inoculation:
   a. Prepare 20 mL of a 50X dilution from overnight *Agrobacterium* culture with liquid plant growth medium (i.e., MS0) and pour into a 9-cm Petri dish.
   b. Inoculate cotyledon explants (30–40 explants) in the *Agrobacterium* suspension (fully submerging all explants) for 3–5 min.
   c. Remove explants from the *Agrobacterium* suspension and blot them on sterile filter paper to remove excess liquid.
3. Explant cocultivation: Cocultivate the inoculated explants (about 30) for 2–3 d on CSR medium. Return cultures to culture room. When early signs of *Agrobacterium* growth appear in the medium near to the culture explants, prepare the culture for the next stage.
4. Regeneration in selection medium: Remove explants from the cocultivation medium and wash off the *Agrobacterium* from them in a Universal bottle with sterile liquid MS0 medium, blot them on a sterile filter paper, and then culture explants onto CSS1 medium. Return cultures to culture room.
5. Maintenance and subculture: Subculture explants into new medium (i.e., CSS1) every 2–3 wk. After about 4–6 wk explants should be free of *Agrobacterium* contamination, when they can be subcultured every 3–4 wk in selection medium containing 50 mg/L Km only and no vancomycin or augmentin (i.e., CSS2 medium). Kanamycin can also be omitted from the medium at later subcultures.

3.3. Transformation of Arabidopsis Root Explants

1. Explant preparation:
   a. Pull aseptically germinated seedlings out of the germination medium, put them in a sterile Petri dish, and, using a sharp scalpel, cut off roots from the rest of the seedling without including any of the adjacent green tissue.
   b. Culture aforementioned root explants in RCI medium for 3–4 d. It is essential that root explants have good contact with the medium to induce growth at all stages of *Arabidopsis* root explant culture.

   c. Remove callus induced root explants from RCI medium and cut them into about 0.5-cm pieces in a sterile Petri dish. These explants are now ready for inoculation with *Agrobacterium*.

2. Explant inoculation: Inoculate callus induced root explants in 50X dilution of an overnight *Agrobacterium* culture for 3–5 mins (as in cotyledon explants inoculation method, *see* Section 3.2., step 2).

3. Explant cocultivation: Cocultivate 40–60 inoculated explants in a 9-cm Petri dish containing RCI medium for 2–3 d in the culture room. When early signs of *Agrobacterium* growth appears in the medium near to the cultured explants, prepare the cultures for the next stage.

4. Regeneration in selection medium:

   a. Remove explants from the cocultivation medium and wash off *Agrobacteria* from them with sterile liquid MS0 medium (this can be done in a sterile small bottle such as Universal bottle), blot them on a sterile filter paper, and then culture them onto RSS1 (shoot regeneration selection medium) for 2–3 wk. After about 3 wk, small green calli form on the root explants that will regenerate shoots.

   b. Subculture root explants to RSS2 medium. After 2–3 wk, transformed shoots will appear on the root explants.

   c. Subculture every 2–3 wk to new medium. Remove vancomycin/ augmentin from RSS2 medium after about 6 wk of exposure to the antibiotic. Presence of antibiotic is not necessary once the *Agrobacterium* is eliminated from the explants. Vancomycin can also induce vitrification in the shoots.

   d. Subculture three to four transformed shoots in MS0 medium to a Petri dish or Magenta GA-7 vessels (Sigma, St. Louis, MO) to grow to maturity for seed set. Plants will set seed in vitro, however, condensation on the Petri dish lids reduces seed set and should be avoided. About 50% of the shoots will form roots in MS0 medium and they can be transferred to potting mix to set seed in vivo.

5. Maintenance and subculture: Subculture explants into new medium (i.e., RSS2) every 2–3 wk. After about 4–6 wk explants should be free from *Agrobacterium* contamination, when they can be subcultured every 3–4 wk in selection medium containing 50 mg/L Km only and no vancomycin or augmentin (i.e., GM + 50 mg/L Km). Kanamycin can also be omitted from the medium at later subcultures.

3.4. Inheritance Test (Germination of F1 Progeny)

1. Collect seeds from the transgenic plants and initially vernalize the sterilized seeds as before (*see* Section 2.2., item 3.) to allow improved and uniform germination. Then germinate sterilized seeds in MS0 medium containing 50 mg/L Km. Record the germination of sensitive wild-type

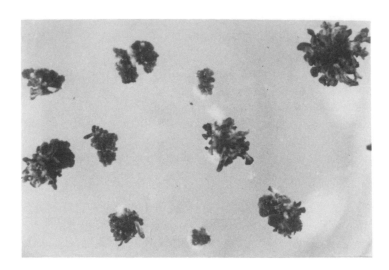

Fig. 1. Shoot regeneration from 4-d-old *Arabidopsis* cotyledon explants in MS medium supplemented with 0.4 mg/L NAA + 1.0 mg/L BAP (4 wk after culture initiation).

(nontransgenic) and resistant (transgenic) germinated seedlings after 10–15 d. The transgenic seedlings will be resistant to Km and will grow normally producing root, green leaves, and shoots, but the sensitive (wild-type) seedlings will not form roots or leaves and will have bleached (white) cotyledons. It should be noted that in vitro germination of seeds in the presence of a selection agent (i.e., kanamycin) cannot always be taken as evidence of the inheritance of kanamycin resistance (i.e., transgenic material) as some large seeds such as pea may germinate in vitro primarily relying on the humidity from the culture jar and the reserves of the seed for initial growth. Also, if the concentration of the selection agent (i.e., Km) is below the level to bleach out the new growth or if Km is not fully absorbed by the new growth of the germinating seeds, they may appear green leading to an error in visual recording. GUS histochemical assay of the germinated seeds can be used to confirm in vitro germination results (*see* Chapter 18).

2. Maintain in vitro produced plants (transformed and wild-type) on MS0 in Magenta GA-7 vessels (Sigma) to grow to maturity for seed set and testing of inheritance of the introduced genes to the progeny.

4. Notes

1. When transferred to culture, cotyledon explants expand rapidly, produce callus at the cut surface within 3–4 d, and then produce multiple shoots soon after (Fig. 1) Transgenic shoot initials are observed in the selection

Table 1
Shoot and Callus Regeneration from Immature Cotyledons
of *Arabidopsis* Ecotype "C24"

Media[a]	Stage of maturity[b]					
	Callus growth			Shoot growth		
	I	II	III	I	II	III
1	100	100	94	95	77	64
2	100	100	96	92	62	70
3	15	35	19	15	32	15
4	35	38	87	4	0	33

[a]Media: MS medium containing 1.0 mg/L BAP plus the following auxins (mg/L): (1) 0.1 NAA, (2) 0.4 NAA, (3) 0.1 IBA, and (4) 0.4 IBA.

[b]Stages of seed maturity: (I) The first three seed pods maturing (browning) at the base of the inflorescence (most mature), (II) The three seed pods above I (moderate mature), (III) The three seed pods above II (least mature).

From ref. *20*.

Table 2
Effect of BAP Concentration on Shoot Regeneration
from Immature Cotyledons[a] of *Arabidopsis* Ecotypes "Landsberg" and "C24"

Growth	BAP, mg/L					
	"Landsberg"			"C24"		
	1	2	4	1	2	4
Callus	100	100	100	100	100	100
Shoot	88	80	85	85	78	68

[a]Stage of cotyledon maturity: Cotyledons from the first three seed pods maturing (browning) at the base of the inflorescence were used. Media: MS medium containing 0.1 mg/L NAA and the above concentrations of BAP.

From ref. *20*.

medium in about 10 d *(18–20)*. It is suggested that rapid shoot regeneration from cotyledon explants may result in the induction of less somaclonal variation than in shoots produced from leaf *(8)* or root *(12)* explants, which take much longer to regenerate.

2. A very large proportion of cotyledon explants produce callus (up to 100%) and shoot regeneration is as high as 95% (Fig. 1, Tables 1 and 2). This applies to a wide range of media *(20)*. A medium containing 0.1–0.4 mg/L NAA and 1–2 mg/L BAP is suitable for regeneration of transgenic shoots (Fig. 2.). Physiological abnormalities such as vitrification and reduced

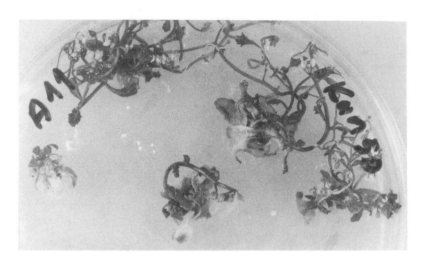

Fig. 2. Transgenic shoot regeneration from *Arabidopsis* cotyledon explants in MS medium supplemented with 0.4 mg/L NAA + 1.0 mg/L BAP + 50 mg/L kanamycin (13 wk after culture initiation; *20*).

shoot elongation is common at higher than 4 mg/L BAP. These abnormalities have been reported to be associated with high concentrations of BAP in shoot initiation medium by Barghchi and Alderson *(23)*.

3. Immature cotyledon explants at the later stages of development can often have better shoot regeneration potential than mature cotyledons. These explants are dissected out of the seed pods (siliques) at the final stage of maturity of the seeds but before desiccation and seed pod break from plants grown in the glasshouse or growth room. Dissection of cotyledons from sterilized seed pods is carried out with the aid of a dissecting microscope at low magnification in a laminar air-flow cabinet. Large numbers can be dissected in a relatively short time but care should be taken not to allow dehydration of the explants.

4. Cotyledons from in vitro germinated mature seeds are easier to work with than immature cotyledons to induce shoot regeneration. Shoot regeneration is reduced dramatically as seedlings grow, and 4–8 d after germination is optimum for shoot regeneration from cotyledons (Table 3). Shoot regeneration frequency from cotyledon explants was found to decline from 80% at 4–8 d postgermination to 8% after 16 d *(20)*.

5. Although a variety of antibiotics can be used to eliminate *Agrobacterium* after the inoculation stage, it is important to monitor their effect on plant regeneration. Cefotaxime and carbenicillin inhibit shoot regeneration in *Arabidopsis* cultures, whereas shoot production is not inhibited in the pres-

Table 3

Effect of Cotyledon Age on Shoot Regeneration
from *Arabidopsis* Explants Ecotype "C24"[a]

Growth response	Days after seed germination, cotyledon age			
	4	8	12	16
Callus induction, %	92	94	92	90
Shoot regeneration, %	80	60	15	3

[a]Medium: MS medium containing 0.4 mg/L NAA and 1.0 mg/L BAP.
From ref. *20*.

Table 4

Transformation Efficiency (%) in the Growth
of Transgenic Callus and Transformed Shoots[a]

Antibiotic (mg/L)	Transgenic regeneration frequency/cultured explant (%)	
	Callus	Shoot[b]
Vancomycin (500)	84	66
Augmentin (400)	86	68

Data were recorded 6 wk after inoculation. Medium: MS medium
containing 0.4 mg/L NAA and 1.0 mg/L BAP. From ref. *20*.

[a]From 4-d-old Cotyledon explants of *Arabidopsis* ecotype "C24"
in the presence of 50 mg/L kanamycin and augmentin or vancomycin.

[b]An average of four transformed shoots per cultured explant
were produced.

ence of vancomycin or augmentin (Table 4). Vancomycin and augmentin
are both satisfactory antibiotics for regeneration, however, augmentin is
more successful in controlling *Agrobacterium* after cocultivation (Table
4). There is even some evidence that augmentin improves shoot regeneration in *Arabidopsis* (M. Barghchi, unpublished data).

6. The origin of the explant and its physiological stage is important for in
vitro regeneration. The nearer the cut is made to the end of the cotyledon
proximal to the main axis, the higher is the production of shoots suggesting that a polar phenomenon affecting morphogenesis exists in *Arabidopsis*
cotyledons. Similar observations are made in apple, soybean, and pea
(14,23–26,27). Green transformed callus starts to grow on *Arabidopsis*
cotyledon explants in selection medium within 6–8 d. The first transformed
shoot appears 2 wk after inoculation and as many as 68% of explants produce at least one transformed shoot within 4 wk (Table 4). An average of
three to four transformed shoots per explant is produced *(20)*.

7. It is suggested that the rapid shoot regeneration from cotyledon explants reported by the present author reduces the potential for increased somaclonal variation owing to long in vitro incubation *(20)*. This high efficiency of transformed shoot production will be reduced dramatically if cotyledon explants dehydrate or wilt during the explant preparation.

8. *Agrobacterium* strain LBA4404(pAL4404) (octopine type) carrying a binary vector pBinl9 was not efficient in producing *Arabidopsis* transformants in our studies.

9. Transient gene expression of GUS activity in the kanamycin resistant transformed shoots and callus can be confirmed using the histochemical GUS assay (*see* Chapter 17). CaMV 35SGUS gene expression is observed in transformed leaves, shoots, stem, and flowers of all *Arabidopsis* transformed regenerants. GUS expression is stronger in the most actively growing regions of the plants and in the younger vascular tissues. The CaMV 35S-GUS gene is often expressed only at the later stages of *Arabidopsis* pollen development. This may indicate that expression of CaMV 35S-GUS gene in *Arabidopsis* exhibits a degree of developmental control.

10. Southern blot hybridization can be used to confirm integration of T-DNA in the nucleus of transformed plants (2, *see* Chapter 11). The results of a Southern blot hybridization assay including two independent transgenic *Arabidopsis* plants regenerated from cotyledon explants is presented in Fig. 3 for a limited discussion *(20)*. Digestion of DNA with *Pst*I releases a single internal T-DNA fragment of 1.4 kb of *npt*II gene (Fig. 3, lanes 1, 3, and 6). Digestion with *Hind*III, which has single restriction site in the T-DNA, confirms integration of T-DNA into the plant genome (Fig. 3, lanes 4 and 7). *Pst*I digesting GUS (+) plants and pBI 121 plasmid DNA reveals 1.4 kb internal fragment of *npt*II gene (Fig. 3, lanes 1, 3, and 6). No hybridization is observed from wild-type *Arabidopsis* plants (Fig. 3, lane 5).

11. The procedure for production of transgenic *Arabidopsis* shoots from cotyledon explants reported in this chapter is simple and has a very high efficiency of transformation in a system that produces a large number of shoots within 2 wk and transformed shoot yields of 66–68% with an average of three to four transformed shoots per explant within 3–4 wk. However, in other protocols, shoot transformation efficiency is much lower and it takes 3–4 mo to produce transformed shoots. Using the method developed by Barghchi et al. *(20)*, the callus phase is very short, thus reducing the risk of somaclonal variation. Reliability and the high production frequency of transformed *Arabidopsis* from cotyledon explants in a very short time pro-

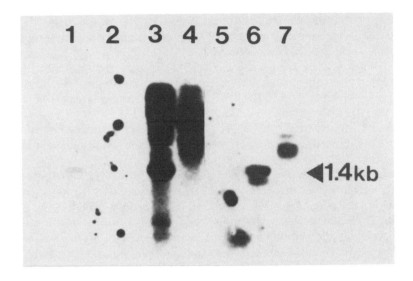

Fig. 3. Southern blot analysis of DNA from transformed *Arabidopsis* plants. Blots were hybridized with a 1.4-kb fragment of *npt*II gene. Lanes: 1. pBI 121, PstI; 2. λ-marker; 3. GUS (+) plant A11, *Pst*I; 4. GUS (+) plant A11, *Hind*III; 5. wild-type plant, *Pst*I; 6. GUS (+) plant A26, *Pst*I; 7. GUS (+) plant A26, *Hind*III. The arrow shows the position of the 1.4 kb *Pst*I fragment of *npt*II gene *(20)*.

vides an efficient transformation system enabling routine and widespread use of *Arabidopsis* transformation in plant molecular genetics and developmental biology.

References

1. Negrutiu, I. and Jacobs, M. (1978) Factors which enhance *in vitro* morphogenesis of *Arabidopsis thaliana. Z. Pflanzenphysiol.* **90,** 423–430.
2. Negrutiu, I., Jacobs, M., and de Gree, F. W. (1978) *In vitro* morphogenesis of *Arabidopsis thaliana:* the origin of the explant. *Z. Pflanzenphysiol.* **90,** 363–372.
3. Goto, N. (1979) *In vitro* organogenesis from leaf explants of some dwarf mutants of *Arabidopsis thaliana* (L) Heynh. *Jpn. J. Genet.* **54,** 303–306.
4. Huang, B. C. and Yeoman, M. M. (1984) Callus proliferation and morphogenesis in tissue culture of *Arabidopsis thaliana. Plant Sci. Letters,* **33,** 353–363.
5. Acedo, G. N. (1986) Regeneration of *Arabidopsis* callus *in vitro. Plant Cell Tissue Org. Cult.* **6,** 109–114.
6. Feldman, K. A. and Marks, M. D. (1986) Rapid and efficient regeneration of plants from explants of *Arabidopsis thaliana. Plant Sci.* **47,** 63–69.
7. Gleddie, S. (1989) Plant regeneration from cell suspension cultures of *Arabidopsis thaliana* heynh. *Plant Cell Rep.* **8,** 1–8.

8. Lloyd, A. M., Barnason, A. R., Rogers, S. G., Byrne, M. C., Fraley, R. T., and Horsch, R. B. (1986) Transformation of *Arabidopsis thaliana* with *Agrobacterium tumefaciens. Science* **234,** 464–466.

9. An, G., Watson, B. D., and Chiang, C. C. (1986) Transformation of tobacco, potato, and *Arabidopsis* using a binary Ti vector system. *Plant Physiol.* **81,** 301–305.

10. Zhang, H. and Somerville, C. R. (1987) Transfer of the maize transposable element MU1 into *Arabidopsis thaliana. Plant Sci.* **48,** 165–173.

11. Feldman, K. A. and Marks, M. D. (1987) *Agrobacterium*-mediated transformation of germinating seeds of *Arabidopsis thaliana:* a non-tissue culture approach. *Mol. Gen. Genet.* **208,** 1–9.

12. Valvekens, D., Van Montagu, M., and Van Lijsebettens, M. (1988) *Agrobacterium tumefaciens*-mediated transformation of *Arabidopsis thaliana* root explant using kanamycin selection. *Proc. Nat. Acad. Sci. USA* **85,** 5536–5540.

13. Damm, B., Schmidt, R., and Willmitzer, L. (1989) Efficient transformations of *Arabidopsis thaliana* direct gene transfer to protoplasts. *Mol. Gen Genet.* **217,** 6–12.

14. Ozcan, S., Barghchi, M., and Draper, J. (1992) High-frequency adventitious shoot regeneration from immature cotyledons of pea (*Pisum sativum* L). *Plant Cell Rep.* **11,** 44–47.

15. Duncan, D. R., Williams, M. E., Zehr, B. E., and Widholm, J. M. (1985) The production of callus capable of plant regeneration from immature embryos of numerous *Zea mays* genotypes. *Planta* **165,** 322–332.

16. Ranch, J. P., Oglesby, L., and Zielinski, A. C. (1985) *In vitro* plant regeneration from embryo-derived tissue cultures of soybeans. *In Vitro Cell Dev. Biol.* **21,** 653–658.

17. Patton, D. and Meinke, D. (1988) High-frequency plant regeneration from cultured cotyledons of *Arabidopsis thaliana. Plant Cell Rep.* **7,** 233–237.

18. Barghchi, M., Turgut, K., Paul, W., Hodge, R., Griffiths, N., Draper, J., and Scott, R. (1990) *Genetic Engineering of* Arabidopsis thaliana, Abstracts VIIth International Congress on Plant Tissue Culture, Amsterdam, p. 46.

19. Barghchi, M., Turgut, K., Griffiths, N., and Draper, J. (1991) Transformation of *Arabidopsis* by *A. tumefaciens. In Vitro Cell Dev. Biol.* **27,** 3.150.

20. Barghchi, M.,Turgut, K., Scott, R., and Draper, J. (1994) High-frequency *Agrobacterium*-mediated transformation of *Arabidopsis thaliana* ecotypes "C24" and "Landsberg *erecta." J. Plant Growth Reg.* **14,** 61–67.

21. Murashige, T. and Skoog, F. (1962) A revised medium for rapid growth and bioassays with tobacco tissue culture. *Physiol. Plant.* **15,** 473–497.

22. Jefferson, R. A., Kavanagh. T. A., and Bevan, M. W. (1987) GUS fusions: β-glucuronidase as a sensitive and versatile gene fusion marke in higher plants. *EMBO J.* **6,** 3301–3307.

23. Barghchi, M. and Alderson, P. G. (1989) Pistachio (*Pistacia vera* L.), in *Biotechnology in Agriculture and Forestry,* vol. II (Bajaj, Y. P. S. ed.), Springer Verlag, Berlin, pp. 68–98.

24. Kouider, M., Korban, S. S., Skirvin, R. M., and Chu, C. M. (1984) Influence of embryonic dominance and polarity on adventitious shoot formation from apple

(*Malus domestica* cultivar Delicious) cotyledons *in vitro. Amer. Soc. Hort. Sci.* **109,** 383–385.

25. Mante, S., Scorza, R., and Cordts, J. (1989) A simple, rapid protocol for adventitious shoot development from mature cotyledons of *Glycine max* cv Bragg. *In Vitro Cell Dev. Biol.* **25,** 385–388.

26. Maniatis, T., Fritsch, E. F., and Sambrook, J. (1982) *Molecular Cloning: A Laboratory Manual.* Cold Spring Harbor Laboratory, Cold Spring Harbor, NY.

27. Ozcan, S., Barghchi, M., Firek, S., and Draper, J. (1993) Efficient adventitious shoot regeneration and somatic embryogenesis in pea. *Plant Cell Tissue Org. Cult.* **34,** 271–277.

CHAPTER 15

Transformation Protocols for Broadleaved Trees

Trevor M. Fenning and Kevan M. A. Gartland

1. Introduction

Plant genetic manipulation by transformation is the insertion of DNA directly into the genome of a plant cell, and the regeneration of whole plants from such cells. This is brought about by human intervention using a combination of plant tissue-culture techniques, molecular biology, and microbiological methods, and allows a chosen plant variety to be modified in a small but highly specific manner.

For tree species, this potential is especially important because the long breeding cycles involved necessarily delay and restrict improvement programs by conventional plant breeding techniques, and require considerable resources *(1)*. In addition, the nonflowering juvenile growth of a tree seedling may last for several years, and its performance during this period is not necessarily an indication of its eventual behavior when mature *(2)*. Therefore, it is preferable, in theory at least, to identify a mature tree already having as many desirable characteristics as possible (the "elite" specimen), and genetically engineer those few that may be lacking.

As specimens of wider scientific interest, trees are problematic subjects, as their distinguishing characteristics ("woody perennial plants, typically having a single stem or trunk growing to a considerable height" *[3]*) are difficult to study and are not unique in the plant kingdom, except in combination. These properties may be studied singly more easily in other plants (all higher plants synthesize lignin, for instance, and many

From: *Methods in Molecular Biology, Vol. 44:* Agrobacterium *Protocols*
Edited by: K. M. A. Gartland and M. R. Davey Humana Press Inc., Totowa, NJ

Table 1
Transformed and Regenerated Tree Species[a]

Species	Source material	Transforming genes	Refs.
Allocasuarina	Seedling explants	Wild-type Ri T-DNA	*(6)*
Apple	Leaf discs of cultured shoots	Markers + CPTI gene + *B. thuringiensis* toxin genes	*(7–10)*
Apricot	Embryos	Markers + plum pox virus coat proteins	*(11)*
Citrus	Embryo protoplasts + seedling explants	Markers only	*(12,13)*
Eucalyptus?[b]	?	Markers only?	*(14)*[b]
Pecan	Embryos	Markers only	*(15)*
Plum	Embryos	Markers only	*(16)*
Poplar	In vitro leaf and internode explants	Numerous	*(17–19)*
Walnut	Embryos	Markers + as apple	*(9,20)*
White spruce	Embryos	Markers + *B. thuringiensis* toxin genes	*(21)*

[a]From refereed publications only.
[b]Private companies may have already transformed *Eucalyptus* spp, but are unlikely to publish a full protocol.

are perennial), or are not yet readily amenable to study by genetic manipulation in the absence of a detailed physiological knowledge. Consequently, no tree species is likely to become a model species in the general sense that *Arabidopsis* or tobacco have become, but instead are worthy of study as valuable and interesting subjects in their own right, to which knowledge gained from the model species can be applied.

For these reasons the genetic manipulation of tree species has attracted increasing interest, but in comparison to the research efforts dedicated to *Arabidopsis* or tobacco, the level of activity is still relatively limited. However, it should be realized that the task of developing a transformation protocol for a tree species is not fundamentally different from that for any other plant species, as reviewed in this volume and elsewhere *(4,5)*, progress being limited mainly by the effort available.

To date, regenerant plants have been obtained from at least nine transformed tree species as described in refereed journals (*see* Table 1), and others are likely in the near future (e.g., alder, silver birch, cherry, elm, willow), but success may be limited to a single amenable variety or seed explants. Embryo transformations are of limited interest because the difficulties in assessing likely mature characteristics are the same as for conventionally derived seedlings. It may be the case, however, that some essential improvement cannot be achieved any other way, such as the transfer of genes for resistance to a disease for which the entire species is susceptible.

Careful thought should be given to the aims of the work before beginning an effort to transform a previously unstudied tree species, as the development costs involved are likely to be high. If the same ends could have been met more simply, quickly, and cheaply by screening a large enough wild population for the desired characteristics, then the effort will have been a high-tech waste of money. Ideally, a genetic manipulation program will be linked to a conventional breeding scheme, or at least to a collection and field trial of wild or other available varieties, enabling a number of elite genotypes to be identified and assessed for their suitability for genetic manipulation. Reliable information from growers, foresters, tree breeders, or even tree pathologists engaged with a particular species may give an indication of which types of problems might be best met by a genetic manipulation approach, and the different strategies that could be successful if adopted. Subsequent to these considerations, a strategy for transforming the chosen tree species may be devised.

2. Materials

These are the materials that are required for most plant tissue-culture and genetic manipulation procedures. Extra materials for a particular method, such as specific plant cultures or media are noted at the start of the relevant method sections.

1. A class II containment bench, standard tissue-culture apparatus (e.g., scalpels, forceps, 9-cm Petri dishes), and sterilizing equipment (autoclave and Bunsen burner) (*see* Note 1).
2. A tissue-culture growth-room maintained at 25°C ± 2°C with an 8 h dark/ 16 h light photoperiod, providing approx 50 µE/m/s photosynthetically active light at shelf level (*see* Note 2).
3. A sterile in vitro shoot culture of the tree species intended for transformation (*see* Note 3).
4. A shaker and incubator kept at 28°C for growing *Agrobacterium tumefaciens* (*see* Note 4).
5. Supplies of filter sterilized antibiotics kanamycin, ampicillin, tetracycline, and cefotaxime (claforan).

6. Bottles of sterile distilled water.
7. Sterile blotting paper.
8. A spectrophotometer capable of taking optical density (OD) readings over the range 420–660 nm.

3. Methods

3.1. Preamble

Not all tree species or varieties will be amenable to the processes required for a genetic manipulation protocol. The differences of behavior in vitro between genotypes even within a single species can be very significant, as trees are highly heterozygous as compared to most crop plants *(2)*. Each variety or genotype of a tree species must therefore be evaluated independently, and if one proves recalcitrant, others should be tested for their suitability. If resources allow, it clearly would be advantageous to conduct feasibility studies on several genotypes simultaneously *(see* Section 3.2.).

The step that is often limiting when attempting to develop a transformation protocol for an established tree variety (as opposed to seed material), is obtaining suitable plant material with which to experiment. An in vitro shoot culture of the chosen variety is the preferred source, enabling a supply of explants already habituated to tissue-culture conditions all year round *(see* Sections 3.2., 3.3., 3.4., and 3.5.). If seed explants have to be used, however, the response of these tissues to the transformation procedures may vary with the length of time the seeds are stored, apart from the inherent genetic variability among such material, making generalizations about methods unreliable. For this reason, no protocol for transforming seed explants is provided in this chapter.

The problems associated with establishing a proliferating culture from the shoots of a mature tree are beyond the scope of this chapter and have been reviewed amply elsewhere *(22,23)*, but there are two problems of principal importance. These are: forcing the excised shoots from the mature tree to proliferate in culture and eradicating microorganisms that normally systemically infest tree tissues under field conditions. This contamination may kill the shoots before they successfully establish as a proliferating culture, or affect the performance of explants under transforming conditions. Once these problems have been resolved the next obstacle is the choice of gene delivery system.

There is now a wide variety of gene delivery systems for plant transformation to choose from including viruses, DNA electroporation into protoplasts, direct DNA uptake into seeds or floral parts, and biolistic approaches *(4)*. The biolistic approach, where microscopic gold or tungsten particles coated with DNA are fired into a plant tissue, is rapidly becoming more popular especially for transient expression studies. The methodologies and equipment used have been extensively described in the *Plant Cell, Tissue and Organ Culture Journal* (**33**, 219–257, 1993). The technique is being increasingly used to produce stable transformants *(21,24)*, but *A. tumefaciens* based transformation protocols still dominate, and are therefore the ones described in this chapter.

The mechanism by which *A. tumefaciens* transforms a plant cell has been extensively reviewed *(25–28, see* Chapter 2), but briefly *A. tumefaciens* contains a large plasmid up to 200 kb in size that largely controls the transformation process. This is called the Ti plasmid (or Ri plasmid for the similar plasmid in *A. rhizogenes*), and has a region of transforming DNA (the T-DNA) for transfer into plant cells, and a virulence region controlling the transfer process.

The important factor enabling what is naturally a disease-causing organism to be adapted for genetic manipulation purposes, is that the genes to be transferred as T-DNA are in a distinct delineated region of the plasmid and independent of the virulence functions. Provided a gene to be transferred is between the so-called border sites on either side of the T-DNA, then it will be transferred into susceptible plant cells when the process is activated. The oncogenic genes of the wild-type *A. tumefaciens* strains (which modify the hormone physiology of transformed cells) may be deleted, creating a "disarmed Ti plasmid," and replaced with genes of interest.

This artificial T-DNA region will activate even if on another plasmid, provided it is in an *A. tumefaciens* host cell containing a Ti or Ri plasmid with a functional virulence region. As plasmids of 200 kb are very fragile when manipulated in vitro, artificial T-DNA regions are often placed into smaller plasmids (called binary plasmids), which are then transferred into suitable (disarmed or wild-type) *A. tumefaciens* or *A. rhizogenes* strains.

For ease of obtaining transformed plants, scorable and selectable marker genes are usually included in the artificial T-DNA region. A scorable marker is a gene that produces an easily detectable product (e.g., β-glucuronidase from the GUS gene) giving an indication of which cells are transformed. A selectable marker is a gene encoding resistance to a particular antibiotic or herbicide (e.g., kanamycin or byalophos), which may then be added to the plant media to encourage transformed cells to grow at the expense of nontransformed cells. Used together, these genetic systems have been responsible for most of the successful plant transformation protocols.

It is the purpose of this chapter to outline a number of repeatable transformation procedures that have been used with apple and poplar varieties, which may be extended to other tree species. Before beginning such work it is first necessary to conduct feasibility studies on the chosen genotype(s) as described next.

3.2. Feasibility Studies

The use of disarmed strains of *A. tumefaciens* with binary vector plasmids containing screenable marker genes such as the GUS gene fusion system *(29)*, theoretically allows transformation events to be assessed on plant material at an early stage even in the absence of the ability to regenerate plants. However, if marker gene activity is not detected it could be owing to any number of biochemical reasons, quite apart from a failure of the chosen *A. tumefaciens* strain to transform the target plant cells under the conditions employed. Equally, nontransformed tissues of some plant species under some conditions may appear to express marker genes, GUS included.

Therefore, it is often instructive to inoculate plants with wild-type *A. tumefaciens* or *A. rhizogenes* strains (as distinct from engineered strains), which will demonstrate transformation of the host plant by the production of visible tumors. These should not only continue to grow in the absence of exogenously supplied growth regulators under in vitro conditions, but should prove positive when analyzed for transformation by Southern blot, and tested for the presence of opines *(30)*, which are synthesized by plant cells transformed with wild-type T-DNA *(25,28)* (*see* Table 2).

Transformation success with a wild-type *Agrobacterium* strain can also give an indication of the physical conditions necessary to transform the plant material in vitro with disarmed strains carrying engineered Ti plasmids. This information is particularly useful if a disarmed version of the wild-type strain to which the chosen tree genotype is susceptible can be obtained. Examples are listed in Table 2.

It is important to remember, however, that the ability of a particular strain of *Agrobacterium* (disarmed or wild-type) to transform a plant cell

Table 2
Selected Wild-Type and Disarmed *Agrobacterium tumefaciens* Strains

Wild-type strain	Ti plasmid	Opine type	Disarmed derivatives	Refs.
A281	pTiBO542	agropine/	EHA101	(31)
		succinamopine	+ ALG1	(32)
82.139[a]	pTi82.139	nopaline	—	(18,33)
Ach5	pTiAch5	octopine	LBA4404	(34)
			(pAL4404)	
B6S3	pTiB6S3	octopine	pGV2260	(35)
C58[b]	pTiC58	nopaline	pMP90	(36)
			pGV3850	(37)

[a]Probably a naturally occurring "shooty mutant" strain, tumors from which readily produce mostly nontransformed shoots. There is no disarmed version available.

[b]Strain C58C1 is C58 cured of its wild-type Ti plasmid, enabling replacement with a disarmed Ti plasmid. It is commonly used as a host for disarmed Ti plasmids from other *Agrobacterium* strains.

is the function of many interacting factors including: the susceptibility of that plant genotype to the *Agrobacterium* strain (i.e., host range of the *Agrobacterium* strain), the susceptibility and physiological condition of the plant tissues exposed to the agrobacteria, the host range of the Ti or Ri plasmid, the physical condition of the *Agrobacterium* culture used, and induction of the *Agrobacterium* virulence mechanism (7,26). The fundamental prerequisite is that the plant tissue inoculated is susceptible to the *Agrobacterium* strain used under the conditions employed.

For these reasons it is important to test several *Agrobacterium* strains on the tree species under investigation, and on various different tissues. Wild-type *A. rhizogenes* strains can even be used to alter the root architecture of susceptible tree varieties, without the need to produce fully transformed plants with disarmed *Agrobacterium* strains (38,39). The base of a shoot growing in vitro or conventional cutting is infected with an appropriate *A. rhizogenes* strain, and many of the roots that emerge will be of the "hairy root" phenotype that are nevertheless functional. This tactic may ease the rooting of cuttings from some hard-to-root tree genotypes and increase the ability of the root system to absorb water and nutrients.

Examples of full transformation protocols are described in the following sections. Before proceeding, however, it is advisable to test the effects of the various antibiotics used in the transformation protocols described on the tree species being investigated under nontransforming conditions. Some species may not tolerate certain antibiotics in the growth or regeneration medium, whereas others may be resistant to the antibiotic needed for a particular selectable marker gene intended for the transformation.

The transformation protocol described in Section 3.3. is an ideal starting point for those beginning this work for the first time, or for where shoot regeneration conditions have not been optimized. This coinoculation transformation procedure relies on the wild-type *A. tumefaciens* strain 82.139, tumors from which readily produce sufficient shoots for it to be used as the regeneration mechanism. If the plant is simultaneously infected with an *Agrobacterium* strain containing an engineered T-DNA sequence, then some of the shoots produced from the 82.139 tumor may be transformed with the engineered genes.

3.3. Coinoculation Transformations

This method is a generalized transformation protocol (developed by Lise Jouanin and colleagues *[18]*) for the *Populus tremula* X *P. alba* genotypes, as well as other recalcitrant tree and crop species *(40)*.

3.3.1. Additional Materials

1. A sterile culture of *P. tremula* X *P. alba* growing on a modified Murashige and Skoog (MS) medium *(41)* with 3% (w/v) sucrose and 0.7% (w/v) agar as setting agent and no growth regulators.
2. The wild-type *A. tumefaciens* strain 82.139 and the disarmed strain C58C1/pMP90 (Table 2), containing a binary vector plasmid such as pBIN19 *(42)* or p35s-GUS/INTRON *(43)*.
3. Grow the *A. tumefaciens* 82.139 in liquid YMB medium *(5)*, (1% [w/v] mannitol, 0.04% [w/v] yeast extract, 0.05% [w/v] K_2HPO_4, 0.02% [w/v] $MgSO_4 \cdot 7H_2O$, and 0.01% [w/v] NaCl, adjusted to pH 7.0) without antibiotic selection, and C58C1/pMP90 with the binary plasmid in liquid YEP medium *(5)* (1% [w/v] bacto-peptone, 1% [w/v] bacto-yeast extract, and 0.5% [w/v] NaCl, adjusted to pH 7.0), supplemented with 50 µg/mL kanamycin *(see* Note 5).
4. Petri dishes of growth regulator free MS medium (as in step 1), but supplemented with 2 g/L activated charcoal, 1000 µg/mL ampicillin, 500 µg/mL cefotaxime, and 50 µg/mL kanamycin.

3.3.2. Procedure

1. Three days prior to the intended transformation experiment, inoculate 10 mL of liquid YMB medium with *A. tumefaciens* 82.139 in a 30-mL Universal

bottle and incubate at 28°C for 2 d in an orbital shaker at 160 rpm. Transfer 1 mL of this culture into another 10 mL of YMB medium and incubate as previously for 24 h to obtain as rapidly growing culture as possible.

2. Two days prior to the intended transformation experiment, inoculate 10 mL of liquid YEP and kanamycin with *A. tumefaciens* C58C1/pMP90/ p35s-GUS/INTRON or similar strain, and incubate as in step 1 for 24 h followed by transfer of 1 mL of this culture to 10 mL of fresh YEP and kanamycin, then leave to grow for another 24 h.

3. Prepare three separate dilutions of *A. tumefaciens* 82.139 in growth medium to OD_{600} 0.1, 0.5, and 1.0 (or undiluted, if the culture will not reach 1.0), and mix each 1:1 with the *Agrobacterium* C58C1/pMP90/p35s-GUS/INTRON culture.

4. Inoculate 40 shoots each on medium free of antibiotics and hormones with the three dilutions of *A. tumefaciens*; use a pointed scalpel or sharp tweezers dipped in the agrobacterial cell suspension to wound the shoots in various internode positions from the growing point downward. Tumors are likely to form preferentially at one internode position compared to others. On repeat experiments inoculate only at the most susceptible internode position on the shoots, and at the most dilute *A. tumefaciens* 82.139 suspension still giving maximum tumor formation.

5. After 4 d, transfer the shoots to fresh medium, still free of growth regulators but containing 200 µg/mL cefotaxime.

6. After 4–6 wk, score the shoots by eye for tumor formation and excise all definite tumors, taking care to exclude any lateral buds. Gently shake the tumors for 8 h in sterile water containing 25 µg/mL tetracycline, before transfer to Petri dishes containing the MS/charcoal/antibiotic medium (as in Section 3.3.1., step 4).

7. Monitor the tumors for shoot production for up to 3 mo (subculture regularly, *see* Note 6), removing shoots as they emerge and transferring them to standard proliferation or rooting medium. Up to 50% of tumors will give rise to shoots with *A. tumefaciens* 82.139 alone, but far fewer when used in combination with an engineered *A. tumefaciens* strain and under selective conditions.

8. Analyze the shoots for transformation (*see* Note 7). These will be of two possible types: shoots transformed with the wild-type T-DNA genes of *A. tumefaciens* 82.139, and consequently demonstrating the altered morphology typical of such plants *(25,28)*; and shoots transformed with the T-DNA and marker genes of the disarmed-binary vector strain. This is achieved by cross feeding between the two transformed plant cell types in the tumor. The chance of an emerging shoot being transformed by both the wild-type and engineered *A. tumefaciens* strains is extremely remote.

3.4. Leaf Disc Transformations

This method is based on the protocol developed by David James and colleagues *(7,8)* for the apple (*Malus pumila* Mill.) cultivar Greensleeves.

3.4.1. Additional Materials

1. A sterile shoot culture of *M. pumila* Mill. cv. Greensleeves (growing on A8; a modified MS medium *(41)* with 3% (w/v) sucrose, 2.0 µg/mL BAP, 0.1 µg/mL IBA, and 1.0 µg/mL gibberellic acid, GA_3). Liquid and solidified supplies (with 0.7% agar as setting agent) of growth medium will be required (*see* Note 3).
2. A 7-mm cork borer.
3. Access to the disarmed *A. tumefaciens* strain LBA4404 *(34)* or C58C1/pGV3850 *(37)* containing a suitable binary vector plasmid, e.g., pBIN19 *(42)* or p35S-GUS/INTRON *(43)*, grown on YEP medium *(5; see* Note 5) supplemented with 50 µg/mL kanamycin.
4. Liquid MS20 medium (MS salts with 2% [w/v] sucrose) supplemented with 0.1 m*M* acetosyringone and 1 m*M* betaine phosphate, and adjusted to pH 5.2 for maximum *A. tumefaciens* virulence induction *(26)*.
5. Centrifuge and appropriate bottles.
6. BNZ511 medium in Petri dishes for the cocultivation of apple leaf discs with *A. tumefaciens*, containing MS salts, 3% (w/v) sucrose, and the growth regulators BAP (5 µg/mL), NAA (1 µg/mL), and thidiazuron (1 µg/mL). To regenerate shoots and eliminate the agrobacteria, BNZ511 is also used but with 25 µg/mL kanamycin and 200 µg/mL cefotaxime added.

3.4.2. Procedure

1. Three days before the intended transformation procedure, inoculate 10 mL of YEP medium containing 50 µg/mL kanamycin with *A. tumefaciens* LBA4404/p35s-GUS/INTRON (or other disarmed strain) and leave shaking at 28°C for 2 d. Inoculate another 10 mL of YEP medium with 1 mL from the previous culture for a further 24 h, to obtain as rapidly growing a culture as possible.
2. Centrifuge the agrobacteria at low speed (2500 rpm), then resuspend and dilute them in liquid MS20 medium (with acetosyringone and betaine) to an OD_{420} of 0.5. Incubate 10 mL of this suspension in a 30-mL Universal bottle at 25°C for another 5 h in an orbital shaker at 160 rpm, to fully induce the virulence mechanism of the agrobacteria.
3. Cut 50 leaf discs with the cork borer from rapidly growing Greensleeves shoots taken directly from proliferation medium and place onto a single 9-cm Petri dish containing antibiotic free BNZ511 apple regeneration medium, to hold the discs until the *A. tumefaciens* induction is complete.

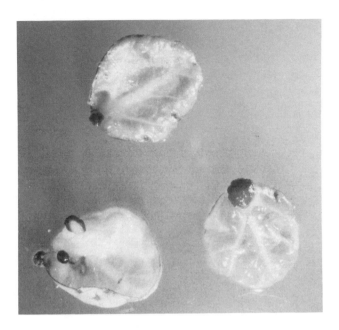

Fig. 1. Leaf discs of the apple cultivar "Greensleeves" on regeneration medium, 1 mo after infection with *A. tumefaciens* C58C1-pGV2260 containing the binary vector plasmid p35s-GUS/INTRON (*see* Section 3.4.). Transformed cells show up as blue areas after development in the GUS enzyme substrate X-GLUC. Courtesy of D. J. James and A. J. Passey of HRI-East Malling. Photo: M. Scutt.

4. When the *A. tumefaciens* induction is complete, pour the bacteria over the leaf discs in the holding plate, and leave for 20–30 min. After this time remove the leaf discs and blot dry on sterile paper, then return to fresh plates of antibiotic-free BNZ511 medium for 3 d cocultivation (adaxial side up) in the dark at 25°C (four discs/plate).

5. After this period, transfer the leaf discs carefully to fresh BNZ511 medium with antibiotics and incubate in the dark at 25°C for up to 5 mo to await the production of transformed shoots. Subculture at regular intervals onto fresh BNZ511 medium with cefotaxime (*see* Note 6). Kanamycin may be omitted after 3 wk as most callus formed by this period will have originated from transformed cells and the continued presence of the antibiotic will inhibit the regeneration of transformed shoots. In trial experiments, a reliable indication of transformation success can be gained by staining sample leaf discs for GUS histochemically with X-GLUC (5-bromo-4-chloro-3-indolyl β-D-glu-curonide cyclohexammonium salt), the blue areas indicating transformed cells (*29*; and as described in Chapter 17, also *see* Note 7).

6. Remove shoots as they appear and transfer to standard proliferation medium (as in Section 3.4.1., step 1) for propagation to provide sufficient material for Southern blot or GUS assay without utilizing all the plant material. Up to 10% of leaf discs are likely to produce transformed shoots over this period, although many more will have transformed callus.

3.5. Internode Transformations

This method is the protocol developed for one amenable genotype of *Populus tremula* X *P. alba*, at the INRA Versailles laboratory of Lise Jouanin *(44)*.

3.5.1. Additional Materials

1. A sterile culture of *P. tremula* X *P. alba* growing on a modified MS medium *(41)* with 3% (w/v) sucrose and 0.7% (w/v) agar as setting agent.
2. A disarmed *A. tumefaciens* strain, such as, C58C1/pMP90 *(36)* containing a suitable binary vector plasmid, e.g., pBIN19 *(42)* or p35s-GUS/INTRON *(43)*, grown on YEP medium *(5; see* Note 5) supplemented with 50 µg/mL kanamycin sulfate.
3. M1 medium for preconditioning the internode explants and cocultivation with *A. tumefaciens*. This is identical to the propagation medium but supplemented with the plant growth regulators NAA (10 µ*M*) and 2iP (5 µ*M*).
4. M2 medium for callusing the infected explants and killing the transforming agrobacteria. This is identical to M1, but with the addition of 250 µg/mL cefotaxime and 500 µg/mL carbenicillin *(see* Note 6).
5. M3 medium for the production of shoots from callused tissue. This is identical to the basic propagation medium, but supplemented with 0.1 µ*M* of the cytokinin-like growth regulator thidiazuron and antibiotics as in M2.
6. Half-strength propagation medium (as in step 1) without hormones, but supplemented with 50 µg/mL kanamycin for rooting transformed shoots.

3.5.2. Procedure

1. Prepare 20 internode segments from a rapidly proliferating *Populus* culture taking care to excise all lateral buds, and cut the segments lengthways before placing them 10 to a Petri dish containing 25 mL M1 medium for 2 d preconditioning at 25°C in the dark.
2. Inoculate 10 mL of YEP medium containing 50 µg/mL kanamycin with *A. tumefaciens* C58C1/pMP90/p35s-GUS/INTRON (or other disarmed strain) and leave shaking at 28°C for 2 d in an orbital shaker at 160 rpm. Inoculate another 10 mL of YEP medium as just mentioned with 1 mL from the previous culture for another 24 h, to obtain as rapidly growing a culture as possible.

3. Dilute an aliquot of this second culture to an OD_{660} of 0.3 with the basic proliferation medium (as in Section 3.5.1., step 1). Add the 40 internode halves and leave gently shaking for 16 h to allow the agrobacteria to attach to the plant cells.

4. Remove the internode explants and blot dry on sterile paper, then transfer to M1 medium (four explants per 9-cm Petri dish with 25 mL of M1), for 48 h cocultivation.

5. Remove the explants and wash free of contaminating agrobacteria in sterile distilled water containing 25 µg/mL tetracycline. Repeat the washes two more times for 5 min each, followed by three 5-min washes in water only, before transferring the explants to M2 medium.

6. Maintain the explants on M2 medium for 2 wk at 25°C in the dark (*see* Note 3), before transfer to M3 medium and standard light conditions for up to 4 mo.

7. Remove putative transformed shoots as they grow out of the callus nodules and transfer to the rooting medium and analyze for transformation. Up to 50% of internode explants are likely to give rise to transformed callus or shoots with this protocol.

4. Notes

1. Bacterial strains or infected plant material should not be handled in a conventional laminar air flow bench, as these blow air toward the operator, as well as spreading any bacteria handled to the surrounding area. Class II containment benches, however, take air in from the front and filter it before passing it over the operating area, and filter it again before venting at the back or top.

2. The quality of light in a tissue-culture growth room is very important, usually needing to be as close to balanced daylight as possible. Many commercially available fluorescent bulbs do not provide this, but it may be achieved by mixing different types of bulbs if necessary. In addition it is important not to provide excess light to plant material in tissue-culture (especially explants in Petri dishes), as photosynthetic mechanisms become decoupled as a result of the external nutrient supply and render the plant tissues more susceptible to photoinhibition of growth.

3. Adapt media and lighting regimes as appropriate for other tree species or genotypes to obtain high levels of shoot proliferation and regeneration from suitable explant sources.

4. Exposure of *A. tumefaciens* or *A. rhizogenes* to temperatures in excess of 28°C even for short periods can result in the loss of their main Ti or Ri plasmid, and hence their ability to transform plant cells. They will continue to grow apparently normally, however, so great care must be taken to ensure that this temperature is not exceeded.

5. The binary vectors pBIN19 and p35s-GUS/INTRON each encode resistance to the antibiotic kanamycin sulfate (from the *npt*II gene) twice—
   under two separate promoters: one in the T-DNA region under a plant
   promoter for expression in transformed plant cells, and one under the original bacterial promoter to render the *A. tumefaciens* host similarly resistant.
   Bacterial strains harboring artificial plasmids should always be maintained
   in the presence of the selective antibiotic (kanamycin for these) as such
   plasmids are readily lost from the host strain otherwise.

6. Once the agrobacteria have been used to transform the plant cells they will
   continue to persist in the tissue and under in vitro conditions can overgrow
   the explants and kill them. For this reason high levels of cefotaxime and
   carbenicillin are used, but this is not always immediately successful as
   these antibiotics deteriorate rapidly and the plant tissue itself can harbor
   the agrobacteria particularly if it is callused. It is therefore necessary to
   closely monitor explants that have been in contact with *A. tumefaciens* and
   subculture them onto media with fresh antibiotics not less than every 2 wk,
   or at the first sign of any bacterial growth. Different authors quote various
   strategies for eliminating agrobacteria from plant cultures after transformation, but that for the poplar internode protocol of Section 3.5. *(44)* is the
   most comprehensive. The antibiotic ticarcillin used in combination with
   cefotaxime has also been reported as effective for eliminating agrobacteria
   from tissue-culture material (W. Broekaert, personal communication, 1994).

7. Plants transformed with a vector such as p35s-GUS/INTRON will express
   the β-glucuronidase enzyme of the GUS (*uid*A) gene and the kanamycin
   resistance gene *npt*II, in addition to any other genes added into the vector
   later. Such plants may be analyzed for GUS activity histochemically or
   fluorimetrically *(29*; and *see* Chapter 17), by ELISA test for *npt*II, and by
   PCR *(45)* or Southern blot at the DNA level. For a few days after transformation with *A. tumefaciens*, GUS activity in plant tissues can represent
   transient expression *(7,46)*, but after 1–2 wk of growth under selection
   GUS expression is derived from stably transformed T-DNA in plant cells.

5. Conclusions

   Most tree species are virtually unimproved in terms of their performance for human use, being little more than selections from open pollinated wild populations with highly variable phenotypes and heterogenous
in the extreme. If similar improvements in productivity can be made for
the main commercial tree species as have been achieved with arable and
vegetable crops in recent decades, then there will at least be the opportunity for commercial plantations to keep pace with the global demand for
timber and other tree products without the logging of virgin forest or

heavy use of pesticides. As genes of agronomic value become available, genetic manipulation will become an important extra tool for achieving this aim if applied with judicious care and in the context of wider improvement programs for the breeding, silviculture, and the conservation of all our trees.

Acknowledgments

The authors wish to acknowledge K. Tobutt, A. J. Passey, and T. J. Riggs for their advice in preparing this manuscript; to L. Jouanin for being willing to provide many of the strains described here; and to the Ministry of Agriculture, Fisheries and Food for funding T. Fenning. Special thanks are also given to the late Gordon Browning for all his encouragement.

References

1. Alston, F. H. and Batlle, I. (1992) Genetic markers in apple breeding. *Phytoparasitica* **20(Suppl.)**, 89–92.
2. Libby, W. J. (1987) Genetic resources, and variation in forest trees, in *Improving Vegetatively Propagated Crops* (Abbott, A. J. and Atkin, R. K., eds), Academic, London, pp. 199–209.
3. *The Oxford English Dictionary* (1993) Clarendon Press, Oxford, UK.
4. Dandekar, A. M. (1992) Transformation, in *Biotechnology of Perennial Fruit Crops* (Hammerschlag, F. A. and Litz, R. E., eds.), CAB International, Wallingford, UK, pp. 141–168.
5. Gelvin, S. B., Schilperoort, R. A., and Verma, D. P. S. (1988) *Plant Molecular Biology Manual*. Kluwer, London.
6. Phelep, M., Petit, A., Martin, L., Duhoux, E., and Tempé, J. (1991) Transformation and regeneration of a nitrogen-fixing tree, *Allocasurina verticillata* Lam. *Biotechnology* **9**, 461–466.
7. James, D. J., Uratsu, S., Cheng, J., Negri, P., Viss, P., and Dandekar, A. M. (1993) Acetosyringone and osmoprotectants like betaine or proline synergistically enhance *Agrobacterium*-mediated transformation of apple. *Plant Cell Rep.* **12**, 559–563.
8. James, D. J., Passey, A. J., Barbara, D. J., and Bevan, M. W. (1989) Genetic transformation of apple (*Malus pumila* Mill.) using a disarmed Ti-binary vector. *Plant Cell Rep.* **7**, 658–661.
9. Dandekar, A. M., McGranahan, G. H., Uratsu, S. L., Leslie, C., Vail, P. V., Tebbets, J. S., et al. (1992) Engineering for apple and walnut resistance to codling moth, in *Brighton Crop Protection Conference–Pests, and Diseases*, vol. 2, BcPc Publishers, Farnham, UK, pp. 741–747.
10. Maheswaran, G., Welander, M., Hutchinson, J. F., Graham, M. W., and Richards, D. (1992) Transformation of apple rootstock M26 with *Agrobacterium tumefaciens*. *J. Plant Physiol.* **139**, 560–568.
11. da Câmara-Machado, M. L., da Câmara-Machado, A., Hanzer, V., Weiss, H., Regner, F., Steinkellner, H., et al. (1992) Regeneration of transgenic plants of *Prunus armeniaca* containing the coat protein gene of plum pox virus. *Plant Cell Rep.* **11**, 25–29.

12. Vardi, A., Bleichman, S., and Aviv, A. (1990) Genetic transformation of *Citrus* protoplasts, and regeneration of transformed plants. *Plant Sci.* **69,** 199–206.
13. Moore, G. A., Jacono, C. C., Neidigh, J. L., Lawrence, S. D., and Cline, K. (1992) *Agrobacterium*-mediated transformation of *Citrus* stem segments and regeneration of transgenic plants. *Plant Cell Rep.* **11,** 238–242.
14. Brackpool, A. L. and Ward, M. R. (1990) Optimisation of adventitious shoot formation for transformation in eucalyptus, in *Abstracts of the VIIth International Congress on Plant Tissue, and Cell Culture*, Kluwer, Dordrecht, The Netherlands, p. 9.
15. McGranahan, G. H., Leslie, C. A., Dandekar, A. M., Uratsu, S. L., and Yates, I. E. (1993) Transformation of pecan, and regeneration of transgenic plants. *Plant Cell Rep.* **12,** 634–638.
16. Mante, S., Morgens, P. H., Scorza, R., Cordts, J. M., and Callahan, A. M. (1991) *Agrobacterium*-mediated transformation of plum (*Prunus domestica* L.) hypocotyl slices and regeneration of transgenic plants. *Biotechnology* **9,** 853–857.
17. Fillatti, J. J., Sellmer, J., McCown, B., Haissig, B., and Comai, L. (1987) *Agrobacterium* mediated transformation, and regeneration of *Populus. Mol. Gen. Genet.* **206,** 192–199.
18. Brasileiro, A. C. M., Leplé, J. C., Muzzin, J., Ounnoughi, D., Michel, M. F., and Jouanin, L. (1991) An alternative approach for gene transfer in trees using wild-type *Agrobacterium* strains. *Plant Mol. Biol.* **17,** 441–452.
19. Brasileiro, A. C. M., Tourneur, C., Leplé, J. C., Combes, V., and Jouanin, L. (1992) Expression of the mutant *Arabidopsis thaliana* acetolactate synthase gene confers chlorosulfuron resistance to transgenic poplar plants. *Transgenic Res.* **1,** 133–141.
20. McGranahan, G. H., Leslie, C. A., Uratsu, S. L., Martin, L. A., and Dandekar, A. M. (1988) *Agrobacterium*-mediated transformation of walnut somatic embryos and regeneration of transgenic plants. *Biotechnology* **6,** 800–804.
21. Ellis, D. D., McCabe, D. E., McInnes, S., Ramachandran, R., Russell, D. R., Wallace, K. M., et al. (1993) Stable transformation of *Picea glauca* by particle acceleration. *Biotechnology* **11,** 84–89.
22. Bonga, J. M. and von Aderkas, P. (1992) *In Vitro Culture of Trees.* Kluwer, London.
23. Bonga, J. M. and Durzan, D. J. (1987) *Cell and Tissue Culture in Forestry.* Martinus Nijhoff, Lancaster, UK.
24. Christou, P. (1992) Genetic transformation of crop plants using microprojectile bombardment. *Plant J.* **2,** 275–281.
25. Gelvin, S. B. (1990) Crown gall disease, and hairy root disease. *Plant Physiol.* **92,** 281–285.
26. Winans, S. C. (1992) Two-way chemical signalling in *Agrobacterium*-plant interactions. *Microbiol. Rev.* **56,** 12–31.
27. van Wordragen, M. F. and Dons, H. J. M. (1992) *Agrobacterium tumefaciens*-mediated transformation of recalcitrant crops. *Plant Mol. Biol. Rep.* **10,** 12–36.
28. Zambryski, P. C. (1988) Basic processes underlying *Agrobacterium*-mediated DNA transfer to plant cells. *Annu. Rev. Genet.* **22,** 1–30.
29. Jefferson, R. A., Kavanagh, T. A., and Bevan, M. W. (1987) GUS fusions: β-glucuronidase as a sensitive and versatile gene fusion marker in higher plants. *EMBO J.* **6,** 3901–3907.

30. Draper, J., Scott, R., Armitage, P., and Walden, R. (1988) *Plant Genetic Transformation and Gene Expression: A Laboratory Manual.* Blackwell Scientific, Oxford, UK.
31. Hood, E. E., Helmer, G. L., Fraley, R. T., and Chilton, M. D. (1986) The hypervirulence region of *Agrobacterium tumefaciens* A281 is encoded in a region outside of T-DNA. *J. Bacteriol.* **168**, 1291–1301.
32. Lazo, G. R., Stein, P. A., and Ludwig, R. A. (1991) A DNA transformation-competent *Arabidopsis* library in *Agrobacterium. Biotechnology* **9**, 963–967.
33. Michel, M. F., Brasileiro, A. C. M., Depierreux, C., Otten, L., Delmotte, F., and Jouanin, L. (1990) Identification of different *Agrobacterium* strains isolated from the same forest nursery. *Appl. Environ. Microbiol.* **56**, 3537–3545.
34. Hoekema, A., Hirsch, P. R., Hooykaas, P. J. J., and Schilperoort, R. A. (1983) A binary plant vector strategy based on separation of *vir-* and T-region of the *Agrobacterium tumefaciens* Ti-plasmid. *Nature* **303**, 179,180.
35. Deblaere, R., Bytebier, B., de Greve, H., Schell, J., Van Montague, M., and Leemans, J. (1985) Efficient octopine Ti plasmid-derived vectors for *Agrobacterium*-mediated gene transfer to plants. *Nucleic Acids Res.* **13**, 4777–4788.
36. Koncz, C. and Schell, J. (1986) The promoter of T_L-DNA gene 5 controls the tissue specific expression of chimaeric genes carried by a novel type of *Agrobacterium* binary vector. *Mol. Gen. Genet.* **204**, 383–396.
37. Zambryski, P., Joos, H., Genetello, C., Leemans, J., Van Montague, M., and Schell, J. (1983) Ti plasmid vector for the introduction of DNA into plant cells without alteration of their normal regeneration capacity by *Agrobacterium tumefaciens. EMBO J.* **2**, 2143–2150.
38. Lambert, C. and Tepfer, D. (1991) Use of *Agrobacterium rhizogenes* to create chimeric apple trees through genetic grafting. *Biotechnology* **9**, 80–83.
39. Macrae, S. and van Staden, J. (1993) *Agrobacterium rhizogenes*-mediated transformation to improve rooting ability in eucalypts. *Tree Physiol.* **12**, 411–418.
40. Béclin, C., Charlot, F., Botton, E., Jouanin, L., and Doré, C. (1993) Potential use of the *aux*2 gene from *Agrobacterium rhizogenes* as a conditional negative marker in transgenic cabbage. *Transgenic Res.* **2**, 48–55.
41. Murashige, T. and Skoog, F. (1962) A revised medium for rapid growth and bioassays with tobacco tissue culture. *Physiol. Plant.* **15**, 473–497.
42. Bevan, M. (1984) Binary *Agrobacterium* vectors for plant transformation. *Nucleic Acids Res.* **12**, 8711–8721.
43. Vancanneyt, G., Schmidt, R., O'Connor-Sanchez, A., Willmitzer, L., and Rocha-Sosa, M. (1990) Construction of an intron containing marker gene: splicing of the intron in transgenic plants and its use in monitoring early events in *Agrobacterium*-mediated plant transformation. *Mol. Gen. Genet.* **220**, 245–250.
44. Leplé, J. C., Brasileiro, A. C. M., Michel, M. F., Delmotte, F., and Jouanin, L. (1992) Transgenic poplars: expression of chimeric genes using four different constructs. *Plant Cell Rep.* **11**, 137–141.
45. Hamill, D. J., Rounsley, S., Spencer, A., Todd, G., and Rhodes, M. J. C. (1991) The use of the polymerase chain reaction in plant transformation studies. *Plant Cell Rep.* **10**, 221–224.
46. Rossi, L., Escudero, J., Hohn, B., and Tinland, B. (1993) Efficient and sensitive assay for T-DNA-dependent transient gene expression. *Plant Mol. Biol. Rep.* **11**, 220–229.

CHAPTER 16

Agrobacterium-Mediated Antibiotic Resistance for Selection of Somatic Hybrids

The Genus Lycopersicon as a Model System

Michael R. Davey, Rajendra S. Patil, Kenneth C. Lowe, and John B. Power

1. Introduction

Somatic hybridization, involving protoplast fusion, is a well-established procedure for overcoming interspecific and intergeneric sexual incompatibilities between plants. Chemical fusion and/or electrofusion of isolated protoplasts present few difficulties. However, the success of somatic hybridization depends, primarily, on an ability to select heterokaryons from homokaryons and unfused protoplasts soon after fusion treatment, or to select heterokaryon-derived cells from which hybrid plants can be regenerated.

Several selection procedures of differing complexity have been reported. These include manual selection of heterokaryons postfusion (1), automated high technology flow cytometry (2), and the use of biochemical complementation systems (3). Chimeric genes for antibiotic resistance can be readily introduced into plants using various delivery methods, including those based on Agrobacterium, direct DNA uptake into protoplasts, and biolistics (4,5). Such chimeric genes have been employed to genetically mark plants prior to the isolation and fusion of their protoplasts. Subsequently, heterokaryon-derived tissues have been

From: *Methods in Molecular Biology, Vol. 44:* Agrobacterium *Protocols*
Edited by: K. M. A. Gartland and M. R. Davey Humana Press Inc., Totowa, NJ

selected by their ability to grow on antibiotic-containing culture medium. For example, somatic hybrids between *Lycopersicon esculentum* and *L. peruvianum* were selected based on the dominant kanamycin resistance of *L. esculentum* and the dominant plant regeneration capacity of *L. peruvianum (6)*. The neomycin phosphotransferase (*npt*II) gene, conferring kanamycin resistance, was introduced into *L. esculentum* by leaf disc transformation with *A. tumefaciens (7)*. A similar approach, but with different plant accessions, was used to routinely select somatic hybrids between *L. esculentum* with *L. peruvianum* or *L. chilense* (Patil, Davey, and Power, unpublished). Likewise, the sexual hybrid *L. esculentum* × *L. pennelli*, was transformed to kanamycin resistance by *A. tumefaciens* prior to somatic hybridization with *Solanum lycopersicoides* or *S. melongena (8)*. Other studies have combined antibiotic selection with biochemical complementation. The *npt*II gene was introduced into *L. hirsutum* by *A. rhizogenes* prior to the fusion of iodoacetamide-treated protoplasts of *L. hirsutum* with kanamycin-sensitive protoplasts of *L. esculentum (9)*. The latter strategy was also previously employed for *Lotus* species *(10)*. Transformed roots were induced on *L. tenuis* by inoculation with *A. rhizogenes* LBA9402 carrying both the wild-type Ri plasmid 1855 and the binary vector pBIN19 with the *npt*II gene *(11)*. Kanamycin-resistant callus derived from transformed roots was used to initiate a cell suspension from which protoplasts were isolated. Such protoplasts were treated with a sublethal dose of sodium iodoacetate prior to electrofusion with seedling cotyledon protoplasts of *L. corniculatus*. Following biochemical complementation, somatic hybrid tissues from which plants were regenerated, were selected on medium containing kanamycin sulfate. Kanamycin resistance, combined with sodium iodoacetate treatment, has also been employed at the intervarietal level to select somatic hybrids in *Daucus carota (12)*.

In general, antibiotic resistance gives "tight" selection of putative somatic hybrid tissues. The disadvantage of this approach is the time involved in genetically marking at least one of the sources of parental protoplasts and the fact that the presence of chimeric genes may be undesirable in somatic hybrid plants and/or their derivatives.

This chapter provides examples of the generation of somatic hybrids between *L. esculentum* cv Pusa Ruby with *L. peruvianum* and *L. chilense*, and between *L. esculentum* cv Roma with *L. hirsutum*, using kanamycin resistance to select somatic hybrid tissues.

2. Materials

2.1. Initiation of Shoot Cultures

1. Seeds of *L. esculentum* cv Pusa Ruby (Mahatma Phule Agricultural University, Rahuri, India), *L. esculentum* cv Roma (National Agricultural Research Centre, Islamabad, Pakistan), *L. chilense* LA2930, *L. hirsutum* LA1353, and *L. peruvianum* LA2744 (C. M. Rick, Tomato Genetic Center, University of California, Davis, CA).
2. "Domestos" bleach (Lever, Kingston-Upon-Thames, UK).
3. MS0 medium: Murashige and Skoog-based medium *(13)*, with macronutrients, micronutrients, and vitamins as in Table 1. This medium can be purchased from Flow Laboratories (Irvine, Scotland), to which is added 30 g/L sucrose and 8 g/L agar (Sigma, St. Louis, MO), pH 5.8. This medium locks growth regulators.
4. Nescofilm: Bando (Kobe, Japan).

2.2. Initiation of Cell Suspensions of L. chilense

MSP1 medium: Murashige and Skoog-based medium *(13)* with 30 g/L sucrose, 2 mg/L α-naphthalene acetic acid (NAA), and 0.5 mg/L 6-benzylaminopurine (BAP), pH 5.8. For semi-solid medium, add 8 g/L Sigma agar.

2.3. Transformation of L. esculentum and L. hirsutum

1. Disarmed *A. tumefaciens* strain GV3Ti11SE containing a modified octopine Ti plasmid (pTiB6S3SE::pMON200) carrying the *npt*II gene *(7)*.
2. *A. rhizogenes* strain R1601 with the *npt*II gene cointegrated into the TL-DNA *(14)*.
3. MSIZ medium: Murashige and Skoog-based medium *(13)* with 30 g/L sucrose, 0.1 mg/L indoleacetic acid (IAA), 1.0 mg/L zeatin, and 8 g/L Sigma agar, pH 5.8.
4. APM medium: 5 g/L yeast extract, 0.5 g/L casamino acids, 8 g/L mannitol, 2 g/L $(NH_4)_2SO_4$, 5 g/L NaCl. For semi-solid medium, add 10 g/L Sigma agar.
5. Kanamycin sulfate: 10 mg/mL stock in distilled water. Filter sterilize through a membrane of pore size 0.2 μm. Store 10-mL aliquots in 30-mL screw-capped Universal glass bottles at –20°C.
6. Streptomycin sulfate: 10 mg/mL stock solution in distilled water. Store as in item 5.
7. Ampicillin: 10 mg/mL stock solution in distilled water. Store as in item 5.
8. Chloramphenicol: 10 mg/mL stock solution in absolute ethanol. Store at 4°C.

Table 1
Formulation of Media Macronutrients, Micronutrients, Vitamins, and other Supplements[a]

Component	Concentration, mg/L			
	MS[b]	TM1[c]	TM2[c]	TM3[c]
Macronutrients				
KH_2PO_4	170	—	170	170
$CaCl_2 2H_2O$	440	150	440	440
KNO_3	1900	2530	1500	1500
NH_4NO_3	1650	320	—	—
$NH_4H_2PO_4$	—	230	—	—
$(NH_4)_2SO_4$	—	134	—	—
$MgSO_4 7H_2O$	370	250	370	370
Micronutrients				
KI	0.83	0.38	0.38	0.38
H_3BO_3	6.20	6.20	6.20	6.20
$MnSO_4 4H_2O$	22.30	22.30	22.30	22.30
$ZnSO_4 7H_2O$	8.60	8.60	8.60	8.60
$Na_2MoO_4 2H_2O$	0.25	0.25	0.25	0.25
$CuSO_4 6H_2O$	0.025	0.025	0.025	0.025
$CoCl_2 6H_2O$	0.025	0.025	0.025	0.025
$FeSO_4 7H_2O$	27.85	13.90	13.90	13.90
Na_2EDTA	37.25	18.50	18.50	18.50
Vitamins				
Nicotinic acid	0.50	2.50	2.50	5.00
Thiamine HCl	0.10	10.00	10.00	0.50
Pyridoxine HCl	0.50	1.00	1.00	0.50
Folic acid	—	0.50	0.50	0.50
Biotin	—	0.05	0.05	0.05
D-Ca pantothenate	—	0.50	0.50	—
Choline chloride	—	0.10	0.10	0.10
Glycine	2.00	0.50	0.50	2.50
Casein hydrolysate	—	50.00	150.00	100.00
L-Cysteine	—	1.00	1.00	—
Malic acid	—	10.00	10.00	—
Ascorbic acid	—	0.50	0.50	—
Adenine sulphate	—	—	40.00	40.00
L-Glutamine	—	—	100.00	100.00
Myo-inositol	100.00	100.00	4600.00	100.00
Riboflavin	—	0.25	0.25	—
Other supplements				
Sucrose	30,000	30,000	68,400	50,000
Mannitol	—	—	4560	—
Xylitol	—	—	3800	—
Sorbitol	—	—	4560	—
2-(N-morpholino)ethane sulphonic acid (MES)	—	—	97,600	97,600
pH	5.8	5.8	5.6	5.8

[a]Growth regulators omitted from these formulations.
[b]After ref. *13*.
[c]After ref. *15*.

9. Cefotaxime (Claforan; Roussel, Wembley, UK): 50 mg/mL stock solution in distilled water. Store as in item 5.
10. Whatman (Maidstone, UK) No. 1 filter papers. Place in a screw-capped glass jar; autoclave.

2.4. Polymerase Chain Reaction (PCR)

1. PCR extraction buffer: 200 mM Tris HCl (pH 7.5), 250 mM NaCl, 25 mM Na$_2$EDTA, 5 g/L sodium dodecyl sulfate (SDS).
2. Isopropanol.
3. TE buffer: 10 mM Tris HCl (pH 8.0), 1 mM Na$_2$EDTA.
4. Reaction buffer (x10): 10 mM Tris HCl (pH 8.0), 1.5 mM MgCl$_2$, 50 mM KCl, 0.1% (v/v) Triton X-100.
5. Thermostable DNA polymerase (Dynazyme; Flowgen, Sittingbourne, Kent, UK).
6. dNTP mixture (Promega, Southampton, UK).
7. Agarose gel electrophoresis: Dissolve 0.16 g of agarose (Sea Kem, FMC, Rockland, MD) in 20 mL of TBE buffer and cast in a 12.5 × 17.5-cm tray (Horizon 58 electrophoresis apparatus [Life Technologies, Paisley, Scotland, UK] or equivalent).
8. TBE buffer: 0.1M Tris base, 0.1M boric acid, 2 mM Na$_2$EDTA.
9. Tracking dye: 0.25% (w/v) bromophenol blue, 0.25% (w/v) xylitol, 40.0% (w/v) sucrose in TE buffer.
10. Ethidium bromide: 0.5 µg/mL aqueous solution.

2.5. Isolation of Protoplasts from Leaves and Cell Suspensions

1. Preincubation medium: Half-strength Murashige and Skoog salts *(13)*, TM1 vitamins *(15; see* Table 1), 0.5 mg/L BAP, 1.0 mg/L 2,4-dichlorophenoxyacetic acid (2,4-D), pH 5.7.
2. Enzyme solution E1: 10 g/L cellulase R10 (Yakult Honsha, Nishinomiya, Japan), 1.5 g/L macerozyme R10 (Yakult Honsha), 100 mg/L 2(*N*-morpholino) ethane sulfonic acid (MES), 137.0 g/L sucrose with cell and proplast wash (CPW) salts *(16)*, pH 5.6.
3. Enzyme solution E2: 3 g/L cellulase RS (Yakult Honsha), 3 mg/L pectolyase Y23 (Seishim, Tokyo, Japan), 5 mM MES, 84 g/L mannitol with CPW salts *(16)*, pH 5.6.
4. CPW salts solution *(16)*: 27.2 mg/L KH$_2$PO$_4$, 101 mg/L KNO$_3$, 246 mg/L MgCl$_2$ · 7H$_2$O, 0.16 mg/L KI, 0.025 mg/L CuSO$_4$ · 5H$_2$O, 1480 mg/L CaCl$_2$ · 2H$_2$O. Add 90 g/L mannitol to produce CPW9M solution, pH 5.8.
5. Flotation solution: 137 g/L sucrose, 100 mg/L MES, CPW salts, pH 5.6.
6. W5 solution: 9 g/L NaCl, 18.38 g/L CaCl$_2$ · 2H$_2$O, 0.37 g/L KCl, 0.9 g/L glucose, pH 5.6.

2.6. Protoplast Fusion, Culture, and Selection of Somatic Hybrids

1. W5 solution: *see* Section 2.5., item 6.
2. Polyethylene glycol (PEG) fusion solution: 300 g/L PEG 4000 (BDH, Poole, UK), 54 g/L glucose, 970 mg/L $CaCl_2 \cdot 2H_2O$, pH 6.0. Filter sterilize before use; store in the dark (4°C).
3. W10 solution: Stock A—72.06 g/L glucose, 0.97 g/L $CaCl_2 \cdot 2H_2O$, 100 mL/L dimethyl sulfoxide. Stock B—54 g/L glucose, 22.52 g/L glycine. Adjust to pH 10.5 with 10M NaOH. Filter sterilize stock solutions A and B. Immediately before use, mix 9 vols of stock A with 1 vol of stock B.
4. Electrofusion solution: 73.5 mg/L $CaCl_2 \cdot 2H_2O$, 84 g/L mannitol, pH 5.7.
5. Iodocetamide: 1 mM aqueous solution. Filter sterilize.
6. TMp medium: As described previously *(6)* with TM2 salts, vitamins, and other supplements *(15*; Table 1), 103 g/L sucrose, 1 mg/L NAA, and 0.5 mg/L BAP, pH 5.6.
7. TMd medium: As for TMp medium (item 6), but with 0.1 mg/L NAA and 67.5 g/L sucrose.
8. TMc medium: As described previously *(6)* with TM3 salts and vitamins *(15*; see Table 1), 36 g/L mannitol, 2.5 g/L sucrose, 0.1 mg/L NAA, 0.5 mg/L BAP, and 8 g/L Sigma agar, pH 5.8.
9. MSZ medium: As for MSIZ medium (*see* Section 2.3., item 3), but lacking IAA.
10. MS0 medium: As in Section 2.1., item 3.
11. Kanamycin sulfate: As in Section 2.3., item 5.

3. Methods

3.1. Initiation of Shoot Cultures of L. esculentum, L. chilense, L. hirsutum, and L. peruvianum

1. Surface sterilize seeds by immersion in 8% (v/v) "Domestos" bleach for 20 min. Wash thoroughly with sterile distilled water (five changes).
2. Place the seeds on agar-solidified MS0 medium (20 seeds on 20 mL medium/9-cm Petri dish). Seal the dishes with Nescofilm.
3. Incubate the cultures at 25°C (30 μmol/m^2/s daylight fluorescent illumination, 16 h photoperiod).
4. After 8–10 d, transfer seedlings to 45-mL aliquots of MS0 agar medium in 175-mL capacity screw-capped glass jars (three seedlings/jar). Incubate as in step 3 for 14 d.
5. Cut the stem of each seedling into three equal-size pieces, each with axillary buds. Place the bases of the explants into MS0 medium. Incubate as in step 3. Excise stem explants and repeat the procedure to maintain a stock of axenic shoots (*see* Note 1).

3.2. Initiation of Cell Suspensions of L. chilense

1. Excise 0.5-cm stem explants from axenic shoots.
2. Place explants horizontally on agar-solidified MSP1 medium in 9-cm Petri dishes (20 mL medium, 9 explants/dish). Seal dishes with Nescofilm.
3. Incubate as in Section 3.1., step 3.
4. Excise explant-derived callus after 28 d. Transfer to fresh MSP1 agar medium in 175-mL capacity screw-capped glass jars (three calli/45 mL medium). Subculture fast growing, green, nodular callus every 28 d for three transfers.
5. Dissect callus into fine pieces and transfer to MSP1 liquid medium (2 g fresh weight of tissue in 50-mL medium in a 250-mL flask). Incubate on a horizontal rotary shaker (20 μmol/m^2/s continuous daylight fluorescent illumination; 75 rpm).
6. After 7 d, filter the suspension through a nylon sieve of pore size 125 μm (*see* Note 2). Add 10 mL of filtrate to 40 mL of MSP1 medium. Incubate as in step 5. Subculture 10 mL of suspension to 40 mL of medium every 7 d. After four to five transfers, subculture 3 mL packed cell volume plus 7 mL of spent medium to 40 mL of MSP1 liquid medium every 7 d (*see* Note 3).

3.3. Transformation of L. esculentum by A. tumefaciens

1. Remove leaves from shoot cultures 14 d after transfer. Excise 0.5–1.0 cm^2 pieces from the leaves using a scalpel.
2. Incubate the cultures for 2 d at 25°C (30 μmol/m^2/s daylight fluorescent illumination, 16 h photoperiod) with their adaxial surfaces in contact with agar-solidified MSIZ medium (20 mL of medium/9-cm Petri dish; five explants/dish).
3. Maintain the disarmed *A. tumefaciens* strain GV3Ti11SE on APM agar medium with 15 μg/mL kanamycin sulfate, 25 μg/mL streptomycin sulfate, and 25 μg/mL chloramphenicol.
4. Transfer a single bacterial colony to 10 mL of APM liquid medium with antibiotic selection in a 100-mL flask and incubate at 28°C in the dark for 24 h.
5. Dilute the bacterial suspension 1:20 (v/v) with liquid MS0 medium.
6. Remove the leaf explants from MSIZ medium and immerse in the bacterial suspension for 30 min. Remove excess bacteria by washing briefly with sterile distilled water; blot the explants between sterile Whatman No. 1 filter papers.
7. Dispense 20-mL aliquots of agar-solidified MSIZ medium into 9-cm Petri dishes. Overlay the medium with a sterile Whatman No. 1 filter paper. Place the leaf explants inoculated with *Agrobacterium* onto the surface of

the filter paper (five explants/dish) (*see* Note 4). Seal the dishes with Nescofilm and incubate for 48 h as in step 2.

8. Transfer the explants to agar-solidified MSIZ medium (without the filter paper) supplemented with 500 µg/mL cefotaxime and 100 µg/mL kanamycin sulfate. Subculture explants to fresh MSIZ medium with the same antibiotics every 10 d.

9. Remove the shoots that differentiate from the explants and that remain green in the presence of kanamycin sulfate following 28–92 d of culture (*see* Note 5).

10. Transfer excised, green shoots to MS0 medium with 500 µg/mL cefotaxime and 100 µg/mL kanamycin sulfate for rooting (three shoots/45 mL medium in 175-mL glass jars).

11. Incubate as in Section 3.1., step 3.

12. Maintain the shoots by transferring 5–6-cm apical explants to fresh MS0 agar medium with 100 µg/mL kanamycin sulfate every 28 d. Progressively reduce the cefotaxime concentration to 400, 200, 100, and 0 µg/mL during subsequent transfers (*see* Note 6).

3.4. Transformation of L. hirsutum *by* A. rhizogenes

1. Maintain *A. rhizogenes* strain R1601 on APM agar medium with 100 µg/mL ampicillin and 100 µg/mL kanamycin sulfate.

2. Transfer a single bacterial colony to 10 mL of APM liquid medium with antibiotic selection in a 100-mL flask and incubate at 28°C in the dark for 24 h.

3. Remove 2.5-cm explants from the stems of cultured shoots of *L. hirsutum*. Decapitate the explants. Insert the explants vertically into MS0 agar medium, with the bases of the explants 0.5 cm below the surface of the medium.

4. Inoculate the top of each explant with one loopful of *A. rhizogenes* strain R1601. Incubate as in Section 3.1., step 3.

5. Excise individual roots that develop from the site of inoculation and transfer to the surface of agar-solidified MS0 medium containing 500 µg/mL cefotaxime and 100 µg/mL kanamycin sulfate (15-mL medium/9-cm Petri dish) (*see* Note 7).

6. Subculture the roots every 21 d by excising 1 cm^2 portions from the mat of actively growing roots and transferring to MS0 agar medium. Reduce the cefotaxime concentration to 200 followed by 100 µg/mL during subsequent transfers. Maintain kanamycin sulfate in the medium at 100 µg/mL.

7. Confirm the synthesis of T-DNA-specific opines in *Agrobacterium*-induced roots as described previously *(17)*.

8. Cut the roots into 0.5–2.5-cm lengths and transfer to agar-solidified MSIZ medium supplemented with 100 µg/mL cefotaxime and 100 µg/mL kanamycin sulfate for shoot induction (*see* Note 8).

9. Excise the regenerated shoots and handle as in Section 3.3., steps 9–12. Omit cefotaxime from the medium.

3.5. Confirmation of the Transgenic Nature of Kanamycin-Resistant Shoots of L. esculentum and L. hirsutum

A rapid assay for *npt*II activity to confirm expression of the *npt*II gene introduced into *L. esculentum* and *L. hirsutum* is described in detail in Chapter 19. The polymerase chain reaction (PCR) provides a rapid assay for the presence of the *npt*II gene in transgenic plants.

3.6. PCR Amplification of a Segment of the nptII Gene Using Specific Primers

Plant genomic DNA is extracted using a simple, rapid method *(18)*. The extraction procedure should be performed in an isolated, clean area in order to reduce the risk of contamination from other sources of DNA.

1. Remove 100 mg fresh weight of leaf material from putatively transformed and nontransformed (control) plants. Place in sterile 1.5-mL capacity Eppendorf tubes and macerate each sample with a plastic rod at 22°C for 15 s.
2. Add 400 µL of PCR extraction buffer. Vortex the tubes (5 min) and centrifuge (13,000 rpm; 60 s).
3. Transfer 300 µL of the supernatants to clean Eppendorf PCR tubes. Add 300 µL of isopropanol (4°C). Centrifuge (13,000 rpm; 5 min). Remove the isopropanol and vacuum dry the DNA pellet. Dissolve the DNA pellet in 100 µL of TE buffer. Store at 4°C.
4. Construct 2 primers specific for the *npt*II gene. The forward primer has a sequence of GTCGCTTGGTCGGTCATTTCG; the sequence of the reverse primer is GTCATCTCACCTTGCTCCTGCC (*see* Note 9).
5. Set up the PCR reaction in sterile 0.5-mL Eppendorf tubes with 2.5 µL of 10X reaction buffer, 0.25 µL (0.5 U) of DNA polymerase, 1 µL of 5 m*M* dNTP mixture, 2.5 µL of each primer, 2.5 µL of plant DNA, and 13.75 µL of distilled water, to give a total volume of 25 µL. One microliter of plasmid DNA, e.g., pCaMVNEO *(19)*, should be used as a control.
6. Add 25 µL of mineral oil to each tube to restrict evaporation. Centrifuge the tubes (13,000 rpm; 30 s).
7. Transfer the Eppendorf tubes to a thermal cycler (e.g., Techne PHC-C Dri-Block Cycler [Techne Cambridge Ltd., Duxford, Cambridge, UK]); run 40 cycles. Each cycle consists of 95°C for 45 s (to denature DNA strands), 58°C for 40 s (to anneal primers), and 72°C for 2 min (for exten-

sion of new DNA strands by DNA polymerase). Terminate the reaction at 5°C for 10 min.

8. Separate the PCR-amplified DNA fragment on a 0.8% (w/v) agarose gel, loading 2 μL of tracking dye with 16 μL of each PCR mixture into the wells of the gel. Also mix 5 μL of *Hin*dIII-digested λ DNA with 1 μL of dye; load 2 μL of the mixture into an outer well of the gel to act as a molecular weight marker.

9. Electrophorese in TBE buffer at 2.85 V/cm for 90 min.

10. Stain the gel with ethidium bromide for 45 min at 22°C; destain the gel in water for 10 min.

11. Visualize the gel under UV illumination. A 540 bp PCR fragment should be present in the plasmid control and in transgenic plants containing the *npt*II gene. The fragment should be absent from nontransformed plants.

3.7. Isolation of Leaf Protoplasts from Kanamycin-Resistant Shoots of L. esculentum and L. hirsutum, and from Nontransformed Shoots of L. peruvianum

1. Transfer shoot cultures (14 d after subculture) to the dark 24 h prior to leaf excision.

2. Excise leaflets and float on the surface of pre-incubation medium (10-mL aliquots in 9-cm Petri dishes) at 4°C in the dark for 24 h.

3. Cut the leaflets into strips (each about 1-mm wide) along both sides of the midrib (*see* Note 10).

4. Incubate statically 1 g fresh weight of tissue in 10 mL enzyme solution E1 at 30°C for 16 h in the dark.

5. Filter the enzyme–protoplast mixture through a nylon sieve of 64-μm pore size. Divide the filtrate into two 5-mL aliquots in 16-mL capacity screw-capped centrifuge tubes.

6. Add 5 mL of flotation solution to each tube, overlaid with 2 mL of W5 solution.

7. Centrifuge (100*g*; 10 min). Remove the protoplasts from the junction of the flotation-W5 solutions (*see* Note 11). Transfer to screw-capped centrifuge tubes.

8. Wash the protoplasts twice by resuspending and centrifuging in W5 solution.

9. Resuspend the protoplasts in a known volume of W5 solution. Count the protoplasts using a hemocytometer.

10. Assess the viability of isolated protoplasts using FDA. Add 2 μL of FDA stock solution to 10 mL of W5 washing solution. Mix 1 mL of this solution with an equal volume of protoplast suspension. Incubate for 5 min before observing the protoplasts under UV illumination (e.g., using a Nikon

Diaphot TMD inverted microscope with high pressure mercury vapor lamp HBO 100 W/2, a B1 FITC exciter filter IF 420–485 nm, dichromic mirror DM510, and eyepiece absorption filter 570). Viable protoplasts fluoresce yellow-green.

3.8. Isolation of Protoplasts from L. chilense Cell Suspensions

1. Harvest cells 3–5 d after transfer. Pass the suspension through a nylon sieve of 250-μm pore size, collecting the filtrate in 14–cm Petri dishes.
2. Remove the culture medium using a Pasteur pipet (*see* Note 12).
3. Weigh the cells; adjust the fresh weight of cells to 5 g/dish.
4. Add 20 mL of CPW9M solution to plasmolyse the cells. Incubate in the dark at 25°C for 1 h.
5. Remove the CPW9M solution with a Pasteur pipet. Add 20 mL of enzyme mixture E2. Seal the dish with Nescofilm. Incubate on a slow horizontal shaker (40 rpm) for 10–14 h at 25°C in the dark.
6. Pass the enzyme–protoplast mixture through a nylon sieve of pore size 64 μm. Transfer the filtrate to screw-capped 16-mL capacity centrifuge tubes. Centrifuge (80*g*, 5 min).
7. Remove the supernatant. Resuspend the protoplasts in 10 mL of W5 solution. Centrifuge (80*g*, 5 min) to pellet the protoplasts.
8. Resuspend the protoplast pellet in 10 mL of flotation solution topped with 2 mL of W5 solution.
9. Centrifuge (100*g*, 10 min). Follow steps given in Section 3.7., steps 7–10.

3.9. Parental Combinations and Selection Systems for the Production of Lycopersicon Somatic Hybrids

The procedures described in this chapter can be employed to generate somatic hybrids in *Lycopersicon* using the following combinations of parental protoplasts:

1. Kanamycin-resistant leaf protoplasts of *L. esculentum* cv. Pusa Ruby with kanamycin-sensitive leaf protoplasts of *L. peruvianum*.
2. Kanamycin-resistant leaf protoplasts of *L. esculentum* cv. Pusa Ruby with kanamycin-sensitive suspension cell protoplasts of *L. chilense*.
3. Kanamycin-sensitive leaf protoplasts of *L. esculentum* cv. Roma with kanamycin-resistant leaf protoplasts of *L. hirsutum*.

Following fusion, in refs. *1* and *2*, heterokaryon-derived tissues are selected on medium containing kanamycin sulfate. Tissues derived from unfused protoplasts and homokaryons of *L. esculentum* are able to grow on antibiotic-containing medium, but do not produce shoots. Tissues derived from unfused protoplasts and homokaryons of *L. peruvianum* or *L. chilense* are sensitive to kanamycin and do not survive selection. Consequently, only somatic hybrid tissues regenerate shoots. In ref. *3*, kanamycin-resistant protoplasts of *L. hirsutum*, capable of plant regeneration, are treated immediately after isolation with the metabolic inhibitor iodoacetamide. This arrests mitotic division of unfused protoplasts and homokaryons of *L. hirsutum* following the fusion treatment. Unfused protoplasts and homokaryons of *L. esculentum* do not survive on selection medium containing kanamycin sulfate. However, heterokaryon-derived cells are kanamycin resistant and capable of division through complementation of the *L. esculentum* genome overcoming the iodoacetamide inhibition of the *L. hirsutum* genome. Heterokaryon-derived somatic hybrid tissues grow on kanamycin-containing medium and regenerate shoots.

3.10. Protoplast Fusion

Lycopersicon protoplasts can be fused chemically and/or electrically. Both procedures are well documented; the chemical method is based on a published procedure *(20)*, with modifications. The electrical method utilizes the apparatus that can be constructed from published plans *(21)*.

3.10.1. PEG-Mediated Chemical Fusion

1. Suspend isolated parental protoplasts in W5 solution at a density of $1.0 \times 10^6/mL$. Mix equal volumes of the two parental suspensions.
2. Place 0.5-mL aliquots of the mixed protoplast suspension in the center of a 5-cm Petri dish (Sterilin, Feltham, UK). Allow the protoplasts to settle for 20 min.
3. Add 0.4 mL of PEG fusion solution dropwise to the peripheral regions of the protoplast drop; allow the droplets of PEG to coalesce around and over the settled protoplasts (*see* Note 13).
4. Incubate for 12.5 min at room temperature.
5. Carefully remove the liquid over the settled protoplasts using a Pasteur pipet; replace with two 0.5-mL aliquots of W10 solution. Leave for 10 min (*see* Note 14).
6. Replace, as in step 5, the W10 solution with W5 solution.
7. Set up suitable controls for each experiment consisting of untreated parental protoplasts, self-fused parental protoplasts, and a mixture of unfused parental protoplasts (*see* Note 15).

3.10.2. Electrofusion of Protoplasts

1. Where necessary, treat isolated protoplasts of one fusion partner (e.g., those isolated from kanamycin-resistant shoots of *L. hirsutum*) with iodoacetamide

for 20 min at 4°C, followed by washing with prechilled (4°C) electrofusion solution (*see* Note 16).

2. Wash the protoplasts twice by centrifugation (100*g*; 5 min) in electrofusion solution. Resuspend protoplasts at 0.5×10^6/mL in electrofusion solution. Mix equal volumes of the protoplast suspensions.

3. Dispense 1-mL aliquots of the mixture of parental protoplasts into each square well of a 5×5 grid dish (Sterilin).

4. Sterilize the multielectrode block by immersion in 70% (v/v) ethanol for 10 min. Allow to dry in the air stream of a laminar flow hood.

5. Place the electrode block into a well containing the protoplasts to be fused. Align the protoplasts in an AC field (187 V/cm at 1 MHz for 10 s) (*see* Note 17). Increase the AC field to 250 V/cm to increase protoplast contact. Immediately apply a single DC pulse (600 V/cm, 0.3 ms) to fuse the protoplasts (*see* Note 18).

6. Move from one well to another well within the dish, repeating step 5.

7. Allow the protoplasts to settle in the electrofusion solution (15–30 min).

8. Replace the electrofusion solution with TMp medium (5 mL/well).

9. Set up suitable controls as in Section 3.8.1., step 7.

3.11. Culture of Fusion-Treated Protoplasts and Selection of Putative Somatic Hybrid Tissues and Shoots

1. Resuspend the fusion-treated protoplasts in TMp liquid medium at 1×10^5/mL.

2. Dispense 3-mL aliquots into 5-cm Petri dishes. Seal the dishes with Nescofilm. Incubate the cultures in the dark at 25°C for 3 d.

3. Expose the cultures on day 4 to low-intensity illumination (20 µmol/m²/s daylight fluorescent illumination, 16 h photoperiod).

4. Add 3.0 mL of TMd medium to the dishes on d 4, 7, 10, 13, 17, and 21 of culture. At d 7, include kanamycin sulfate in the dilution medium to give a final antibiotic concentration of 25 µg/mL.

5. Assess the protoplast plating efficiency (number of dividing protoplasts expressed as a percentage of the number of protoplasts plated) at d 7 of culture (*see* Note 19).

6. Increase the concentration of kanamycin sulfate in the medium to 50 µg/mL at d 10.

7. On d 13 and thereafter, transfer protoplast-derived colonies (1–2 mm in diameter) to the surface of agar-solidified TMc callus induction medium with 50 µg/mL of kanamycin sulfate (20 colonies/20 mL of medium/9-cm Petri dish) (*see* Note 20). Incubate the cultures under continuous daylight fluorescent illumination (40 µmol/m²/s) at 25°C.

8. After 28 d, transfer green, protoplast-derived tissues (1–2 cm diameter) to agar-solidified MSIZ medium with 100 µg/mL kanamycin sulfate (three tissues/45 mL medium in 175-mL jars). Incubate as in step 7, but with a 16 h photoperiod.

9. Subculture protoplast-derived tissues every 28 d to MSZ medium containing 100 µg/mL kanamycin sulfate (*see* Note 21). Shoots (putative somatic hybrids) should appear 28–112 d after protoplast fusion, depending on the source of parental protoplasts.

10. Excise developing shoots when 4–5 cm in height. Root the shoots by transfer to MS0 agar medium containing 100 µg/mL kanamycin sulfate.

11. Micropropagate the shoots as described in Section 3.1., step 5; maintain some shoots in both the presence and absence of kanamycin sulfate.

3.12. Confirmation of the Hybridity of Plants Regenerated Following Protoplast Fusion and Kanamycin Selection

Assessments can be performed that, collectively, confirm the hybridity of plants selected following protoplast fusion. Morphologically, regenerated shoots can be scored in vitro for characteristics such as stem and leaf shape and pigmentation. Additionally, floral characteristics, fertility, and fruit color and shape can also be determined following transfer of plants from culture to the growth chamber or glasshouse. Since the morphology of somatic hybrids of tomato has a tendency toward the morphology of the wild tomato parent, it is essential to perform as many diagnostic tests as possible.

The number of chloroplasts per leaf guard cell gives an indication of ploidy *(6)*, which can be confirmed by mitotic chromosome complements of root-tip preparations. Flow-cytometric analysis of isolated nuclei also provides a rapid indication of ploidy *(22)*. Protein markers are useful in confirming hybridity. For example, somatic hybrids generally exhibit isozyme bands characteristic of the parental plants, together with additional bands. The isozymes most suitable for a specific species combination may have to be determined empirically. For example, acid phosphatase bands have been used to confirm hybridity between *L. esculentum* and *L. hirsutum* or *L. chilense*, whereas leucine aminopeptidase is more suitable for *L. esculentum* (+) *L. peruvianum*. Demonstration of *npt*II activity in somatic hybrids confirms the presence of genetic material from a transgenic, kanamycin-resistant parent. This is particularly important in confirming the presence of genetic material from *L. esculentum* in hybrids between cultivated tomato and wild *Lycopersicon* species, because:

1. The phenotype of the somatic hybrid is dominated by the phenotype of the wild parent;
2. Regeneration of somatic hybrid shoots is dependent on the wild parent; and
3. Sometimes protein banding patterns may be dominated by those of the wild parent.

DNA markers are becoming more widely utilized for characterizing somatic hybrids, through the use of PCR and Random Amplified Polymorphic DNA (RAPD) analyses and Southern hybridization.

4. Notes

1. Once established, axenic shoots can be propagated to provide a large stock of experimental material.
2. Inexpensive, nylon sieves in a range of pore sizes may be obtained from Wilson Sieves (Nottingham, UK).
3. Initiate callus and cell suspensions every 3 mo to maintain a supply of totipotent cultures for protoplast isolation.
4. The filter paper prevents the surface of the agar medium becoming overgrown with *Agrobacterium*.
5. Shoots that remain green in the presence of antibiotic selection should be putative transformants.
6. Cefotaxime is very effective in eliminating *Agrobacterium* from cultures.
7. Cefotaxime will eliminate the bacteria; transformed roots should express the *npt*II gene and be resistant to kanamycin sulfate.
8. Dissecting the roots into explants normally stimulates shoot regeneration.
9. Several laboratories and commercial companies offer a primer construction service.
10. Use a new scalpel blade to ensure cutting rather than tearing and bruising of the leaf material.
11. Remove the protoplasts in as small a volume of liquid as possible.
12. Carefully withdraw the medium until the cells appear "dry."
13. Avoid spreading the protoplast suspension and PEG droplets over a wide area of the dish.
14. Approximately 25% of *Lycopersicon* protoplasts are fused by PEG treatment, with approx 50% of these being heterokaryons.
15. These controls are essential to confirm the efficiency of the selection procedure.
16. Handle under the conditions appropriate for toxic compounds.
17. At this stage, the protoplasts should align at right angles to the electrodes to form "pearl chains."
18. Approximately 10% of *Lycopersicon* protoplasts are fused by electrical treatment, predominantly on a 1:1 basis (i.e., binucleate heterokaryons).

19. Count at least 200 protoplasts from randomly selected fields of view. The plating efficiency of fusion-treated protoplasts is 60–70% following PEG treatment, and 40–60% following electrofusion.
20. Transfer the colonies using either a Pasteur pipet from which the end has been removed to increase the size of the orifice, or fine jewellers forceps.
21. Fusion of a total of 2×10^6 *Lycopersicon* protoplasts results in 10–12 kanamycin resistant putative somatic hybrid tissues.

References

1. Mendis, M. H., Power, J. B., and Davey, M. R. (1991) Somatic hybrids of the forage legumes *Medicago sativa* L. and *M. falcata* L. *J. Exp. Bot.* **42,** 1565–1573.
2. Hammatt, N., Lister, A., Blackhall, N. W., Gartland, J., Ghose, T. K., Gilmour, D. M., et al. (1990) Selection of plant heterokaryons from diverse origins by flow cytometry. *Protoplasma* **154,** 34–44.
3. Davey, M. R. and Kumar, A. (1983) Plant protoplasts: retrospect and prospect. *International Review of Cytology, Supplement 16, Plant Protoplasts* (Giles, K. L., ed.), Academic, New York, pp. 219–299.
4. Davey, M. R., Rech, E. L., and Mulligan, B. J. (1989) Direct DNA transfer to plant cells. *Plant Mol. Biol.* **13,** 273–285.
5. Davey, M. R., Power, J. B., and Rech, E. L. (1994) Direct gene transfer to plant protoplasts, cells and tissues, in *Plant Cell Culture and Its Applications* (Charlwood, B., Dixon, R., and Torres, J. M., eds.), Edward Arnold, London, in press.
6. Wijbrandi, J., Van Capelle, W., Hanhart, C. J., Van Loenen Martinet-Schuringa, E. P., and Koornneef, M. (1990) Selection and characterization of somatic hybrids between *Lycopersicon esculentum* and *Lycopersicon peruvianum. Plant Sci.* **70,** 197–208.
7. Horsch, R. B., Fry, J. E., Hoffmann, N. L., Eichholtz, D., Rogers, S. G., and Fraley, R. T. (1985) A simple and general method for transferring genes into plants. *Science* **227,** 1229–1231.
8. Guri, A., Dunbar, L. J., and Sink, K. C. (1991) Somatic hybridisation between selected *Lycopersicon* and *Solanum* species. *Plant Cell Rep.* **10,** 76–80.
9. Latif, M. (1993) *Genetic improvement in* Lycopersicon esculentum *Mill. through Somatic Hybridisation and Transformation.* Doctoral dissertation, *University of Nottingham*, Nottingham, UK.
10. Aziz, M. A., Chand, P. K., Power, J. B., and Davey, M. R. (1990) Somatic hybrids between the forage legumes *Lotus corniculatus* L. and *L. tenuis* Waldst et Kit. *J. Exp. Bot.* **41,** 471–479.
11. Hamill, J. D., Prescott, A., and Martin, C. (1987) Assessment of the efficiency of co-transformation of the T-DNA of disarmed binary vectors derived from *Agrobacterium tumefaciens* and the T-DNA of *A. rhizogenes. Plant Mol. Biol.* **9,** 573–584.
12. Ichikawa, H. and Immaura, J. (1990) A highly efficient selection method for somatic hybrids which uses an introduced dominant selectable marker combined with iodoacetamide treatment. *Plant Sci.* **67,** 227–235.
13. Murashige, T. and Skoog, F. (1962) A revised medium for rapid growth and bioassays with tobacco tissue cultures. *Physiol. Plant* **15,** 473–497.

14. Pythoud, F., Sinkar, V. P., Nester, E. W., and Gordon, M. P. (1987) Increased virulence of *Agrobacterium rhizogenes* conferred by the *vir* region of pTiBo542: application of genetic engineering of poplar. *Bio/Technol.* **5,** 1323–1327.
15. Shahin, E. A. (1985) Totipotency of tomato protoplasts. *Theor. Appl. Genet.* **69,** 235–240.
16. Frearson, E. M., Power, J. B., and Cocking, E. C. (1973) The isolation and regeneration of *Petunia* leaf protoplasts. *Dev. Biol.* **33,** 130–137.
17. Golds, T. J., Davey, M. R., Rech, E. L. and Power, J. B. (1990) Methods of gene transfer and analysis in higher plants, in *Methods in Molecular Biology, Vol. 6, Plant Cell and Tissue Culture* (Pollard, J. W. and Walker, J. M., eds.), Humana, Clifton, NJ, pp. 341–371.
18. Edwards, K., Johnstone, C., and Thompson, C. (1991) A simple and rapid method for the preparation of plant genomic DNA for PCR analysis. *Nucleic Acids Res.* **19,** 1349.
19. Fromm, M. E., Taylor, L. P., and Walbot, V. (1986) Stable transformation of maize after gene transfer by electroporation. *Nature* **319,** 791–793.
20. Menczel, L., Nagy, F., Kiss, Z. R., and Maliga, P. (1981) Streptomycin resistant and sensitive somatic hybrids of *Nicotiana tabacum* + *Nicotiana knightiana*: correlation of resistance to *N. tabacum* plastids. *Theor. Appl. Genet.* **59,** 191–195.
21. Jones, B., Lynch, P. T., Handley, G. J., Malaure, R. S., Blackhall, N. W., Hammatt, N., Power, J. B., Cocking, E. C., and Davey, M. R. (1994) Equipment for the large-scale electromanipulation of plant protoplasts. *BioTechniques*, **16,** 312–321.
22. Hammatt, N., Blackhall, N. W. and Davey, M. R. (1991) Variation in the DNA content of *Glycine* species. *J. Exp. Bot.* **42,** 659–665.

CHAPTER 17

Histochemical GUS Analysis

Stanislav Vitha, Karel Beneš, Julian P. Phillips, and Kevan M. A. Gartland

1. Introduction
1.1. General Remarks

Patterns in the localization of compounds, metabolic processes, and regulatory machineries at the cellular and subcellular levels are studied by scientific disciplines, called cytochemistry (biochemistry of the cell, cellular topochemistry) or histochemistry, if the tissue level is concerned. In order to reach these goals, a whole range of procedures have been developed, which may be divided in two groups: the in vitro techniques, based on homogenate fractionation and the *in situ* techniques, realized on sections. Only *in situ* techniques are considered in this chapter *(1–5)*.

Jefferson *(6–9)* stated the possibilities existing for the localization of β-glucuronidase (GUS) in transgenic plants, bearing the *uid*A gene from *Escherichia coli*. In plant tissues GUS is present only very rarely, and even in such cases *(10,11)*, intrinsic GUS activity usually can be distinguished from the introduced one. The procedures proposed by Jefferson have been used by many authors to study such issues as cell and tissue specific expression of marker genes driven by different promoters. Gene expression is intended here to mean the production of functional protein, not only mRNA.

It should be stressed that any *in situ* assay of enzymes—including GUS—must follow appropriate steps to avoid false-positive or false-negative results, and false localization as well *(12,13)*. In this chapter,

From: *Methods in Molecular Biology, Vol. 44:* Agrobacterium *Protocols*
Edited by: K. M. A. Gartland and M. R. Davey Humana Press Inc., Totowa, NJ

therefore, procedures for the study of GUS localization, which proved competent in our laboratories, are provided and possible sources of errors are highlighted.

1.2. Characteristics of the Procedure

There are two possible approaches to *in situ* localization of enzymes in cells and tissues. The first approach is based on the macromolecular (mostly antigenic) nature of the enzyme, the second is based on its catalytic activity. Immunocytochemistry is the best known example of the first approach. Although the antibody against *E. coli* GUS is available (Clontech, Palo Alto, CA), this technique is not commonly used so far. In this chapter, the enzymatic activity basis of GUS assay is considered in depth.

When assaying enzyme activity, three stages can be identified:

1. Processing of the object under study;
2. The incubation itself; and
3. Postincubation handling.

Special attention, however, should be paid to possible sources of error and their elimination.

1.2.1. Processing of the Objects

The common techniques for intrinsic glycosidases of plant tissues studied in our laboratory *(14)* failed in the case of transgenic plants bearing GUS of prokaryotic origin. Usually, sections from fixed material are preferred (*see* Note 1). Fixation decreases the enzyme activity (the important question is whether proportionally in all allozymes), but it minimizes the problems caused on the one hand by diffusion of the enzyme out of the tissue and on the other hand by capricious penetration of the substrate and/or other ingredients of the medium into the object. In the case of GUS assays in sections from transgenic plants, only faint or no staining is sometimes obtained. That is why the *in toto* incubation is preferred, very often in combination with the addition of detergents into the medium and with vacuum infiltration (*see* Note 2). The objects are either live or gently fixed (for a short time, in an ice bath).

Some authors incubate an unfixed object *in toto*, then postfix, and finally prepare sections, or even section the live-incubated object after the incubation and then postfix. Which procedure has been used must always be precisely stated. This concerns numerous combinations—to fix vs not to fix, to section vs not to section—in various orders. For example, the sectioning of *in toto* incubated objects is a rather doubtful procedure. The fixation employed must always be stated explicitly as well. Most often, a buffered formaldehyde or glutaraldehyde solution is used, which can be prepared according to different prescriptions *(4)*. It is also necessary to define the sectioning method (freehand sections, cold knife techniques, cryostat, and so on; *15,16*).

1.2.2. Incubation of the Object

In the incubation itself it is necessary to distinguish between the enzymatic reaction and the visualization reaction. In the enzymatic reaction the substrate is hydrolyzed and the primary product is formed, which is then processed in the visualization reaction. Finally, a colored product is obtained, the localization of which should correspond to the site within the tissue, where the enzyme is present. Various types of reactions can be distinguished. For GUS, three of them should be taken into consideration: the oxidation-chelation, the indigogenic, and the azo coupling reactions.

The oxidation-chelation method is based on hydrolysis of 8-hydroxyquinoline β-D-glucuronide and on the formation of Prussian blue as a final reaction product *(17,18)*. This method is capricious and therefore is not recommended.

The most common method for localization of GUS activity in transgenic plants is the indigogenic one. Variously halogenated substrates can be used, most often 5-bromo-4-chloro-3-indolyl β-D-glucuronide. The presence of ferri- and ferrocyanide in the incubation medium is a critical point for this procedure. Their presence minimizes the diffusion of primary reaction product and protects the formed indigo from further oxidation. On the other hand, it concurrently inhibits the enzyme activity, though in our case the inhibition usually does not exceed 60% (unpublished results). These facts must be taken into consideration in each particular case. Principles derived from the works on carboxylesterases *(19,20)* are useful for all indigogenic procedures.

In animal tissues, the azocoupling techniques are common for the localization of intrinsic β-glucuronidase *(21)*. Jefferson *(7–9)* mentioned the possibility of using procedures of this type, but the first positive results with this technique in transgenic plants were published only recently *(22)*. The postcoupling procedure, with naphthol ASBI β-D-glucuronide and

Fast Blue BB salt as a coupler, proved competent. Although the diffusion of (apparently) primary product has been observed, this substantially cheaper technique could be used in many instances with satisfactory results.

Recently a procedure with Sudan II-glucuronide has been described *(23)*. Detailed comparative experiments are necessary before the usefulness of this technique is evaluated.

1.2.3. Postincubation Handling

The nature of postincubation handling should be considered in cases where permanent preparations are intended or if the treated material is to be stored for very long, where the crystallization of final product could occur. This usually does not affect the results at the tissue level. In the case of permanent preparations, the final reaction product is exposed to the actions of dehydrating, clearing, and mounting media that could dissolve it or cause its redistribution. Fading of the resulting color on exposure to light can be neglected.

1.2.4. Possible Errors and their Elimination

In order to minimize the possibility of false-negative results, in vitro and *in situ* findings should be compared. For in vitro assays, the fluorometric technique with 4-methylumbelliferyl β-D-glucuronide as a substrate is usually used *(7–9,* and *see* Chapter 18). The procedure based on hydrolysis of *p*-nitrophenyl β-D-glucuronide is less sensitive. As data on substrate specifity of GUS from *E. coli* are lacking, it is difficult to give any guidance for conflict situations.

There are several possibilities for obtaining false-positive results. Although the intrinsic GUS activity is negligible in many plant materials, it can be significant in some objects and cause false-positive results. It applies namely to pollen grains *(10,11,24)*. That is why it is important to perform a parallel assay with either nontransgenic material or material transformed in the same way but by a construct lacking the *uid*A gene. Various methods *(24–27)* of eliminating the intrinsic GUS activity (addition of methanol, elevated pH, increased incubation temperature; *see* Note 3) have been described.

Besides the intrinsic activity, false-positive results can be caused by the presence of *uid*A-bearing *Agrobacterium* that survived the action of antibiotics. This danger is avoided by the use of constructs with an intron inserted in the *uid*A gene *(28,29)*. Tör *(30)* reported false-positive results caused by endogenous *Corynebacterium*, possessing GUS activity.

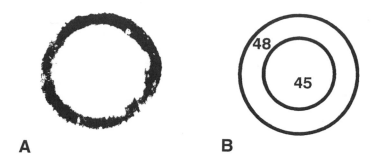

A **B**

Fig. 1. Localization artifact caused by *in toto* incubation. Two leaf disks (7 mm in diameter) from the same leaf of transgenic, *uid*A-bearing tobacco were fixed and washed (*see* Sections 2.1. and 3.1.). (**A**) One disk was incubated *in toto* (*see* Section 3.1., vacuum infiltration not performed). Only the tissue close to the margin stained blue. (**B**) The central and peripheral parts of the second disk were separated using corkborer and GUS activity measured quantitatively (fluorogenic in vitro assay, *see* ref. *33* and Chapter 19). The activity (units of fluorescence/mg of fresh weight tissue) is equal both in the center and in the periphery.

In the indigogenic reaction, false-positive results could also appear owing to the formation of colored product from ferri/ferrocyanide. In the azocoupling reactions, tissue staining as a result of the reaction of phenolic compounds with diazonium salt and the absorption of decomposition products of the medium must be considered. In both types of reaction the use of without-substrate controls is therefore recommended. In the simultaneous azocoupling technique, the diazonium salt breaks down during prolonged incubations. The postcoupling procedure should therefore be used.

The danger of false localization has already been mentioned, namely when tissue processing was discussed. It is always necessary to consider the limits of the technique used. If the *in toto* incubation is performed and if the postcoupling procedure is used, the results cannot be assessed at the cellular or even subcellular levels. It is very difficult to disprove an objection that the enzyme pattern revealed after the *in toto* incubation (Fig. 1) is rather a consequence of uneven penetration of the medium into the tissue than the result of differential gene expression *(31)*.

2. Materials

2.1. Indigogenic GUS Histochemistry

1. Buffer: phosphate buffer pH 7.0 (made of $0.05M$ NaH_2PO_4 and $0.05M$ Na_2HPO_4) or citrate-HCl buffer pH 7.0 (made of $0.05M$ sodium citrate and $0.1N$ hydrochloric acid.
2. Fixative: 4% formaldehyde (formalin diluted 1:10) in phosphate buffer.

3. Incubation medium: Dissolve 1 mg 5-bromo-4-chloro-3-indolyl β-D-glucuronide (X-gluc) in 0.1 mL methanol (*see* Note 4), add 1.87 mL phosphate or citrate buffer (*see* Note 5), 10 μL 0.1M potassium ferrocyanide, 10 μL 0.1M potassium ferricyanide, and 10 μL 10% (w/v) solution of Triton X-100.
4. 70% (v/v) Ethanol.
5. 50% (v/v) and 100% (v/v) Glycerol.
6. Vacuum pump and desiccator.

2.2. Azocoupling GUS Histochemistry

1. Buffer: phosphate buffer pH 7.0 (made of 0.05M NaH$_2$PO$_4$ and 0.05M Na$_2$HPO$_4$) or citrate-HCl buffer pH 7.0 (made of 0.05M sodium citrate and 0.1N hydrochloric acid.
2. Fixative: 4% formaldehyde (formalin diluted 1:10) in phosphate buffer.
3. Incubation medium: Dissolve 2.5 mg of naphthol ASBI β-D-glucuronic acid in 0.2 mL of methanol (*see* Note 4), add 9.8 mL of phosphate or citrate buffer (*see* Note 5). Prepare fresh before use.
4. Diazonium salt solution: 10 mg Fast Blue BB salt dissolved (*see* Note 6) in 10 mL of phosphate or citrate buffer. Prepare immediately before use.
5. Vacuum pump and desiccator.

3. Methods

3.1. Indigogenic GUS Histochemistry

1. Fix for 30 min in ice-cold fixative, shaking occasionally.
2. Wash for 30–60 min in several changes of ice-cold distilled water.
3. Vacuum infiltrate the incubation medium into the objects.
4. Incubate in darkness at room temperature (*see* Note 7) in the incubation medium (several hours or overnight) until distinct blue staining appears.
5. Rinse in distilled water.
6. Clear green objects in 70% ethanol for several hours until the chlorophyll is removed, then transfer to distilled water again.
7. Place in 50% glycerol (*see* Note 8), leave for several minutes, transfer to pure glycerol, then leave for another several minutes.
8. Mount objects in 100% glycerol on microscope slides and examine under a microscope.

3.2. Azocoupling GUS Histochemistry

1. Repeat steps 1–3 of Section 3.1.
2. Incubate overnight in darkness at room temperature (*see* Note 7) in the incubation medium.
3. Wash 30 s in distilled water at room temperature.
4. Transfer objects to freshly prepared diazonium salt solution and leave for 10 min at room temperature.

5. Wash 5 min in water at room temperature.
6. Mount in distilled water on microscope slides and examine under a microscope. The sites with GUS activity have a blue-violet color (*see* Note 9).

4. Notes

1. The use of sections is preferred. Try different ways of sectioning of fixed or unfixed material. If the results are unsatisfactory, then *in toto* incubation is the only way. If sections are used, the fixation and washing steps can be shortened and vacuum infiltration omitted.
2. For *in toto* incubation the objects should be as small as possible to avoid localization artifacts caused by limited penetration of the substrate.
3. If problems with intrinsic GUS activity are encountered, it is recommended to use incubation medium with an increased concentration of methanol (20% [v/v]) to elevate pH to 7.5 or to raise temperature to 60°C *(24–27)*.
4. In the original indigogenic procedure of Jefferson *(7–9)*, *N,N*-dimethyl-formamide is used to dissolve the substrate. As dimethylformamide inhibits GUS activity *(25)*, the use of methanol is preferred *(32)*.
5. In citrate buffer the GUS activity is somewhat higher than in the phosphate one. This was confirmed by in vitro experiments with X-gluc (unpublished results) and with the fluorogenic substrate *(33)*.
6. If in the azocoupling procedure the diazonium salt does not dissolve completely, it may be necessary to filter the solution.
7. No differences were found between results after incubation at the room temperature or at 37°C.
8. Mounting in glycerol is not necessary; the objects can be examined as water mounts (Section 3.1., steps 7 and 8 omitted). Glycerol mounting, however, provides better preparation optical quality.
9. Besides Fast Blue BB salt, many other diazonium salts are available. The color of the final reaction product varies with the salt used.

References

1. Burstone, M. S. (1962) *Enzyme Histochemistry.* Academic, London.
2. Chayen, J., Bitensky L., Butcher, R. G., and Poulter, L. W. (1969) *A Guide to Practical Histochemistry.* Oliver & Boyd, Edinburgh, UK.
3. Lojda, Z., Gossrau, R., and Schiebler, T. H. (1976) *Enzymhistochemische Methoden.* Springer-Verlag, Berlin.
4. Pearse, A. G. E. (1991) *Histochemistry: Theoretical and Applied,* 4th ed. Churchill Livingstone, Edinburgh.
5. Gahan, P. B. (1984) *Plant Histochemistry and Cytochemistry.* Academic, London.
6. Jefferson, R. A., Burgess, S. M., and Hirsh, D. (1986) β-glucuronidase from *Escherichia coli* as a gene-fusion marker. *Proc. Natl. Acad. Sci. USA* **83,** 8447–8451.

7. Jefferson, R. A., Kavanagh, T. A., and Bevan, M. W. (1987) GUS fusions: β-glucuronidase as a sensitive and versatile gene fusion marker in higher plants. *EMBO J.* **6,** 3901–3907.

8. Jefferson, R. A. (1987) Assaying chimeric genes in plants: the GUS gene fusion system. *Plant Mol. Biol. Rep.* **5,** 387–405.

9. Jefferson, R. A. (1988) Plant reporter genes: the GUS gene fusion system, in *Genetic Engineering*, vol. 10 (Setlow, J. K., ed.), Plenum, New York, pp. 247–263.

10. Plegt, L. and Bino, R. J. (1989) β-glucuronidase activity during development of the male gametophyte from transgenic and non-transgenic plants. *Mol. Gen. Genet.* **216,** 321–327.

11. Hu, C. Y., Chee, P. P., Chesney, R. H., Zhou, J. H., Miller, P. D., and O'Brien, W. T. (1990) Intrinsic GUS-like activities in seed plants. *Plant Cell Rep.* **9,** 1–5.

12. Holt, S. J. (1959) Factors governing the validity of staining methods for enzymes, and their bearing on the Gomori acid phosphatase technique. *Exp. Cell Res.* **7,** 1–27.

13. van Duijn, P. (1973) Fundamental aspects of enzyme cytochemistry, in *Electron Microscopy and Cytochemistry* (Wisse, E., Daems, W. Th., Molenaar, I., and van Duijn, P., eds.), North Holland, Amsterdam, The Netherlands, pp. 3–23.

14. Beneš, K., Ivanov, V. B., and Hadačová, V. (1981) Glycosidases in the root tip, in *Structure and Function of Plant Roots* (Brouwer, R., Gašparíková, O., Kolek, J., and Loughman, B. C., eds.), Nijhoff, The Hague, The Netherlands, pp. 137–139.

15. Beneš, K. (1973) On the media improving freeze-sectioning of plant material. *Biol. Plant.* **15,** 50–56

16. O'Brien, T. P. and McCully, M. E. (1981) *The Study of Plant Structure: Principles and Selected Methods.* Termarcarphi Ltd., Melbourne, Australia.

17. Fishman, W. H. and Baker, J. R. (1956) Cellular localization of β-glucuronidase in rat tissues. *J. Histochem. Cytochem.* **4,** 570–587.

18. Fishman, W. S., Goldman, S. S., and Green, S. (1964) Several biochemical criteria for evaluating β-glucuronidase localization. *J. Histochem. Cytochem.* **12,** 239–251.

19. Holt, S. J. (1958) Indigogenic staining methods for esterases, in *General Cytochemical Methods,* vol. 1 (Danielli, J. F., ed.), Academic, New York, pp. 375–397.

20. Beneš, K. (1977) Histochemistry of carboxyl esterases in the broad bean root tip with indoxyl substrates. *Histochemistry* **53,** 79–87.

21. Fishman, W. H. and Goldman, S. S. (1965) A postcoupling technique for β-glucuronidase employing the substrate, naphthol AS-BI-β-D-glucosiduronic acid. *J. Histochem. Cytochem.* **13,** 441–447.

22. Beneš, K., Vitha, S., Gartland, K. M. A., and Elliott, M. C. (1995) The localization of β-glucuronidase in roots of transgenic sugar beet by means of azocoupling procedure, in *Developments in Plant and Soil Sciences,* vol. 58 (Baluška, F., Barlow, P. W., Gašparíková, O., and Čiamporová, M., eds.), Kluwer, Dordrecht, The Netherlands, pp. 87–89.

23. Terryn, N., Brito-Arias, M., Engler, G., Tiré, C., Villarroel, R., Van Montagu, M., and Inzé, D. (1993) *rha*1, a gene encoding a small GTP binding protein from *Arabidopsis,* is expressed primarily in developing guard cells. *Plant Cell* **5,** 1761–1769.

24. Nishihara, M., Ito, M., Tanaka, I., Kyo, M., Ono, K., Irifune, K., and Morikawa, H. (1993) Expression of the β-glucuronidase gene in pollen of lily (*Lilium longiflorum*),

tobacco (*Nicotiana tabacum*), *Nicotiana rustica*, and peony (*Paeonia lactiflora*) by particle bombardment. *Plant Physiol.* **102,** 357–361.

25. Kosugi, S., Ohashi, Y., Nakajima, K., and Arai, Y. (1990) An improved assay for β-glucuronidase in transformed cells: methanol almost completely suppresses a putative endogenous β-glucuronidase activity. *Plant Sci.* **70,** 133–140.

26. Hodal, L., Bochhardt, A., Nielsen, J. E., Mattson, O., and Okkels, F. T. (1992) Detection, expression and specific elimination of endogenous β-glucuronidase activity in transgenic and non-transgenic plants. *Plant Sci.* **87,** 115–122.

27. Gallagher, S. R., ed. (1992) *GUS Protocols: Using the GUS Gene as a Reporter of Gene Expression.* Academic, San Diego, CA.

28. Vancanneyt, G., Schmidt, R., O'Connor-Sanchez, A., Willmitzer, L., and Rocha-Sosa, M. (1990) Construction of an intron-containing marker gene: splicing of the intron in transgenic plants and its use in monitoring early events in *Agrobacterium*-mediated plant transformation. *Mol. Gen. Genet.* **220,** 245–250.

29. Ohta, S., Mita, S., Hattori, T., and Nakamura, K. (1990) Construction and expression in tobacco of a beta-glucuronidase (GUS) reporter gene containing an intron within the coding sequence. *Plant Cell Physiol.* **31,** 805–813.

30. Tör, M., Mantell, S. H., and Ainsworth, C. (1992) Endophytic bacteria expressing β-glucuronidase cause false positives in transformation of *Dioscorea* species. *Plant Cell Rep.* **11,** 452–456.

31. Beneš, K., Voldánová-Michalová, M., and Vitha, S. (1993) Localization artifacts caused by *in toto* incubation to reveal GUS activity in transgenic plants *Histochem. J.* **25,** 894.

32. Vitha, S., Beneš, K., Gartland, K. M. A., and Elliott, M. C. (1994) β-glucuronidase in situ localization in roots of transgenic plants: improvements of the indigogenic procedure. *Histochem. J.* **26,** 888.

33. Vitha, S., Beneš, K., Michalová, M., and Ondrej, M. (1993) Quantitative β-glucuronidase assay in transgenic plants. *Biol. Plant.* **35,** 151–155.

CHAPTER 18

Fluorometric GUS Analysis for Transformed Plant Material

Kevan M. A. Gartland, Julian P. Phillips, Stanislav Vitha, and Karel Beneš

1. Introduction

The use of reporter genes in transgenic plants provides an excellent opportunity to investigate the ways in which promoters and other regulatory elements regulate gene expression. Neomycin phosphotransferase II *(1)*, chloramphenicol acetyltransferase *(2)*, luciferase *(3)*, and β-glucuronidase (GUS; *4*) genes may each be used to provide some indications of the extent and sites of gene expression. Each reporter gene system has particular requirements for assaying gene expression and distinctive features. The neomycin phosphotransferase and chloramphenicol acetyltransferase systems require the use of radioisotopes or HPLC, whereas the luciferase system requires a luminometer or darkroom facilities. The GUS reporter gene system, in contrast, is quick, easy to use, sensitive, does not require radioisotopes, and is relatively inexpensive. Plant biotechnologists use the *Escherichia coli GUS*A gene in their assessments of reporter gene activity. The *E. coli* GUS has a monomeric mol wt of 68 kDa, and exists as a tetramer in vivo *(5)*.

GUS activity can be assessed both in terms of the sites of gene expression, via histochemical analysis (*see* Chapter 17), and in quantitative terms via fluorometric or spectrophotometric analysis. The use of a fluorimeter to assess GUS reporter gene expression provides increased sensitivity and wider dynamic range than is available with GUS spectro-

From: *Methods in Molecular Biology, Vol. 44:* Agrobacterium *Protocols*
Edited by: K. M. A. Gartland and M. R. Davey Humana Press Inc., Totowa, NJ

photometry. Dynamic range may be defined as the ratio of the highest quantifiable amount of fluorescent product to the lowest amount of product detectable. With fluorometric GUS analysis, this dynamic range is approximately 100-fold greater than for spectrophotometric analysis *(6)*.

The most commonly used substrate for fluorometric GUS analysis is 4-methyl-umbelliferyl-β-D-glucuronide (MUG). This compound is cleaved by GUS to release methyl umbelliferone and glucuronic acid. MUG fluoresces and may be quantified using excitation and emission wavelengths of 365 and 455 nm, respectively, by reference to the fluorescence produced with known amounts of sodium methyl umbelliferone.

2. Materials

1. GUS extraction buffer: 50 mM sodium phosphate pH 7.0 (*see* Note 1), 10 mM DTT, 10 mM Na$_2$EDTA, 0.1% (w/v) sodium lauryl sarcosine (*see* Note 2), and 0.1% (v/v) Triton X-100 (*see* Note 3). (*See* Note 4 for alternative components.)
2. GUS assay buffer: GUS extraction buffer + 1 mM MUG. Only make up in small batches. Stable at 4°C for a few days (*see* Note 5).
3. Stop buffer: 0.2M sodium carbonate. This can be kept at room temperature. The stop buffer ends the reaction by raising the pH, and enhances fluorescence.
4. 1.5-mL Eppendorf tubes.
5. Disposable plastic rods.
6. Fluorimeter capable of excitation at 365 nm and emission measurement at 455 nm.
7. Disposable fluorimeter cuvets.
8. 37°C Water bath.
9. 1 μM Sodium methyl umbelliferone (mol wt 198). Dissolve in stop buffer

3. Method

The protocol described here is based on the procedures of Jefferson *(7,8)*.

1. Grind 100 mg from each tissue to be assayed, including nontransformed samples, in Eppendorf tubes containing 500 μL extraction buffer and a pinch of sand, for example, with a plastic rod.
2. Microfuge 10 min, 12,000 rpm.
3. Transfer the supernatant to a new tube.
4. Extracts may be stored at –70°C, but *not* at –20°C, where significant loss of activity can occur overnight (*see* Note 6).
5. Prewarm 500-μL aliquots of GUS assay buffer to 37°C.

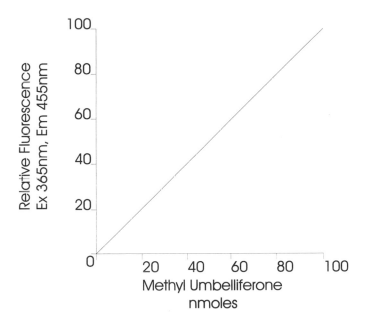

Fig. 1. Calibration graph for fluorescence of methyl umbelliferone.

6. Add 50-µL extract to each 500-µL aliquot, and incubate at 37°C. It is important to include reagent blanks, without plant extracts, to monitor spontaneous substrate degradation.
7. Remove 100-µL aliquots from the reaction mixture at time 0, and four other intervals (c.g., 30, 60, 90, and 120 min). Transfer the aliquots removed from the reaction mixture to an Eppendorf tube containing 900 µL stop buffer.
8. Calibrate the fluorimeter using triplicate 10–100 nM sodium methyl umbelliferone in stop buffer standards, further diluted with stop buffer as for the reaction mix samples. Calibration data can be regressed. Set the fluorimeter so that a known amount (e.g., 100 nmol) of methyl umbelliferone = 100 relative fluorescence units (*see* Fig. 1).
9. Dilute stopped samples further with stop buffer if necessary, store in the dark, and read on a fluorimeter within 1 h of removal (excitation wavelength 365 nm, emission 455 nm). A typical cuvet holds 3 mL (*see* Note 7). Reagent blank values should be subtracted from all other readings before calculating activities. If available, this can be done using an auto-background facility directly on the fluorimeter.

Fluorometric assessments of GUS activity are usually expressed in terms of picomoles methyl umbelliferone per milligram protein per

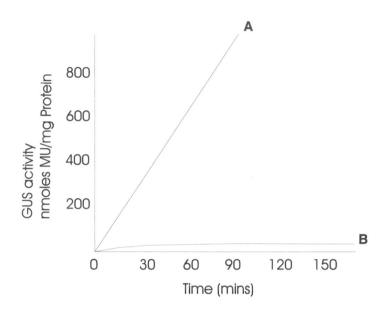

Fig. 2. GUS activity graphs of MUnmoles/mg protein vs time for GUS transformed (A) and nontransformed (B) samples.

minute. Alternatively, picogram DNA or cell number can be used instead
of protein content. Typical sodium methyl umbelliferone-fluorescence
calibration and GUS activity graphs are shown in Figs. 1 and 2.

Statistically valid comparisons between different promoter-GUS constructs, or between transformants can be made if a sufficient degree of
replication is built into the assay program *(9)*. Such comparisons permit
hierarchies of promoter strength to be drawn up. When combined with
the results of histochemical analysis using X-glucuronide, these comparisons may be valuable in deciding on regulatory elements to obtain a
particular degree of expression in specific parts of a plant.

4. Notes

1. Use a mixture of Na_2HPO_4 (mol wt 142) and NaH_2PO_4 (mol wt 120).
2. Sarcosine is sometimes difficult to dissolve and may need gentle heating.
3. The Triton X-100 should be added from a 10% (v/v) stock.
4. β-Mercaptoethanol (0.07% [v/v]) may be used instead of DTT.
5. When removing MUG powder from freezer storage, leave the bottle to
   warm up for a few minutes, in order to reduce condensation, before opening.
6. Extracts should be stored for the minimum duration necessary, because
   even at −70°C, some activity loss can occur after prolonged storage *(9)*.

7. In some tissues, intrinsic fluorescence may interfere with the use of MUG as a substrate for GUS analysis. If this occurs with material known not to be transformed with the GUS gene, background activity can be removed by treating the extract with polyclar and Sephadex G-25 spun column treatment. Alternatively, use a substrate, such as 4-trifluoromethyl umbelliferyl β-D-glucuronic acid, with different optimum excitation and emission wavelengths *(10)*.

References

1. Chen, W.-H., Gartland, K. M. A., Davey, M. R., Sotak, R., Gartland, J. S., Mulligan, B. J., et al. (1987) Transformation of sugarcane protoplasts by direct uptake of a selectable chimaeric gene. *Plant Cell Rep.* **6**, 297–301.
2. Ward, A., Etessami, P., and Stanley J. (1988) Expression of a bacterial gene in plants mediated by infectious geminivirus DNA. *EMBO J.* **7**, 1583–1587.
3. de Wet, J. R., Wood, K. V., DeLuca, M., Helinski, D. R., and Subramani S. (1987) Firefly luciferase gene: structure and expression in mammalian cells. *Mol. Cell Biol.* **7**, 725–737.
4. Jefferson, R. A. (1987) Assaying chimeric gene expression in plants: the GUS gene fusion system. *Plant Mol. Biol. Rep.* **5**, 387–405.
5. Wilson, K. J., Giller, K. E., and Jefferson, R. A. (1991) Beta-Glucuronidase (GUS) operon fusions as a tool for studying plant–microbe interactions, in *Advances in Molecular Genetics of Plant–Microbe Interactions* (Hennecke, H. and Verma, D. P. S., eds.), Kluwer, The Netherlands, pp. 226–229.
6. Gallagher, S. R. (ed.) (1992) *GUS Protocols,* Academic, London.
7. Jefferson, R. A. (1989) The GUS reporter gene system. *Nature* **342**, 837,838.
8. Jefferson, R. A., Kavanagh, T. A., and Bevan, M. W. (1987) GUS fusions: beta-glucuronidase as a sensitive and versatile gene fusion marker in higher plants. *EMBO J.* **6**, 3901–3907.
9. Vitha, S., Beneš, K., Michalová, M., and Ondrej, M. (1993) Quantitative β-glucuronidase assay in transgenic plants. *Biol. Plant* **35**, 151–155.
10. Phillips, J. P., Xing, T., Gartland, J. S., Elliott, M. C., and Gartland, K. M. A. (1992) Variation in β-glucuronidase activity in transgenic sugar beet roots. *Plant Growth Reg.* **11**, 319–325.

CHAPTER 19

The Detection of Neomycin Phosphotransferase Activity in Plant Extracts

Ortrun Mittelsten Scheid and Gabriele Neuhaus-Url

1. Introduction

The selectable marker gene so far most frequently used in plant transformation experiments is the well-known *npt*II or neomycin phosphotransferase II gene, also referred to as *aph*(3')II or aminoglycoside phosphotransferase II *(1)*. Derived from the bacterial transposon Tn5, the gene product (E.C.2.7.1.95) transfers the γ phosphate group of ATP to a specific hydroxyl group of aminoglycosidic antibiotics. In this way, it detoxifies neomycin, kanamycin, G418, as well as paromomycin. Plants expressing the enzyme can tolerate certain concentrations of these antibiotics which lead to bleaching and growth inhibition in nontransformed plants. In many cases, kanamycin sulfate is a suitable additive for selective plant media, but which of the selective agents results in the most reliable distinction between sensitive and resistant cells must be evaluated for the specific plant material under investigation. Although quantification of gene expression and its tissue specificity in plants are routinely determined by the use of the β-glucuronidase (GUS) reporter gene, there are still some applications where the level of gene expression originating from the selectable marker is of interest. Several protocols for in vitro assays of

From: *Methods in Molecular Biology, Vol. 44:* Agrobacterium *Protocols*
Edited by: K. M. A. Gartland and M. R. Davey Humana Press Inc., Totowa, NJ

NPTII activity have been published *(2–11)*. All are based on the transfer of the ³²P-labeled γ phosphate group from ATP to kanamycin *(2)*, with the drawback that other phosphorylation reactions result in additional labeled products. Therefore, assays specific for neomycin phosphotransferase require a protein separation prior to the reaction and/or a separation of the reaction products. Protein separation has been achieved by gel filtration *(3)*, nondenaturing gel electrophoresis *(4)*, phenol extraction *(9)*, or immunoaffinity purification *(10)*. Reaction products can be distinguished by paper chromatography *(3)*, thin-layer chromatography *(7)*, or by the strong binding of the positively charged kanamycin to phosphocellulose ion-exchange paper *(2,4,5,8–11)* in combination with passage through nitrocellulose *(6)*. As usual, each procedure encompasses a certain combination of benefits and obstacles. Immunopurification of the NPTII enzyme *(10)* makes the assay sensitive and specific, but requires some additional handling and expensive chemicals. If the molecular weight of the NPTII protein is of importance, e.g., in translational fusion markers *(12)*, the sensitive assay using electrophoretic separation followed by an *in situ* reaction and subsequent blotting of the reaction products *(4)* can be applied, but this requires time and substantial amounts of radioactivity and limits the sample number. A recently published protocol describes a qualitative assay for NPTII activity without the necessity to prepare plant extracts and with the potential to analyze GUS activity in parallel *(11)*.

The following protocol is entirely based on ref. *5*, with a few additional comments. It describes a fast and easy assay that can be performed simultaneously for a large number of samples, at low costs, and using low amounts of radioactivity. Unspecific phosphorylation is lowered by the addition of cold ATP, and degradation of the phosphorylated kanamycin is decreased by the addition of fluoride ions inhibiting phosphatases. Unspecific binding to the phosphocellulose is reduced by the pretreatment of the ion-exchange paper with pyrophosphate. However, it is not the most sensitive method, being susceptible to interfering substances in plant extracts that limit the reliability of quantification. In addition, these problems vary with the plant species and tissue source. It is therefore recommended that the investigator begins with the following basic protocol *(5)*. If necessary, the method can be modified according to the description in several other publications *(4,6–10)*.

2. Materials

2.1. Solutions

1. Extraction buffer (2X): 125 mM Tris-HCl, pH 6.8, 20% (v/v) glycerol, 10% (v/v) mercaptoethanol, 0.2% (w/v) sodium dodecyl sulfate (SDS) (store in aliquots at –20°C).
2. Reaction buffer (5X): 335 mM Tris-maleate pH 7.1, 210 mM MgCl$_2$, 2M NH$_4$Cl (adjust pH with saturated solution of maleic acid in distilled water, store in aliquots at –20°C).
3. Kanamycin stock: 22 mM kanamycin sulfate (store in aliquots at –20°C).
4. Cold ATP stock: 10 mM ATP (store in aliquots at –20°C).
5. Fluoride stock: 1M NaF (store in aliquots at –20°C).
6. Radiolabeled ATP: 3000 Ci/mmol γ-^{32}P-ATP; 10 μCi/μL.
7. Saturation solution: 20 mM ATP, 100 mM sodium pyrophosphate (prepare freshly for each paper pretreatment).
8. Phosphate buffer (50X): 500 mM Na$_2$HPO$_4$, 500 mM NaH$_2$PO$_4$, pH 7.5 (store at room temperature after autoclaving).
9. Protein reagent: BioRad cat. no. 500-0006 (store at 4°C).
10. Bovine serum albumin (BSA) stock: 20 mg/mL BSA (store in aliquots at –20°C).

2.2. Other Equipment

1. Mortar and pestle.
2. Liquid nitrogen.
3. Eppendorf tubes (1.5 and 2.2 mL).
4. Benchtop centrifuge (high-speed microfuge).
5. Gloves.
6. Ice box.
7. Vortex mixer.
8. Dot-blot chamber (e.g., Minifold I [Schleicher and Schüll, Dassel, Germany], Hybrid Dot [Gibco BRL, Gaithersburg, MD]).
9. PE 81 paper (Whatman [Maidstone, UK], cut according to size of dot-blot chamber).
10. Filter paper (e.g., Whatman 3MM, cut according to size of dot-blot chamber).
11. Heat block or water bath.
12. Photometer and disposable cuvets.
13. X-ray film and cassette.
14. Plastic bags.

3. Methods

3.1. Pretreatment of Ion-Exchange Paper

Soak several PE81 papers in the saturation solution for 30 min at room temperature and then let them dry completely. The papers may be stored at room temperature for several months. Handle only with gloves to avoid background artifacts.

3.2. Preparation of Extracts

1. Take 0.2–1 g plant material (e.g., leaves or callus) (*see* Note 1).
2. Freeze in a mortar with liquid nitrogen and homogenize to a fine powder (*see* Note 2).
3. Transfer the frozen powder with a spatula to 2.2-mL Eppendorf tubes.
4. Add extraction buffer (1 mL/g fresh weight) (*see* Note 3).
5. Vortex thoroughly until all the material is thawed (*see* Note 4).
6. Centrifuge (5 min at 4°C) at full speed in a benchtop centrifuge (14,000 rpm).
7. Transfer the supernatant to a fresh tube.
8. Repeat the centrifugation. Keep the supernatant on ice (*see* Note 5).

3.3. Determination of Protein Concentration (13) (see Note 6)

1. Prepare 7 standard samples by mixing the following: 0, 40, 80, 200, 400, 600, and 800 µL of BSA stock; 800, 760, 720, 600, 400, 200, and 0 µL of distilled water. The resulting concentrations are 0, 1, 2, 5, 10, 15, and 20 µg/mL, respectively.
2. Dilute plant extracts 1:100, 1:500, and 1:1000 (v/v) with distilled water in a total volume of 800 µL.
3. Add 200 µL protein reagent to each sample, including standard samples, and mix well.
4. Read the absorption at 595 nm within 1 h of mixing. Use disposable cuvets, because the dye binds strongly to the cuvet surface. If glass cuvets are used, they have to be rinsed with ethanol. Calculate protein content of plant extracts from absorption of those dilutions that are in the range of the calibration values.
5. Adjust all samples to the same protein content (if possible with extract from nontransformed tissue).

3.4. Enzyme Reaction and Detection (see Note 7)

1. Prepare a reaction mix containing 200 µL of reaction buffer (5X), 787 µL distilled water, 1.4 µL kanamycin stock, 1 µL cold ATP stock, 10 µL fluoride stock and a volume of radiolabeled ATP containing 3 µCi (normally 0.3–0.4 µL). This mix is sufficient for 60 samples and can also be used the next day, if stored at –20°C.
2. Mix 15 µL of each extract with 15 µL of the reaction mix (*see* Notes 8 and 9) and incubate 30 min at 37°C.
3. Assemble the dot-blot chamber with one layer of pretreated PE81 paper on the top of one layer of filter paper.

4. Centrifuge the samples for 2 min at full speed at room temperature.
5. Apply 20 µL of each supernatant to one well of the dot blot chamber (without vacuum).
6. Wait 5 min after application of the last sample.
7. Open the dot-blot chamber and let the PE81 paper dry completely.
8. Preheat 1X phosphate buffer to 70–80°C.
9. Wash the PE81 paper in 1X phosphate buffer at 80°C for 5 min, followed by three washes in 1X phosphate buffer at room temperature for 10 min each.
10. Dry the paper, insert it into a plastic bag, and expose to X-ray film overnight at –70°C (*see* Notes 10 and 11).

4. Notes

1. Extracts should be prepared from young and viable tissue only.
2. If only a limited amount of material is available, grinding can be done in a 1.5-mL tube with a fitting pestle.
3. For any particular plant material, it may be necessary to include protease inhibitors and BSA to the extraction buffer.
4. It is recommended that the extracts are prepared in the coldroom or are kept on ice to minimize inactivation of the enzyme prior to the assay.
5. Plant extracts can be kept at 4°C for several weeks. However, quantitative evaluation should always be performed with fresh extracts.
6. The protein quantification can be omitted if only qualitative evaluation is required. Protein contents vary between 1 and 7 mg/mL for tobacco extracts and are fairly similar in the same series of preparations from the same species.
7. Although the amount of radioactivity is low, normal precautions for work with isotopes should be taken.
8. Extracts from some species may have inhibitory effects on NPTII activity. Controls are possible by mixing unknown extracts with extracts from plants known to be positive.
9. Background activity is normally very low, even after prolonged exposure. Nevertheless, ambiguous results can be clarified by including incubations without kanamycin in the reaction mix.
10. Quantification can be achieved by liquid scintillation counting of single spots cut from the PE81 paper or by image processing using the photometer.
11. NPTII activity may originate from gene expression in bacterial cells, if the plant tissue was not grown in axenic culture (e.g., if the material is from the glasshouse or in the early stages of transformation following incubation with *Agrobacterium*). If the analysis under these conditions needs to be restricted to plant cell-derived enzyme activity, then the plasmid constructions used for transformation should be considered. The *npt*II gene tolerates intron insertions without reduction of the resistance levels

achieved in tobacco *(14)*. Such intron-containing genes are not expressed in bacterial cells.

References

1. Davies, J. and Smith, D. I. (1978) Plasmid-determined resistance to antimicrobial agents. *Ann. Rev. Microbiol.* **32,** 469–518.
2. Ozanne, B., Benveniste, R., Tipper, D., and Davies, J. (1969) Aminoglycoside antibiotics: inactivation by phosphorylation in *Escherichia coli* carrying R factors. *J. Bacteriol.* **100,** 1144–1146.
3. Herrera-Estrella, L., De Block, M., Messens, E., Hernalsteens, J.-P., Van Montagu, M., and Schell, J. (1983) Chimeric genes as dominant selectable markers in plant cells. *EMBO J.* **2,** 987–995.
4. Reiss, B., Sprengel, R., Will, H., and Schaller, H. (1984) A new sensitive method for qualitative and quantitative assay of neomycin phosphotransferase in crude cell extracts. *Gene* **30,** 211–218.
5. Mc Donnell, R. E., Clark, R. D., Smith, W. A., and Hinchee M. A. (1987) A simplified method for the detection of neomycin phosphotransferase II activity in transformed plant tissues. *Plant Mol. Biol. Rep.* **5,** 380–386.
6. Platt, S. G. and Yang, N.-S. (1987) Dot assay for neomycin phosphotransferase activity in crude cell extracts. *Anal. Biochem.* **162,** 529–535.
7. Cabanes-Bastos, E., Day, A. G., and Lichtenstein, C. P. (1989) A sensitive and simple assay for neomycin phosphotransferase II activity in transgenic tissue. *Gene* **77,** 169–176.
8. Staebell, M., Tomes, D., Weissinger, A., Maddock, S., Marsh, W., Huffman, G., Bauer, R., Ross, M., and Howard, J. (1990) A quantitative assay for neomycin phosphotransferase activity in plants. *Anal. Biochem.* **185,** 319–323.
9. Ramesh, N. and Osborne, W. R. A. (1991) Assay of neomycin phosphotransferase activity in cell extracts. *Anal. Biochem.* **193,** 316–318.
10. Henderson, L., Rao, A. G., and Howard, J. (1991) An immunoaffinity immobilized enzyme assay for neomycin phosphotransferase II in crude cell extracts. *Anal. Biochem.* **194,** 64–68.
11. Peng, J., Wen, F., and Hodges, T. K. (1993) A rapid method for qualitative assay of both neomycin phosphotransferase II and β-glucuronidase activities in transgenic plants. *Plant Mol. Biol. Rep.* **11,** 38–47.
12. Datla, R. S. S., Hammerlindl, J. K., Pelcher, L. E., Crosby, W. L., and Selvaraj, G. (1991) A bifunctional fusion between β-glucuronidase and neomycin phosphotransferase: a broad-spectrum marker enzyme for plants. *Gene* **101,** 239–246.
13. Bradford, M. M. (1976) A rapid and sensitive method for the quantitation of microgram quantities of protein utilizing the principle of protein-dye binding. *Anal. Biochem.* **72,** 248–254.
14. Paszkowski, J., Peterhans, A., Bilang, R., and Filipowicz, W. (1992) Expression in transgenic tobacco of the bacterial neomycin phosphotransferase gene modified by intron insertions of various size. *Plant Mol. Biol.* **19,** 825–836.

CHAPTER 20

The Plant Oncogenes *rol*A, B, and C from *Agrobacterium rhizogenes*

Effects on Morphology, Development, and Hormone Metabolism

Tony Michael and Angelo Spena

1. Introduction

The soil pathogen *Agrobacterium rhizogenes* is the etiological agent of hairy root disease and can incite tumor formation on many dicotyledonous plants *(1)*. The disease is so-called because abundant fine roots that resemble hair develop at the site of infection. A segment of the large Ri plasmid, the T-DNA or transferred DNA, is mobilized from the bacterium into the plant genome, thereby initiating the disease *(2–4)*. The T-DNA may consist of one region (e.g., Ri plasmid 8196) or two separate regions, termed the TL-DNA and the TR-DNA. Axenic growth of transformed roots in liquid culture is typically fast, highly branched, and hormone independent.

In tobacco and a number of other species, whole plants can be regenerated from transformed roots *(5,6)*. Such transformed plants exhibit altered phenotypes, including wrinkled leaves, short internodes, reduced apical dominance, reduced root geotropism, delayed flowering, reduced pollen, and reduced seed set. Several loci on the TL-DNA of Ri plasmid A4 (the *rol* genes, for root locus) have been shown to be implicated in the induction of roots on *Kalanchoë daigremontiana (7)*. Three of the *rol* genes, *rol*A, B, and C, which correspond to open reading frames 10, 11, and 12 on the TL-DNA sequence *(8)*, are found on a contiguous EcoRI frag-

From: *Methods in Molecular Biology, Vol. 44:* Agrobacterium *Protocols*
Edited by: K. M. A. Gartland and M. R. Davey Humana Press Inc., Totowa, NJ

ment. These three genes together were found to be capable of directing the development of the hairy root phenotype in regenerated tobacco plants *(9)* and of inducing hairy root formation *(10)*.

To determine the effect of expressing individual *rol* genes on plant growth and development, single *rol* genes were introduced into tobacco. Each *rol* gene was expressed either from its own promoter or from the stronger, constitutive 35S RNA promoter from cauliflower mosaic virus *(11)*. As a result of the nature of the hairy root disease and of the biological assay used to define the *rol* loci, one of the physiological attributes of the transformed plants most keenly studied was the ability of transformed leaf discs to induce root formation on culture medium in the absence of exogenous hormones. Furthermore, plants transgenic for the *rol* genes presented an excellent opportunity to study plant development.

2. Effects of the *rol* Genes on Morphological Development

2.1. rolC

A fragment of the TL-DNA containing the *rol*C gene with its own promoter was used by Oono et al. *(12)* to transform tobacco. The expression of *rol*C caused dwarfism and reduced the corolla size of the flowers. Adventitious roots were not induced from leaf discs when inoculated with *A. tumefaciens* containing the *rol*C gene. Using the same fragment of the TL-DNA containing *rol*C, Spena et al. *(10)* showed that *A. tumefaciens* containing *rol*C was incapable of inducing root formation on tobacco leaf discs and *K. diagremontiana* leaves. Regenerated tobacco plants expressing the *rol*C gene exhibited a dwarf phenotype with reduced apical dominance, altered leaf morphology, small flowers, and reduced seed production *(13)*. The *Agrobacterium* vector and final plasmid constructs used are described in detail elsewhere *(10)*.

When the *rol*C promoter was replaced by the 35S promoter, new developmental abnormalities were produced (Fig.1.1), including pale-green and lanceolate leaves, drastically reduced apical dominance and internodal length, and small, male sterile flowers *(13)*. The reduction of apical dominance by the overexpression of *rol*C (Fig.1.2) was even more pronounced in potato *(14)*. Roots transgenic for the *rol*C gene are highly branched *(10)*. Overexpression of *rol*C in the short day photoperiodic mutant of tobacco, Maryland Mammouth, caused flowering under long day conditions, whereas control plants did not flower (Michael et al., in preparation).

Fig. 1. (**1**) A control tobacco plant on the left and a 35S-*rol*C transformed tobacco on the right exhibiting dwarfism and altered leaf shape. (**2**) A control potato plant on the left and a 35S-*rol*C transformed potato plant on the right exhibiting severe reduction of apical dominance. (**3**) A 35S-*rol*B transformed tobacco plant exhibiting leaf necrosis. (**4**) On the far left a control tobacco flower; the other flowers are *tap*1-*rol*B flowers expressing the *rol*B gene specifically in the tapetum.

By placing the Ac transposon of maize between the 35S promoter and the *rol*C coding sequence, expression of *rol*C is blocked unless an Ac excision event takes place in the transgenic plant. When the Ac transposon jumps, there is a localized expression of *rol*C, visible in leaves as a sharply defined pale-green sector owing to reduced chlorophyll content *(15)*.

2.2. rolB

In inoculation experiments with *A. tumefaciens* on *Kalanchoë* leaves and tobacco leaf discs, the *rol*B gene by itself was able to induce roots *(10)*. Some confusion about the ability of *rol*B to induce root formation existed because of the different *rol*B promoter fragments used to express the gene. A 300-bp promoter fragment was able to produce a weak root-inducing response, whereas a 1100-bp fragment produced a strong root inducing response on tobacco leaf discs and *Kalanchoë* leaves *(10)*. Some reports on the inability of *rol*B to induce roots on tobacco leaf discs used the less effective 300-bp promoter fragment. Replacing the 1100-bp *rol*B promoter with the 35S promoter reduced the root-inducing response to a level similar to the 300-bp *rol*B promoter *(10)*, a result suggesting that the tissue specificity of *rol*B expression is important in *rol*B-induced rhizogenesis.

When *rol*B was expressed by its endogenous promoter in tobacco, relatively subtle changes in phenotype of regenerated plants occurred, such as altered leaf and flower morphology, heterostyly, and increased formation of adventitious roots on stems *(13)*. Overexpression of *rol*B by the 35S promoter resulted in new developmental abnormalities (Fig. 1.3), including rounder leaves and a dramatic leaf necrosis *(13)*. Anther-specific expression of *rol*B in tapetal cells impaired pollen production and caused a reduction in the growth (Fig. 1.4) of the corolla, pedicel, and peduncle, conferring a more condensed appearance to the whole inflorescence *(16)*. Roots transgenic for *rol*B grew straight with very reduced branching *(10)*.

2.3. rolA

A. tumefaciens harboring a fragment of the TL-DNA containing *rol*A with only 638 bp of its own promoter was able to induce root formation on tobacco leaf discs *(10,17)* but not on *Kalanchoë* leaves *(10)*. Regenerated tobacco plants expressing *rol*A from its own promoter exhibited a phenotype consisting of dwarfism caused by reduced growth, internode distance, and wrinkled leaves *(13,18)*. The leaf-wrinkling seen in *rol*A plants is probably caused by differential growth of the vascular bundles and the lamina of the leaf blade so that the lamina tissue puffs out between the more slowly growing veins.

High level overexpression of *rol*A, by employing a 35S promoter with double enhancer, produced an extreme phenotype, including severe leaf wrinkling (Fig. 2.1), and delayed flowering with the number of nodes produced before flowering increasing from approx 38 for the control to 170 (Fig. 2.2) in one 35S-*rol*A transformant (Michael et al., in preparation). The flowers of 35S-*rol*A plants were short with an even shorter

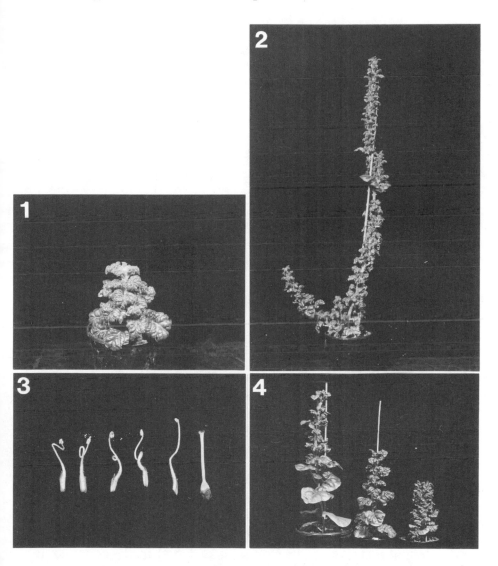

Fig. 2. (**1**) A 35S-*rol*A tobacco plant exhibiting severe leaf-wrinkling. (**2**) A 35S-*rol*A tobacco plant showing greatly increased number of nodes before flowering that is correlated with a delay in flowering. (**3**) The curled stamens and shortened style from a 35S-*rol*A tobacco flower. (**4**) Left, a tobacco plant expressing 35S-*rol*A and 35S-*rol*B; center, 35S-*rol*A only; and right, expressing 35S-*rol*A and 35S-*rol*C.

style and the stamens were curled around the style (Fig. 2.3). The curled stamens again are explained probably by a differential growth of the single central vein of the stamen and the surrounding tissue. Tobacco plants expressing *rol*A from their own promoter exhibit revertant branches associated with a reduction in the level of the *rol*A transcript *(19)*, whereas the 35S-*rol*A plants did not exhibit reversion of phenotype. Crossing the 35S-*rol*A plants to 35S-*rol*C plants produced a more extreme phenotype with greatly reduced internodes and a more erect leaf habit. Flowering was accelerated relative to *rol*A, although all flowers aborted (Michael et al., in preparation). When the 35S-*rol*A plants were crossed to 35S-*rol*B plants, the *rol*A phenotype was attenuated, the tip of each leaf was unwrinkled, and dwarfism was reduced, however, very few flowers were produced (Fig. 2.4).

3. Synergistic Interactions Between the *rol* Genes in Root Induction

An important aspect of the ability of the *rol* genes to induce root formation is their synergistic interaction. Each pairwise combination of *rol* genes (expressed from their own promoter) produces more efficient root induction than any single gene, and all three genes together produce the greatest induction of root formation *(10)*. When leaf discs of transgenic tobacco were placed on solid culture medium lacking hormones, *rol*B was the most efficient at inducing root formation, followed by *rol*A and *rol*C *(13)*.

The root system of 35S-*rol*A plants was inhibited and very poorly developed with fine, crinkled roots, the roots of 35S-*rol*B plants were thicker and relatively unbranched, whereas the roots of 35S-*rol*C plants were abundant and highly branched.

4. Effects of *rol* Gene Expression on Plant Physiology

A fragment of the TL-DNA containing *rol*A, B, and C conferred greater auxin sensitivity to transformed tobacco leaves when measured as the root-inducing response of leaf discs to different auxin concentrations *(20)*. The TL-DNA also confers a 100–1000-fold increase in auxin sensitivity to *Lotus comiculatus* transformed protoplasts when measured as a membrane depolarization response to auxin concentration *(21)*. Protoplasts of transformed tobacco expressing individual *rol* genes also have been analyzed for auxin sensitivity *(22)* by measuring membrane depolarization. The *rol*B gene was found to render protoplasts up to 10,000 times more sensitive to auxin, the *rol*A gene up to 1000 times, and the *rol*C gene up to 30 times. However, the *rol* genes did not act synergistically in increasing auxin sensitivity because protoplasts of tobacco plants expressing all three *rol* genes were only as sensitive to auxin as protoplasts expressing *rol*A alone.

The activity of the *rol* genes has also been associated with a change in polyamine metabolism and in hairy root plants containing the TL-DNA, the delay in flowering is associated with a delayed and decreased accumulation of free and conjugated polyamines *(23)*. Sun et al. *(24)* observed a decreased accumulation of water-insoluble polyamine conjugates in male-sterile tobacco plants that also express *rol*A.

5. *rol* Gene Functions

5.1. *Cellular Localization of the* rolC *Polypeptide in Transgenic Plants*

Several indicators suggested that the activity of *rol*C was cell autonomous: roots induced by *rol*C were always transformed *(25)*, and genetic mosaics for the *rol*C gene possessed variegated leaves with sharp borders *(15)*. Of several bacterially derived hormonal genes tested (i.e., *iaa*M, involved in auxin synthesis; *iaa*L, responsible for conjugating auxin to lysine and ornithine; *ipt*, isopentyltransferase, which synthesizes cytokinin; and *rol*A, *rol*B, *rol*C) as genetic mosaics, only the *ipt* and *rol*C genes gave localized, albeit different, effects on leaves (unpublished). Of the five major classes of phytohormones, only cytokinins are known to elicit localized effects on leaves *(26)*. Furthermore, White et al. *(7)*, showed that one or more of the loci present on the TL-DNA of Ri plasmid A4 could complement an *ipt*-deficient *A. tumefaciens* strain, indicating that one or more of the loci present in the TL-DNA has a cytokinin-like activity. A polyclonal antibody raised against the *rol*C polypeptide overexpressed in *E. coli* was used to demonstrate by Western blotting of subcellular fractions of *rol*C transgenic leaves, that the *rol*C polypeptide was a cytosolic protein *(27)*. By determining the cellular localization of *rol*C, it was possible to eliminate certain hypotheses for the *rol*C function. It was then reasonable to assume that *rol*C did not produce a diffusable growth factor and was unlikely to be involved in the *de novo* synthesis of phytohormones. The cytosolic localization of the *rol*C protein was compatible with models envisaging that its oncogenic activity in plants might be caused by the biochemical modification of endogenous plant growth factors.

5.2. rolC Is a Cytokinin-N-glucosidase

Genetic evidence *(7,13)* had already indicated a cytokinin-like function for the *rol*C gene. Moreover, several but not all of the alterations in plant growth and development in *rol*C transgenic plants are reminiscent of the effects of higher cytokinin levels *(28,29)*. A simple and attractive hypothesis for the function of *rol*C was the cell-specific and reversible release of active forms of growth substances from inactive conjugates *(30,31)*, which would also explain the cell-autonomous behavior of *rol*C. To test this hypothesis, *rol*C protein produced in *E. coli* was used in an in vitro assay for the hydrolysis of cytokinin conjugates *(32)*. The *rol*C protein was able to hydrolyze two cytokinin-*N*-glucosides, 6-benzylaminopurine-9-(β-D-glucoside), a synthetic conjugate, and zeatin-9-glucoside, a natural conjugate. Whereas *O*-glycosylation of cytokinins is known to be a reversible reaction, *N*-glycosylation is considered to be not freely reversible in plants *(30)*. Hydrolysis of *N*-glucosyl linkages to release free cytokinins should have an important effect on cytokinin level and/or activity. The cytokinin-β-glucosidase activity of the *rol*C protein was specifically blocked by anti-*rol*C affinity-purified antibody. The *rol*C protein was also able to hydrolyze cytokinin-*O*-glucosides, which are hydrolyzed by a variety of β-glucosidases, whereas *N*-glucosides are resistant to nonspecific β-glucosidases *(33)*. Recently, a range of cytokinin-*N*7 and *N*9-glucosides have been reported to be hydrolyzed in vitro by the *rol*C enzyme *(34)*.

The *rol*C protein was immunopurified from transgenic plants expressing *rol*C from the 35S promoter. Purified *rol*C protein from tobacco plants released cytokinin aglycones from their glucosides when provided with vascular sap that facilitated the reaction *(32)*. Thus, the data presented here indicate that *rol*C alters plant growth and development by releasing free active cytokinins from their inactive and/or storage conjugates. A short stretch of amino acid homology has been found between the *rol*C protein and a β-glucosidase from maize *(35)*.

5.3. rolB Is an Indole-β-glucosidase

Some of the biological effects caused by the expression of *rol*B in transgenic tobacco are suggestive of auxin-induced effects. Indeed, increased root induction in the lower part of the stem of tobacco plants expressing *rol*B, leaf necrosis in 35S-*rol*B plants *(13)*, and increased auxin sensitivity of *rol*B tobacco protoplasts *(22,36)* can be considered indicative of auxin biological activity. Similar phenotypes have been observed in plants transgenic for transposon split constructions of the auxin-synthesizing *iaa*M gene (Spena, unpublished observations). In addition, the effect of *rol*B overexpression on tobacco development appeared to be counteractive to the effect of *rol*C *(13)*; whereas overexpression of *rol*C led to a protracted lifespan of tobacco plants, similar expression of *rol*B resulted in cellular death as seen in the necrosis of leaves and callus. Plants that overexpress both *rol*C and *rol*A have a phenotype that is more extreme than the phenotype of 35S-*rol*A plants (Fig. 2.4) whereas overexpression of both *rol*A and *rol*B attenuates the phenotype of the 35S-*rol*A plants (Michael et al., in preparation). Anther-specific expression of *rol*B correlates with an increase in the level of free *(16)* and bound (Spena, unpublished) indole acetic acid (IAA). In vitro, the *rol*B peptide is not able to hydrolyze IAA-glucose or IAA-myoinositol, but is able to hydrolyze indoxyl glucosides *(37)*. The hydrolysis of indoxyl glucosides is inhibited by anti-*rol*B antibody; indole and, to a lesser extent, IAA, act as inhibitors of the glucosidase activity *(37)*. The level of indoxyl glucosides in the anthers of plants expressing *rol*B with a tapetum-specific promoter is reduced from 400 pmol/g/fr wt for the control plants to 30 and 170 pmol in two independent transgenic clones (unpublished).

Although *rol*B protein can hydrolyze in vitro cytokinin-*O*-glucosides, free cytokinin (i.e., benzylaminopurine) is not an inhibitor of *rol*B activity (Estruch et al., unpublished); the protein sequence of *rol*B has been found to possess a discrete region of homology with a glucosidase from the fungal pathogen that causes the "take-all" disease of wheat (P. Bowyer, Sainsbury Laboratory, Norwich, personal communication). The *rol*B protein is localized in the cytosol (Estruch et al., unpublished). Experiments are in progress to find out how *rol*B gene expression could alter auxin metabolism.

5.4. Is rolA *Another Hormonal Gene?*

The *rol*A phenotype is the most dramatic of the three *rol* genes, but its function remains unresolved. Beside the morphological changes in plants expressing *rol*A, there are physiological perturbations as well. Alterations in auxin sensitivity *(22)* and in polyamine metabolism *(24)* have been reported in plants expressing *rol*A from its own promoter. Although the biochemical function of *rol*A is undetermined, several hypotheses have been proposed for its activity, including a protein able

```
126  AGNCGAHVVSVGDITLVTKSHFEALNSIKLNVLLGVPS  163 iaaL
     |  .   |. ||.||:.      ..::.|.  ..|.|  ..||
     AKKRKAKRVSPGDVPPDQVAELDDLSVTPLAVTSPGPS        rolA
         |  |
         W  L
```

Fig. 3. Comparison of part of the protein sequences of the *iaaL* gene of *P. savastanoi* and of the *rol*A gene of *A. rhizogenes* 8196. The letters in bold below the sequence represent the new residues found in independent ethylmethyl sulphonated (EMS) mutants that suppress *rol*A activity in *rol*A transformed plants. The *rol*A transcript has recently been shown to be spliced in *Arabidopsis* and contains an intron in the untranslated leader *(48)*.

to alter directly or indirectly polyamine metabolism *(24)* and also a nucleic acid binding protein *(38)*. The *rol*A gene product is a rather small protein of a predicted mol wt of approx 11.5 kDa and an apparent mol wt of 14.5 kDa on PAGE (unpublished). Owing to its small monomeric size, it is conceivable that its biochemical function would be uncomplicated. In particular, transformed phenotypes similar to those observed in *rol*A transgenic tobacco have been reported and pointed out by Romano et al. *(39)* in tobacco plants expressing the *iaaL* gene of *Pseudomonas savastanoi*. Phenotypic alterations in plants expressing the *iaaL* gene included leaf wrinkling and an inhibition of root development *(39,40)*. The *iaaL* gene product conjugates IAA to lysine and ornithine, thereby decreasing free IAA levels. The phenotype is attenuated when plants transgenic for the *iaaL* gene are crossed to plants expressing the *iaa*M gene (which produces IAA) from the 19S promoter, i.e., the phenotype caused by the *iaaL* gene consequently could be interpreted as mainly owing to decreased levels of free IAA. In this regard it is interesting to note that the *iaaL* enzyme and the *rol*A peptide (e.g., strain 8196) show a discrete region of homology (Fig. 3). Moreover, two mutations that inactivate the *rol*A gene *(41)* map in this region (Fig. 3). Furthermore, the phenotype of 35S-*rol*A plants is attenuated by crossing to 35S-*rol*B plants and exacerbated by crossing to 35S-*rol*C plants (Fig. 2.4). This observation can be interpreted as indicating that the *rol*A gene has an anti-auxin effect that is contrasted by the auxin-like effect of the *rol*B gene. An anti-auxin effect of the *rol*A gene has also been reported by Walden et al. *(42)*. In a transient expression system, an auxin-inducible promoter is not induced in tobacco protoplasts transformed with the *rol*A gene. On the basis of this evidence, the phenotype of the *rol*A plants might then appear to be caused by a primary defect of IAA content and/ or activity. Future research will test this and other possible hypotheses.

6. *rol* Homologues

The *rol* homologous genes have been found and sequenced from Ri plasmid 8196 *(43)* and from the genome of *Nicotiana glauca (44)*. Comparisons are shown in Fig. 4. It is interesting to note that the identity between the *rol*A from A4 and the putative homologous *rol*A gene from 8196 is limited to the amino terminal half of the protein and to the last amino acids of the C-terminal. In contrast, the identity among the *rol*C and *rol*B proteins is more evenly distributed. Although no data on single genes have been reported, the *rol* genes of Ri plasmid 8196 are assumed functional. On the contrary, the *rol* genes of *N. glauca* are considered inactive. Transcripts homologous to the *rol*B and *rol*C genes have been detected in genetic tumors *(45)*, however, we are not aware of data on the biological activity of the *rol*B and *rol*C homologs of *N. glauca*.

7. Discussion

The elucidation of the biochemical activity of *rol*B and *rol*C furnished relatively simple interpretations of very complex developmental effects. Considering the bewildering numbers of interactive effects caused by alterations in phytohormone metabolism, it is easy to envisage that a modification in the content and activity of one or a few hormones will trigger alterations in several other physiological parameters leading to the complex morphological and developmental alterations observed in plants transgenic for the *rol* genes. Furthermore, the biological role(s) of glucosidic conjugates in general, and of phytohormone conjugates in particular, still eludes our understanding. Nevertheless, the fact that two genes involved in the pathogenesis of the hairy root disease code for β-glucosidases, whereas a gene able to conjugate IAA with lysine and ornithine (i.e., the *iaa*L gene) is involved in the formation of galls caused by *P. syringae* subsp. savastanoi, indicate that bacterial pathogens have exploited systems of conjugation and hydrolysis to alter plant hormone metabolism and therefore to manipulate plant growth to their advantage.

For some time, glycosylated compounds have been considered either excretion products of plant metabolism or inactive storage forms of biologically active aglycones. However, other possible functions have been suggested. For example, Gilchrist and Kosuge *(46)*, on the basis of work published by Singh and Widholm *(47)*, pointed out that glucosidic con-

```
              1                                                          50
rolA A4       MELAGLNVAG  MAQTFGVLSL  VCSKLVRRAK  A.KRKAKRVS  PGE..RDHLAEPA
rolA 8196     MELAGINVAG  MAQTFGEVSL  VLSDLVRRAK  AKKRKAKRVS  PGDVPPDQVAELD
Consensus     MELAG-NVAG  MAQTFG--SL  V-S-LVRRAK  A-KRKAKRVS  PG----D--AE--


                                                                       100
rolA A4       NLSTTPL AMTSQARPGR  STTRELLRRD  PLSPDVKIQT  YGINTHFETNLRD
rolA 8196     DLSVTPL AVTSPGPSVM  VNDHRIATRG  RFVGGPK... .......NSNLRD
Consensus     -LS-TPL A-TS------  --------R-----  --K--- ---------NLRD



              1                                                          50
rolB A4       MDPKLLFLPR  FQPVDLTPAW  SQINLFEGIR  FAFAIYSRDY  SKPLLHFQKR
rolB 8196     MPTHMHFLPR  FYPRNLTQAW  NQINLCEEIR  FAFITYSQVY  GKTLMEYQKG
rolB Ng       MASQSQFHPR  FQPRNLTPAG  KQINLSKEIQ  SAFMTYSEVY  SKTLLDYQKR
Consensus     M-----F-PR  F-P--LT-A-  -QINL---I-  -AF--YS--Y  -K-LL-YQK-


              51                                                        100
rolB A4       WALAVLDLKE  NSPPIYILKQ  LAELLKNKVC  YHPPMFVSQP  DLARERENDQHV
rolB 8196     WARDVLELEE  NSPPVNILKQ  LAQILKDQLC  YHRPMFVSQP  DLARERERQHV
rolB Ng       WADVIFDLEE  KSLRMDILKQ  LAELLKNKIC  YHPPMFVEQP  DLARERDQRV
Consensus     WA--VLDL-E  -S----ILKQ  LA--LK---C  YH-PMFV-QP  DLARE-DQ-V


              101                                                       150
rolB A4       FVYLSREKMQ  KVLKEQSITF  GMEAVLATTI  QPYRSELALQ  EMLRVHNLAW
rolB 8196     FVYLSREKMQ  KVLREQSITF  GMEAVVATTI  QPYRCDLSVK  ALLHAHNLAW
rolB Ng       FIYLSREKMQ  KVLEEQSITV  GMEAVLATTI  QPYRSDLAVQ  EMLRVHNRAW
Consensus     FVYLSREKMQ  KVL-EQSIT-  GMEAV-ATTI  QPYR-DL---  -ML--HN-AW


              151                                                      .200
rolB A4       PHSRTEEPDL  ECFIAIFASS  LFIHLLELKV  TNVYGREVAC  TFFLRRGTEN
rolB 8196     PLRRMVKSDL  ECFIAIFAST  LFVHLLEAKL  TNLYGREAPC  AFFVRLGTEN
rolB Ng       PHRRMEERDL  ECFIAIFAST  LFIHLTTLKV  TNLYGREVDC  TFFVRRASTN
Consensus     P--R----DL  ECFIAIFAS-  LFIHL---K-  TN-YGRE--C  -FF-R----N


              201                                                       250
rolB A4       RPYDVVACGT  TQFTKNALGI  SRPAASSPEP  DLTLRLSGPD  QEGEEGVMKP
rolB 8196     RRYDVIACGL  TKFDESDCVV  PPPAAAQPDL  NLRLSVQSEE  DVMKPEIVYP
rolB Ng       RPYDVVAFGT  T.........  ..........  ..........  ..........
Consensus     R-YDVVA-G-  T-F-------  --PAA--P--  -L-L------  ---------P


              251
rolB A4       AAVNLKKEA
rolB 8196     -----KNEA
rolB Ng       .........
Consensus     -----K-EA



              1                                                          50
rolC A4       MAEDDLCSLF  FKLKVEDVTS  SDELARHMKN  ASNERKPLIE  PGENQSMDID
rolC 8196     MAEVDLCALF  SNLRVKDVAS  SDELMKHIQS  VSDERVSLIE  LGENPSMDID
rolC Ng       MAEVDLCALF  FNLRVKDVTS  SDELKKHILS  ASDERNPLTE  PQENQSMDVD
Consensus     MAE-DLC-LF  --L-V-DV-S  SDEL--H---  -S-ER--L-E  --EN-SMDID


              51                                                        100
rolC A4       EEGGSVGHGL  LYLYVDCPTM  MLCFYGGSLP  YNWMQGALLT  NLPPYQHDVT
rolC 8196     EEHPPQTPET  LFLYVDCPTM  MQCFYGGWLP  YNSTHGALLT  NLPPYQKNVS
rolC Ng       EEGGTRDPGI  LYLYVDCPTM  MQCFYGTSFP  YNSRHGALLT  NLPPYQKDVS
Consensus     EE--------  LYLYVDCPTM  M-CFYG--LP  YN---GALLT  NLPPYQ--V-


              101                                                       150
rolC A4       LDEVNRGLRQ  ASGFFGYADP  MRSAYFAAFS  FPGRVIKLNE  QMELTSTKGK
rolC 8196     FNEVNRGLRE  ASGFVGYEDP  IRSAYFAALS  FPGHVAKLDE  QLRLTSTDGE
rolC Ng       LSEVSRGLRQ  ASGFFGYEDP  IRSAYFAALS  FPGHVAKLDE  QMELTSTNGE
Consensus     L-EV-RGLR-  ASGF-GY-DP  -RSAYFAALS  FPG-V-KL-E  QM-LTST-G-


              151                              180
rolC A4       CLTFDLYAST  QLRFEPGELV  RHGECKFAIG
rolC 8196     TLIFDLYATR  RHELDRDKVV  SHGECMFG..
rolC Ng       SLTFDLYASD  QLRLEPGAWV  RHGECKFGMD
Consensus     -L-FDLYA--  ---LE----V  -HGEC-F---
```

Fig. 4. *rol*A A4 has 58.8% identity and 74.4% similarity with the putative *rol*A protein from plasmid 8196. The two *rol*C proteins share a 63.4% identity and 75.2% similarity, whereas the two *rol*B proteins show 65.3% identity and 77.9% similarity. For *rol*B and *rol*C the sequence of the *N. glauca* genomic fragment is also compared.

jugates could also function as enzyme inhibitors, and consequently they could *per se* have a regulatory role in biosynthetic pathways. The *rol* genes, along with other genes able to modify hormonal metabolism, are of importance in the general goal of understanding the regulatory mechanisms controlling the synthesis, catabolism, and modification of phytohormones. Moreover, such genes, in conjunction with organ and tissue-specific promoters, will provide tools for altering plant development to produce tailor-made plants for different purposes, an application that could have commercial uses.

Acknowledgment

The authors thank M. Fladung for providing the figure of the 35S-*rol*C potato plant and Jeff Schell for support.

References

1. De Cleene, M. and De Ley, J. (1981) The host range of infectious hairy root. *Bot. Rev.* **47,** 147–194.
2. Chilton, M.-D., Tepfer, D., Petit, A., David, C., Casse-Delbart, F., and Tempe, J. (1982) *Agrobacterium rhizogenes* inserts T DNA into the genomes of host plant root cells. *Nature* **295,** 432–434.
3. White, F. F., Ghidossi, G., Gordon, M. P., and Nester, E. (1982) Tumor induction by *Agrobacterium rhizogenes* involves the transfer of plasmid DNA to the plant genome. *Proc. Natl. Acad. Sci. USA* **79,** 3193–3197.
4. Willmitzer, L., Sanchez-Serrano, J., Bushfeld, E., and Schell, J. (1982) DNA from *Agrobacterium rhizogenes* is transferred to and expressed in axenic hairy root tissue. *Mol. Gen. Genet.* **186,** 16–22.
5. Ackermann, C. (1977) Pflanzen aus *Agrobacterium rhizogenes* Tumoren an Nicotiana tabacum. *Plant Sci. Lett.* **8,** 23–30.
6. Tepfer, D. (1984) Transformation of several species of higher plants by *Agrobacterium rhizogenes*: sexual transmission of the transformed genotype and phenotype. *Cell* **47,** 959–967.
7. White, F. F., Taylor, B. B., Huffman, G. A., Gordon, M. P., and Nester, E. (1985) Molecular and genetic analysis of the transferred DNA regions of the root inducing plasmid of *Agrobacterium rhizogenes. J. Bacteriol.* **164,** 33–44.
8. Slightom, J., Durand-Tardif, M., Jouanin, L., and Tepfer, D. (1986) Nucleotide sequence analysis of the TL-DNA of *Agrobacterium rhizogenes* agropine type plasmid. *J. Biol. Chem.* **261,** 731–744.
9. Jouanin, L., Vilaine, F., Tourneur, J., Tourneur, C., Pautot, V., Muller, J. F., and Caboche, M. (1987) Transfer of a 4.3 kb fragment of the TL-DNA of *Agrobacterium rhizogenes* strain A4 confers the hairy root phenotype to regenerated tobacco plants. *Plant Sci.* **53,** 53–63.

10. Spena, A., Schmülling, T., Koncz, C., and Schell, J. (1987) Independent and synergistic activity of *rolA*, *B* and *C* loci in stimulating abnormal growth in plants. *EMBO J.* **6**, 3891–3899.

11. Odell, J. T., Nagy, F., and Chua, N.-H. (1985) Identification of DNA sequences required for the activity of the cauliflower mosaic virus 35S promoter. *Nature* **313**, 810–812.

12. Oono, Y., Handa, T., Kanaya, K., and Uchimiya, H. (1987) The TL-DNA gene of Ri plasmids responsible for dwarfness of tobacco plants. *Jpn. J. Genet.* **62**, 501–505.

13. Schmulling, T., Schell, J., and Spena, A. (1988) Single genes from *Agrobacterium rhizogenes* influence plant development. *EMBO J.* **7**, 2621–2629.

14. Fladung, M. (1990) Transformation of diploid and tetraploid potato clones with the *rolC* gene of *Agrobacterium rhizogenes* and characterisation of transgenic plants. *Plant Breeding* **104**, 295–304.

15. Spena, A., Aalen, R. D., and Schulze, S. C. (1989) Cell-autonomous behaviour of the *rolC* gene of *Agrobacterium rhizogenes* during leaf development: a visual assay for transposon excision in transgenic plants. *Plant Cell* **1**, 1157–1164.

16. Spena, A., Estruch, J. J., Prinsen, E., Nacken, W., Van Onckelen, H., and Sommer, H. (1992) Anther-specific expression of the *rolB* gene of *Agrobacterium rhizogenes* increases IAA content in anthers and alters anther development and whole flower growth. *Theor. Appl. Genet.* **84**, 520–527.

17. Vilaine, F., Charbonnier, C., and Casse-Delbart, F. (1987) Further insight concerning the TL-DNA region of the Ri plasmid of *Agrobacterium rhizogenes* strain A4: transfer of a 1.9 kb fragment is sufficient to induce transformed roots on tobacco leaf fragments. *Mol. Gen. Genet.* **280**, 111–115.

18. Sinkar, V. P., White, F. F., Furner, I. J., Abrahamsen, M., Pythoud, F., and Gordon, M. P. (1988) Reversion of aberrant plants transformed with *Agrobacterium rhizogenes* is associated with the transcriptional inactivation of the TL-DNA genes. *Plant Physiol.* **86**, 584–590.

19. Sinkar, V. P., Pythoud, F., White, F., Nestor, E., and Gordon, M. (1988) *rolA* locus of the Ri plasmid directs developmental abnormalities in transgenic plants. *Genes Dev.* **2**, 688–698.

20. Spanò, L., Mariotti, D., Cardarelli, M., Branca, C., and Costantino, P. (1988) Morphogenesis and auxin sensitivity of transgenic tobacco with different complements of Ri T-DNA. *Plant Physiol.* **87**, 479–483.

21. Shen, W. H., Petit, A., Guern, J., and Tempé, J. (1988) Hairy roots are more sensitive to auxin than normal roots. *Proc. Natl. Acad. Sci. USA* **85**, 3417–3421.

22. Maurel, C., Barbier-Brygoo, H., Brevet, J., Spena, A., Tempé, J., and Guern, J. (1991) *Agrobacterium rhizogenes* T-DNA genes and sensitivity of plant protoplasts to auxins, in *Advances in Molecular Genetics of Plant–Microbe Interactions* (Hennecke, H. and Verma, D. P. S., eds.), Kluwer, Dordrecht, The Netherlands, pp. 343–351.

23. Martin-Tanguy, J., Tepfer, D., Paynot, M., Burtin, D., Heisler, L., and Martin, C. (1990) Inverse relationship between polyamine levels and the degree of phenotypic alteration induced by the root inducing left-hand transferred DNA from *Agrobacterium rhizogenes*. *Plant Physiol.* **92**, 912–918.

24. Sun, L. Y., Monneuse, M.-O., Martin-Tanguy, J., and Tepfer, D. (1991) Changes in flowering and accumulation of polyamines and hydroxycinnamic acid-polyamine conjugates in tobacco plants transformed by the *rolA* locus from the Ri TL-DNA of *Agrobacterium rhizogenes*. *Plant Sci.* **80,** 145–156.

25. Schmülling, T. (1988) *Studien zum Einfluss der rolA, B und C Gene der TL-DNA von* Agrobacterium rhizogenes *auf die Pflazenentwicklung*. Unpublished doctoral dissertation, Universität zu Köln, Köln, Germany.

26. Moore, T. C. (1979) *Biochemistry and physiology of plant hormones,* 2nd ed., Springer, New York.

27. Estruch, J. J., Parets-Soler, T., Schmülling, T., and Spena, A. (1991) Cytosolic localisation in transgenic plants of the rolC peptide from *Agrobacterium rhizogenes*. *Plant Mol. Biol.* **17,** 547–550.

28. Matthysse, A. G. and Scott, T. K. (1984) Functions of hormones at the whole plant level of organisation, in *Encyclopedia of Plant Physiology*, vol. 10 (Scott, T. K., ed.), Springer, Berlin, pp. 219–243.

29. Smigocki, A. C. and Owens, L. D. (1988) Cytokinin gene fused with a strong promoter enhances shoot organogenesis and zeatin levels in transformed plant cells. *Proc. Natl. Acad. Sci. USA* **85,** 5131–5135.

30. Letham, D. S. and Palni, L. M. S. (1983) The biosynthesis and metabolism of cytokinins. *Annu. Rev. Plant Physiol.* **34,** 163–197.

31. Reinecke, D. M. and Bandurski, R. S. (1987) Auxin biosynthesis and metabolism, in *Plant Hormones and Their Role in Plant Growth and Development* (Davies, P. J., ed.), Martinus Nijhoff, Dordrecht, pp. 24–42.

32. Estruch, J., Chriqui, D., Grossmann, K., Schell, J., and Spena, A. (1991) The plant oncogene *rolC* is responsible for the release of cytokinins from glucoside conjugates. *EMBO J.* **10,** 2889–2895.

33. Van Staden, J. and Papaphillopou, A. P. (1977) Biological activity of *O*-β-D-glucopyranosylzeatin. *Plant Physiol.* **60,** 649,650.

34. Spena, A., Estruch, J. J., Hansen, G., Langenkemper, K., Berger, S., and Schell, J. (1992) The *rhizogenes* tale: modification of plant growth and physiology by an enzymatic system of hydrolysis of phytohormone conjugates, in *Advances in Molecular Genetics of Plant–Microbe Interactions* (Nester, E. W. and Verma, D. P. S., eds.), Kluwer, Dordrecht, The Netherlands, pp. 109–124.

35. Campos, N., Bako, L., Feldwisch, J., Schell J., and Palme, K. (1992) A protein from maize labeled with azido-IAA has novel β-glucosidase activity. *Plant J.* **2,** 675–684.

36. Barbier-Brygoo, H., Guern, J., Ephritikhine, G., Shen, W. H., Maurel, C., and Klämbt, D. (1990) The sensitivity of plant protoplasts to auxin: modulation of receptors at the plasmalemma, in *Plant Gene Transfer, UCLA Symposia on Molecular and Cellular Biology* (New Series), vol. 129 (Lamb, C. J. and Beachy, R. N., eds.), Liss, New York, pp. 165–173.

37. Estruch, J., Schell, J., and Spena, A. (1991) The protein encoded by the *rolB* plant oncogene hydrolyses indole glucosides. *EMBO J.* **10,** 3125–3128.

38. Levesque, H., Delepelaire, P., Rouzé, P., Slightom, J., and Tepfer, D. (1988) Common evolutionary origin of the central portions of the Ri TL-DNA of *Agrobac-*

terium rhizogenes and the Ti T-DNA of *Agrobacterium tumefaciens. Plant Mol. Biol.* **11,** 731–744.

39. Romano, C. P., Hein, M. B., and Klee, H. J (1991) Inactivation of auxin in tobacco transformed with the indoleacetic acid-lysine synthetase gene of *Pseudomonas savastanoi. Genes Dev.* **5,** 438–446.

40. Spena, A., Prinsen, E., Fladung, M., Schulze, S., and Van Onckelen, H. (1991) The indoleacetic acid-lysine synthetase gene of *Pseudomonas syringae* subsp. *savastanoi* induces developmental alterations in transgenic tobacco and potato plants. *Mol. Gen. Genet.* **227,** 205–212.

41. Dehio, C. (1992) *Isolierung und Charakterisierung von Suppressormutanten der Phänotypen* rolA *und* rolB *transgener* Arabidopsis thaliana *Pflanzen.* Unpublished doctoral dissertation, Universität zu Köln, Köln, Germany.

42. Walden, R., Czaja, I., Schmülling, T., and Schell, J. (1993) *Rol* genes alter hormonal requirements for protoplast growth and modify the expression of an auxin responsive promoter. *Plant Cell Rep.* **12,** 551–555.

43. Hansen, G., Larribe, M., Vaubert, D., Tempé, J., Biermann, B., Montoya, A., Chilton, M.-D., and Brevet J. (1991) *Agrobacterium rhizogenes* pRi8196 T-DNA: mapping and DNA sequence of functions involved in mannopine synthesis and hairy root differentiation. *Proc. Natl. Acad. Sci. USA* **88,** 7763–7767.

44. Furner, I., Huffman, G. A., Amasino, R. M., Garfinkel, D. J., Gordon, M. P., and Nestor, E. W. (1986) An *Agrobacterium* transformation in the evolution of the genus *Nicotiana. Nature* **329,** 424–427.

45. Ichikawa, T., Ozeki, Y., and Syono, K. (1990) Evidence for the expression of the *rol* genes of *Nicotiana glauca* in genetic tumours of *N. glauca* x *N. langsdorffi. Mol. Gen. Genet.* **220,** 177–180.

46. Gilchrist, D. and Kosuge, T. (1980) Aromatic amino acid synthesis, in *The Biochemistry of Plants* (Miflin, B. J., ed.), Academic, New York, pp. 507–531.

47. Singh, M. and Widholm, J. M. (1975) *Biochem. Genet.* **13,** 357–367.

48. Magrelli, A., Langenkemper, K., Dehio, C., Schell, J., and Spena, A. (1994) Splicing of the *rol*A transcript of *Agrobacterium rhizogenes* in *Arabidosis. Science* **266,** 1986–1988.

CHAPTER 21

Quantifying Polyamines
in *Agrobacterium rhizogenes* Strains
and in Ri Plasmid Transformed Cells

Nello Bagni, Marisa Mengoli,
and Shigeru Matsuzaki

1. Introduction

Aliphatic polyamines are ubiquitous compounds classified as plant growth substances (*1*). They act mainly in processes based on cell division. It is known that exogenous polyamines can induce cell division in plant tissues temporarily lacking in polyamines (*1*). Many authors have suggested the possible utilization of these compounds as tumor markers because in general they increase during the early stage of the disease, but more detailed studies demonstrate that this increase is positively related in plant and animal tissues to growth rate rather than tumorigenesis *per se*, both in normal and tumor tissues. Even though polyamines alone seem not specifically related with morphogenic effects (i.e., embryogenesis and organogenesis), in association with other plant hormones (notably auxins and cytokinins) they can modify and/or regulate this phenomenon. Their involvement has been reported in the growth of "crown gall" caused by *Agrobacterium tumefaciens* (*2*), and more recently in the genetic transformation induced by *Agrobacterium rhizogenes*, the so-called "hairy root disease"; in this context the big interest about polyamines concerns the transfer and expression of genetic information from Ri plasmid to host plant cell. Results obtained up to now are in favor of changes in polyamine levels (both free and conjugated forms, trichloroacetic acid

From: *Methods in Molecular Biology, Vol. 44:* Agrobacterium *Protocols*
Edited by: K. M. A. Gartland and M. R. Davey Humana Press Inc., Totowa, NJ

[TCA]-soluble and -insoluble) during the growth cycle of transgenic tobacco plants both in vivo and in vitro *(3)*, and of isolated hairy roots in culture *(4)*. In particular, changes in free polyamine levels and in the activities of related enzymes seem to be linked to different growth kinetics (cell division and/or root elongation) depending on the different degree of expression of T-DNA genes. In addition, because some "unusual" polyamines can serve as taxonomic markers in microorganisms, and also in the different species of the genus *Agrobacterium (5)*, it is of interest to analyze polyamine pattern and distribution in the *A. rhizogenes* strains *(6)*, also in relation to the different virulence properties.

The most sensitive and widely used methods of free and bound polyamine analysis, summarized in this chapter, are based on the dansylation of the amino groups under alkaline conditions, separation of the dansyl derivatives by chromatographic techniques (including thin layer chromatography [TLC], ion exchange chromatography [IEC], and high-performance liquid chromatography [HPLC]), and quantification by fluorimetry.

2. Materials

2.1. Free and Bound Polyamine Analysis in A. rhizogenes *Strains by HPLC*

1. Yeast-mannitol medium for *A. rhizogenes* culture *(7)*: 5 g/L yeast extract, 8 g/L mannitol, 5 g/L NaCl, 2 g/L $(NH_4)_2SO_4$, 20 g/L Bacto agar (Difco, Detroit, MI) (only for solid medium).
2. Rotary shaker.
3. Spectrophotometer (Jasco [Tokyo, Japan] 7800) for 595 nm optical density readings.
4. Distilled water.
5. Lyophilization apparatus (Edwards Mini Fast 680).
6. 0.5N Perchloric acid (PCA).
7. Glass-glass homogenizer.
8. Centrifuge (4°C, 17,500g).
9. Freezer –20°C.
10. 6 and 1N HCl.
11. Rotary evaporator and vacuum apparatus.
12. Dowex-50 W (strongly acidic cation exchanger resin; dry mesh: 200–400; Sigma, St. Louis, MO) for chromatographic columns.
13. HPLC apparatus (K-101 AS, Kyowa Seimitsu, Saitama, Japan) with a short column (4.6 × 80 mm) of cation-exchange resin (62210FK, Kyowa Seimitsu); this kind of column is usually used for amino acid analysis. Resins with a narrow range of bead diameter below 10 μm are adopted.

14. Autosampler KSST-60 (Kyowa Seimitsu).
15. Spectrofluorimeter (FP110, Jasco).
16. 1,6-Hexanediamine [$NH_2(CH_2)_6NH_2$] (Sigma).
17. Minisart P filters (pore size 0.45 μm).
18. Citrate buffers containing 2% (v/v) *n*-propanol and different concentrations of KCl: pH 5.55 buffer, 0.5*M* KCl; pH 5.63 buffer, 2.1*M* KCl; pH 5.73 buffer, 2.4*M* KCl (*see* Note 1).
19. Polyamine derivatization solution: 800 mg of *o*-phthalaldehyde (OPA) are dissolved in 10 mL of 95% (v/v) ethanol, and then 2 mL of mercaptoethanol are added. The solution is made up to 1 L by adding 2.5% (v/v) borate buffer, pH 10.4, containing 0.09% (v/v) Brij-35 (Merck, Darmstadt, Germany).
20. Standard polyamines (putrescine, cadaverine, spermidine, and spermine) are purchased from Sigma; homospermidine is synthesized according to Hamana et al. *(8)*.
21. 0.5*N* KOH.
22. Spectrophotometer and spectrofluorimeter cuvets.

2.2. Free Polyamine Analysis in Ri Transformed Plant Tissues by HPLC

1. Chilled mortar and pestle.
2. 0.2*N* Perchloric acid (PCA)
3. Standard putrescine, spermidine, spermine, as their hydrochlorides, obtained from Sigma (1 m*M* stock solutions of these compounds in 0.01*N* HCl are stable for at least 1 mo at –20°C).
4. 1,6-Hexanediamine [$NH_2(CH_2)_6NH_2$] (Sigma).
5. Microcentrifuge (at 4°C).
6. Na_2CO_3 saturated solution.
7. Methanol, acetone, and toluene of HPLC grade.
8. Dansyl chloride (Sigma): solution in acetone (7.5 mg/mL).
9. Proline (Sigma): solution in distilled water (100 mg/mL).
10. Vortex mixer.
11. Nylon membranes (0.2 μm pore, Rainin, Woburn, MA).
12. Rotary evaporator and vacuum apparatus.
13. Freezer –20°C.
14. HPLC apparatus consisting of two solvent metering pumps (Altex, model 110A) programmed with a microprocessor controller (Altex, model 420).
15. HPLC columns including ODS^2 hypersil (5-μm particle diameter, 5 × 250 mm, Shandon, Cheshire, England) and ODS ultrasphere (5-μm particle diameter, 4.6 × 250 mm, Altex, Berkeley, CA).
16. Fluorescence spectrophotometer (equipped with an 8-μL flow through cell, model 650-10 LC, Perkin Elmer, Norwalk, CT) and integrator (Hewlett Packard, model 3390A), both attached to HPLC apparatus.

2.3. Hydroxycinnamoyl Amide Analysis in Transformed Tobacco Tissues by HPLC

1. Homogenizer.
2. Centrifuge (up to 20,000g).
3. Weakly acidic cation exchanger resin column (Fractogel TSK CM-650 [M] Na$^+$, 0.045–0.09 mm, Merck).
4. 8M Acetic acid.
5. Methanol of HPLC grade.
6. Nylon membranes (0.2-μm pore size).
7. p-Coumaroylputrescine (pCP), feruloylputrescine (FP), and diferuloylputrescine (DiFP) are a gift from Ciba Geigy (Basel, Switzerland), and caffeoylputrescine (CP) from G. Leubner (Ruhr-Universität, Germany).
8. Reversed-phase HPLC apparatus using two LC-pumps, and provided by a controller (T-414 Anacomp 220, Kontron Instruments) and an autosampler (MSI 660T, Kontron Instruments).
9. Nucleosil C-18 column (250 × 4.6 mm, particle size 5 μm, Stagroma, Wallisellen, Switzerland), fitted with a guard column.
10. Oven 830 (Kontron), to perform HPLC separation at 30°C.
11. 25 mM Sodium acetate buffer, pH 4.5.
12. Fluorescence spectrophotometer (Spectroflow 980 and 757, Kratos Analytical, Ramsey, NJ).
13. Interface (760 series, Nelson-Stadler, Switzerland).
14. Computer (HP 9000, Nelson-Stadler).

2.4. Free and Bound Polyamine Analysis in Ri Transformed Plant Tissues by TLC

1. Mortar and pestle.
2. 5% (v/v) TCA.
3. 0.1 and 12N HCl.
4. Standard polyamines, as their hydrochlorides, obtained from Sigma; 5-mM stock solutions in 0.1 N HCl are stable for at least 1 mo at –20°C.
5. Acetone (Merck) of analytical grade.
6. Dansyl chloride (Sigma) solution in acetone (3 mg/mL).
7. Centrifuge.
8. Special glass ampules for acid hydrolysis.
9. Apparatus for flame sealing glass ampules.
10. Dry oven at 110°C.
11. Rotary evaporator and vacuum apparatus.
12. NaHCO$_3$ (Merck).
13. Vortex mixer.
14. Proline (Sigma) solution in distilled water (15 mg/mL).

15. Benzene (Merck).
16. Silicagel 60 precoated plates (Merck) with concentrating zone.
17. Chromatographic tank.
18. Cyclohexane (Merck).
19. Ethylacetate (Merck).
20. UV lamp.
21. Spectrofluorimeter (Jasco FP 770) with cuvets.

3. Methods

3.1. Free and Bound Polyamine Analysis in A. rhizogenes Strains by HPLC

3.1.1. Bacterial Cultures

1. Culture *A. rhizogenes* strains 1855 (agropine type) or 8196 (mannopine type) in the dark at $29 \pm 1°C$, in flasks containing 30 mL of yeast-mannitol medium.
2. Place the bacterial liquid cultures on a rotary shaker (80 rpm), and at regular intervals measure the optical density of the culture at 595 nm to determine the growth cycle.
3. Harvest the bacterial cultures during the different growth phases and centrifuge them at 17,500g for 15 min at 4°C.
4. After elimination of liquid culture medium, rinse the pellet twice by resuspension in distilled water (1 mL/10 mL of bacterial culture) and centrifugation (17,500g for 15 min at 4°C).
5. Resuspend the pellets in 1 mL of distilled water in Eppendorf tubes, and put them open in a lyophilizer for 6–8 h at –40°C.

3.1.2. Extraction Procedures

1. Extract polyamines (*see* Note 2) in 4 vol of ice-cold 0.5*N* PCA, by using a glass homogenizer.
2. Collect the supernatants after centrifugation at 3000 rpm for 10 min at 4°C.
3. Homogenize the precipitates in 2.5 vol of 0.5*N* PCA and centrifuge as in step 2.
4. Repeat this procedure again and pool the three supernatants (*see* Note 3).
5. In order to determine polyamines covalently bound in the PCA-soluble fraction, heat the samples at 110°C for 24 h in the presence of 6*N* HCl, evaporate them to dryness, and reconstitute the original volume with appropriate amounts of distilled water.
6. Apply the samples to a Dowex 50W column to remove amino acids and to concentrate polyamines.
7. After washing the columns with 1*N* HCl, evaporate whole polyamines to dryness under reduced pressure (29 cm Hg) at 60°C.

8. Reconstitute the original volume of the samples with appropriate amounts of distilled water.

3.1.3. HPLC Analysis

1. Set the spectrofluorimeter recorder at 1 and 10 mV for 0 and 100% relative fluorescence.
2. In order to determine the recovery rate, a known amount of 1,6-hexanediamine in a preliminary experiment is added to the homogenates (*see* Note 4).
3. To adjust the pHs of the samples, add 2–5 volumes of pH 5.55 citrate buffer and filter the mixture through Millipore filters with pore size of 0.45 µm.
4. Preset the volume of the sample (depending on the concentration of polyamines) and the duration of the assay cycle in the autosampler, and apply automatically 10–50 µL of the filtrates to the column, equilibrated with the pH 5.55 citrate buffer for 10 min.
5. Regulate the buffer change by a sequence controller.
6. The pH 5.55 citrate buffer elutes almost all amino acids, including arginine and homoarginine, within 10 min.
7. The pH 5.63 citrate buffer elutes putrescine, cadaverine, spermidine, and homospermidine within 20 min.
8. The pH 5.73 citrate buffer elutes spermine and thermospermine within 10 min. The chromatogram in Fig. 1 shows the separation of free polyamines from the *A. rhizogenes* strains 1855 and 8196 (*see* Note 5).
9. After completion of the elution, wash the column with 0.5*N* KOH for 3 min and equilibrate with the pH 5.55 citrate buffer for 10 min.
10. The duration of a complete cycle is 53 min; after each 20 samples, known amounts (usually 100 pmol each) of standard polyamines are run.
11. Follow the elution patterns by post-column derivatization with OPA: Samples are mixed with the OPA solution (2:1 [v/v]), according to the flowrates of the buffer (0.5 mL/min) and of the OPA solution (0.25 mL/min); the column is kept at 60°C in a temperature-controlled box.
12. Assay polyamines by fluorescence measurement. The wavelengths of excitation and emission are 338 and 455 nm, respectively. The lower limit of detection is about 1 pmol for each polyamine. Polyamine concentrations in *A. rhizogenes* are expressed as nanomoles per milligram wet or dry weight.

3.2. Free Polyamine Analysis in Ri Transformed Plant Tissues by HPLC

3.2.1. Extraction of Tissues

1. Homogenize plant tissues in a mortar in ice-cold 0.2*N* PCA in the order of 100 mg tissue/mL acid (*see* Note 6).
2. Add to the extracts hexanediamine (at 1 µmol/g fresh wt of tissue) as an internal standard (*see* Note 7).

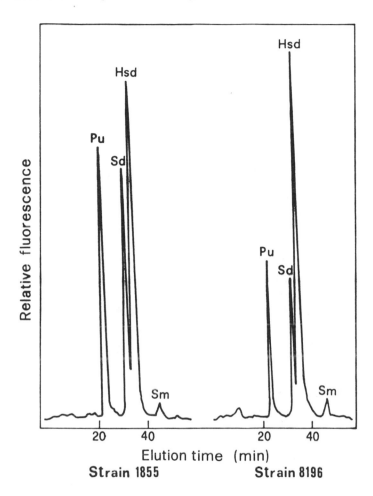

Fig. 1. HPLC elution profile of polyamines from *A. rhizogenes* strains 1855 and 8196. Pu = putrescine; Sd = spermidine; Hsd = homospermidine; Sm = spermine.

3. Centrifuge the homogenates at 4°C in a microcentrifuge (11,000 rpm).
4. Use the supernatants for the free polyamine analysis (*see* Note 8).

3.2.2. Dansylation

1. To 50–100 µL of the extracts, add 200 µL of saturated Na_2CO_3 and 400 µL of dansyl chloride in acetone.
2. After 1 h incubation in the dark at 60°C, in 5-mL tapered reaction vials, add to the mixture 100 µL of proline to remove the excess of dansyl chloride.
3. After 1.5 h incubation, extract polyamines with 500 µL of toluene by vigorous vortexing for 30 s.

4. Remove and discard the lower aqueous phase.
5. Dry completely under nitrogen the organic phase containing the polyamines, and dissolve the residue in 1 mL of methanol, ultrafiltered through nylon membrane.
6. Assay immediately the samples or store them at –20°C (no more than 1 wk).

3.2.3. HPLC Analysis

1. Dilute (5–20-fold) aliquots of the samples before HPLC analysis performed at room temperature.
2. Inject the samples into a fixed 20-µL loop for loading onto the column.
3. Elute the samples from the column with a programmed water:methanol (v/v) solvent gradient, changing from 60 to 95% in 23 min, at a flowrate of 1 mL/min.
4. In these experimental conditions, dansyl polyamines are completely eluted by 27 min: The elution times in minutes from injection are the following: putrescine, 18.1; cadaverine, 18.8; hexanediamine, 19.7; spermidine, 23.5; spermine, 26.8.
5. Wash the column with 100% methanol for 5 min and re-equilibrate it with 60% (v/v) methanol for 5 min before the next sample is injected.

3.2.4. Detection and Quantitation

1. For the detection of the eluates from the column, the system is provided by an attached fluorescence spectrophotometer (excitation and emission wavelengths are 365 and 510 nm, respectively).
2. Eluant peaks with their areas and retention times are recorded by an attached integrator. The scheme for free polyamine extraction and HPLC assay is shown in Fig. 2.

3.3. Hydroxycinnamoyl Amide (HCA) Analysis in Transformed Tobacco Tissues by HPLC

3.3.1. Extraction of Tissues

1. Extract the plant tissues (previously lyophilized) with water:methanol (1:1 [v/v]; 40 mL/g dry wt of tissue) using an homogenizer (*see* Note 9).
2. After 2 h incubation at room temperature, centrifuge the extracts for 20 min at 20,000g.
3. Load the supernatants onto a weakly acidic cation-exchanger column.
4. Elute bound HCAs with 8M acetic acid/methanol (1:1 [v/v]).
5. Lyophilize the eluted extracts, and dissolve them in water/methanol (1:1 [v/v]).

3.3.2. HPLC Analysis

1. Filter the samples prior to injection into the HPLC reverse phase apparatus.
2. Apply the samples through a 20-µL loop onto the column.

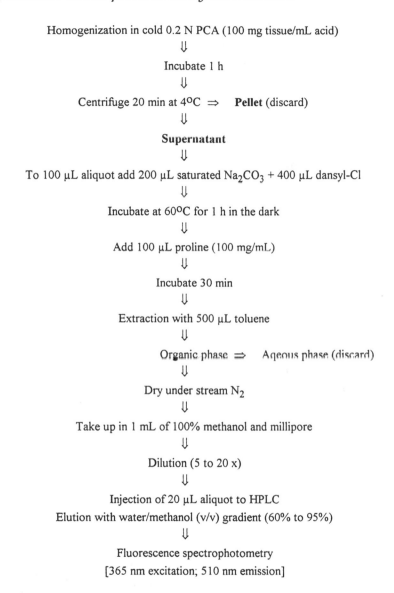

Homogenization in cold 0.2 N PCA (100 mg tissue/mL acid)
⇓
Incubate 1 h
⇓
Centrifuge 20 min at 4°C ⇒ **Pellet** (discard)
⇓
Supernatant
⇓
To 100 μL aliquot add 200 μL saturated Na_2CO_3 + 400 μL dansyl-Cl
⇓
Incubate at 60°C for 1 h in the dark
⇓
Add 100 μL proline (100 mg/mL)
⇓
Incubate 30 min
⇓
Extraction with 500 μL toluene
⇓
Organic phase ⇒ Aqeous phase (discard)
⇓
Dry under stream N_2
⇓
Take up in 1 mL of 100% methanol and millipore
⇓
Dilution (5 to 20 x)
⇓
Injection of 20 μL aliquot to HPLC
Elution with water/methanol (v/v) gradient (60% to 95%)
⇓
Fluorescence spectrophotometry
[365 nm excitation; 510 nm emission]

Fig. 2. Scheme for free polyamines extraction by using HPLC.

3. Perform the separation at 30°C, using a gradient of 25 mM sodium acetate pH 4.5 (solution A) to acetonitrile (solution B) at a flowrate of 1 mL/min. Acetonitrile gradient: 0–13 min, 13%; 13–18 min, 13–35%; 18–28 min, 35%; 28–33 min, 35–100%; 33–37 min, 100%.

4. Detect the compounds by fluorescence (320-nm excitation, >389-nm emission) and absorption (306 nm). Data points are sampled by an interface and integrated with a computer.
5. Calculate hydroxycinnamoyl amide content using standard curves measured for each substance.

3.4. Free and Bound Polyamine Analysis in Transformed Plant Tissues by TLC

3.4.1. Extraction of Tissues and Acid Hydrolysis

1. Homogenize plant tissues in a mortar in 5% cold TCA at a ratio of approx 100 mg tissue/0.3 mL acid (*see* Note 10). A scheme of the assay procedures is reported in Fig. 3.
2. After extraction, centrifuge the samples at 1500*g* for 15 min, to allow the recovery of a supernatant (SN), containing the "free" and the "TCA-soluble bound" polyamine fractions, and a pellet (PT), containing the "TCA-insoluble bound" polyamine fraction.
3. Resuspend the PT in 5% TCA, and repeat this washing procedure twice, followed by centrifugation.
4. Resuspended again the pellet in a final volume of TCA corresponding to that of the supernatant.
5. Place replicates (0.2 mL) of the resuspended PT and of the SN in glass ampules containing the same vol of HCl 12*N*.
6. Incubate the flame-sealed ampules at 110°C for 18 h to allow the acid hydrolysis of covalent linkages between polyamines and other molecules (*see* Note 11).
7. Take to dryness the hydrolysates and resuspend them in 0.2 mL TCA.

3.4.2. Dansylation

1. Mix aliquots (0.1 mL) of the SN, hydrolyzed SN, and hydrolyzed resuspended PT with 200 µL of dansyl chloride in acetone (3 mg/mL) and 50 mg of NaHCO$_3$.
2. Process standard polyamines in the same way, dansylating 50 nmol for each (alone or in combination with other polyamines).
3. After brief vortexing, incubate the assay mixture overnight in darkness at room temperature.
4. Remove the excess of dansyl reagent by reaction with 100 µL of added proline.
5. After 30-min incubation at room temperature, extract dansyl polyamines in 0.5 mL benzene, by vortexing for 30 s.

3.4.3. TLC Analysis

1. Load 25–200-µL of dansylated extracts on the concentrating zone of silicagel 60 plates.

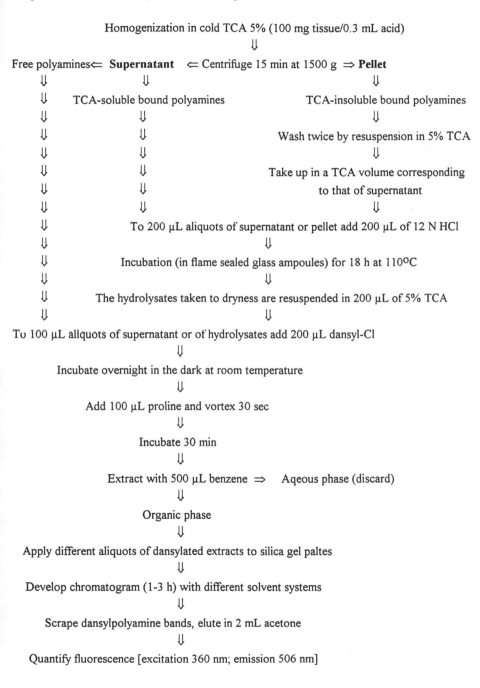

Homogenization in cold TCA 5% (100 mg tissue/0.3 mL acid)
⇓

Free polyamines ⇐ **Supernatant** ⇐ Centrifuge 15 min at 1500 g ⇒ **Pellet**
⇓ ⇓ ⇓

TCA-soluble bound polyamines   TCA-insoluble bound polyamines
⇓ ⇓ ⇓

Wash twice by resuspension in 5% TCA
⇓

Take up in a TCA volume corresponding
to that of supernatant
⇓

To 200 µL aliquots of supernatant or pellet add 200 µL of 12 N HCl
⇓

Incubation (in flame sealed glass ampoules) for 18 h at 110°C
⇓

The hydrolysates taken to dryness are resuspended in 200 µL of 5% TCA
⇓

To 100 µL aliquots of supernatant or of hydrolysates add 200 µL dansyl-Cl
⇓

Incubate overnight in the dark at room temperature
⇓

Add 100 µL proline and vortex 30 sec
⇓

Incubate 30 min
⇓

Extract with 500 µL benzene ⇒ Aqeous phase (discard)
⇓

Organic phase
⇓

Apply different aliquots of dansylated extracts to silica gel paltes
⇓

Develop chromatogram (1-3 h) with different solvent systems
⇓

Scrape dansylpolyamine bands, elute in 2 mL acetone
⇓

Quantify fluorescence [excitation 360 nm; emission 506 nm]

Fig. 3. Scheme for the extraction and assay of free and conjugated polyamines by using TLC.

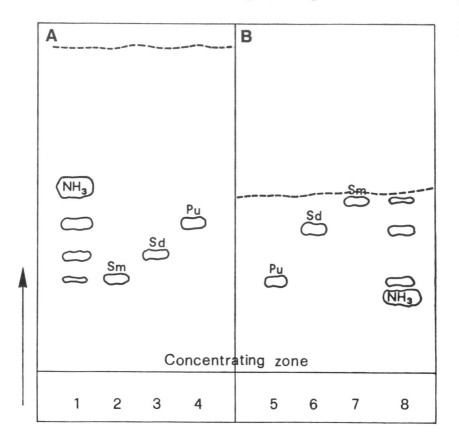

Fig. 4. TLC separation of dansyl polyamines. The solvent systems used to develop chromatogram are: (A) cyclohexane:ethylacetate, 3:2 (v/v); (B) chlorophorm:triethylamine, 25:1 (v/v). Lanes 1 and 8: Ri transformed tobacco leaves; lanes 2–4 and 5–7: standard polyamines (1 nmol each). The arrow indicates the direction of front migration. The dotted line represents the front of migration. Pu = putrescine; Sd = spermidine; Sm = spermine.

2. Put the plate in the tank containing cyclohexane:ethylacetate (3:2 [v/v]); a well-resolved chromatogram can be obtained in 2–3 h (*see* Note 12).
3. Identify the different bands (visualized as spots under UV light) by comparing chromatographic parameters of separation of standard polyamines in the same solvent system (*see* Fig. 4).
4. Scrape the spots from the plate.
5. Elute in 2 mL anhydrous acetone and measure immediately the fluorescence (because of its rapid fading); excitation and emission wavelengths are 360 and 505.5 nm, respectively.

4. Notes

1. Freshly redistilled water is used to prepare all citrate buffers, which are filtered through Minisart P (pore size 0.45 µm) before use.
2. The method described is according to Matsuzaki et al. *(9)*.
3. Polyamines extracted in PCA solutions can be stored at –20°C for several mo without appreciable degradation.
4. In the routine analysis method the recovery rate is around 100%, and therefore generally polyamines are assayed without adding any internal standard. However, when several purification procedures are applied, the addition of an internal standard is essential to calculate the recovery rate.
5. No appreciable amounts of thermospermine and of acetylated polyamines are present in the *A. rhizogenes* strains examined.
6. The method described is according to Smith and Davies *(10)*.
7. Hexanediamine appears to be an appropriate internal standard, as it resolves well from other substances in the mixture, and elutes close to the naturally occurring amines.
8. Samples are derivatized immediately or stored for no more than 2 wk at –20°C.
9. The scheme reported for extracting and assaying hydroxycinnamoyl amides in tobacco tissues follows the method of Wiss-Benz et al. *(11)*. Conjugates of the polyamines with different hydroxycinnamic acids (mainly *p*-coumaric, ferulic, and caffeic acids) are widely distributed in the plant kingdom, and seem to be linked to flowering processes, but their physiological role is far from clear.
10. The method described is according to Smith and Best *(12)* and Torrigiani et al. *(13)*.
11. The partner molecules covalently bound to polyamines consist mainly of hydroxycinnamic acids in the SN, and of proteins in the PT.
12. Different solvent systems can be used, i.e., chlorophorm:triethylamine (18:1 or 25:1 [v/v]), with different resolution patterns.

Acknowledgment

We thank S. Scaramagli for her kind assistance in preparing this chapter.

References

1. Bagni, N. (1989) Polyamines in plant growth and development, in *The Physiology of Polyamines*, vol. 2 (Bachrach, U. and Heimer, Y. M., eds.), CRC Press, Boca Raton, FL, pp. 107–120.
2. Speranza, A. and Bagni, N. (1977) Putrescine biosynthesis in *Agrobacterium tumefaciens* and in normal and crown gall tissues of *Scorzonera hispanica* L. Z. *Pflanzenphysiol.* **81,** 226–233.
3. Mengoli, M., Chriqui, D., and Bagni, N. (1992) Protein, free amino acid and polyamine contents during development of hairy root *Nicotiana tabacum* plants. *J. Plant Physiol.* **139,** 697–702.

4. Mengoli, M., Ghelli, A., Chriqui, D., and Bagni, N. (1992) Growth kinetics, polyamine pattern and biosynthesis in hairy root lines of *Nicotiana tabacum*. *Physiol. Plant.* **85,** 697–703.

5. Hamana, K., Matsuzaki, S., Niitsu, M., and Samejima, K. (1989) Polyamine distribution and the potential to form novel polyamines in phytopathogenic agrobacteria. *FEMS Microbiol. Lett.* **65,** 269–274.

6. Mengoli, M., Matsuzaki, S., and Bagni, N. (1992) Culture and characterization of *Agrobacterium rhizogenes* strains in relation to polyamine pattern and biosynthesis. *Microbios* **69,** 233–242.

7. Petit, A., David, C., Dahl, G. A., Ellis, J. G., Guyon, P., Casse-Delbart, T., and Tempé, J. (1983) Further extension of the opine concept: plasmids of *Agrobacterium rhizogenes* cooperate for opine degradation. *Mol. Gen. Genet.* **190,** 204–214.

8. Hamana, K., Miyagawa, K., and Matsuzaki, S. (1983) Occurrence of *sym*-homospermidine as the major polyamine in nitrogen-fixing cyanobacteria. *Biochem. Biophys. Res. Commun.* **112,** 606–613.

9. Matsuzaki, S., Hamana, K., Imai, K., and Matsuura, K. (1982) Occurrence in high concentrations of *N*-acetylspermidine and *sym*-homospermidine in the hamster epididymis. *Biochem. Biophys. Res. Commun.* **107,** 307–313.

10. Smith, M. A. and Davies, P. (1985) Separation and quantitation of polyamines in plant tissue by high performance liquid chromatography of their dansyl derivatives. *Plant Physiol.* **78,** 89–91.

11. Wiss-Benz, M., Streit, L., and Ebert, E. (1988) Hydroxycinnamoyl amides in stem explants from flowering and non-flowering *Nicotiana tabacum*. *Physiol. Plant.* **74,** 294–298.

12. Smith, T. A. and Best, G. R. (1977) Polyamines in barley seedlings. *Phytochemistry* **16,** 841–843.

13. Torrigiani, P., Altamura, M. M., Capitani, F., Serafini-Fracassini, D., and Bagni, N. (1989) De novo root formation in thin cell layers of tobacco: changes in free and bound polyamines. *Physiol. Plant.* **77,** 294–301.

CHAPTER 22

IAA Analysis in Transgenic Plants

John F. Hall, Sarah J. Brown, and Kevan M. A. Gartland

1. Introduction

The availability of transgenic plant material has provided an important new tool for unraveling the complex mechanisms by which auxins regulate plant development *(1–4)*. The accurate estimation of indole-3-acetic acid (IAA) and IAA-conjugates remains a vital part of the analysis if we are to understand auxin action. Several methods are available utilizing immunoassays, GC-MS, high-performance liquid chromatography (HPLC), and fluorescence. One reliable method for IAA analysis in transgenic material is based on the method of Blakesley et al. *(5)*, which uses a combination of HPLC and fluorimetric detection of the derivatized IAA *(see* Fig. 1). The straightforward method described has been used successfully for the analysis of IAA and indole-3-acetamide (IAM) in *Solanum* species and sugar beet hairy roots *(1,6)* and should be suitable for most plant tissues, although additional purification steps may be required, as explained in the Notes section. This protocol is intended as an introduction to the methodology and the reader is referred to refs. *7* and *8* and Chapter 23 in this volume, which cover the analysis of other metabolites and the methods used.

2. Materials

1. All glassware must be washed thoroughly with Decon or equivalent detergent prior to steeping in chromic acid solution for 12 h (usually overnight). The acid is rinsed at least four times with warm water and then at least

From: *Methods in Molecular Biology, Vol. 44:* Agrobacterium *Protocols*
Edited by: K. M. A. Gartland and M. R. Davey Humana Press Inc., Totowa, NJ

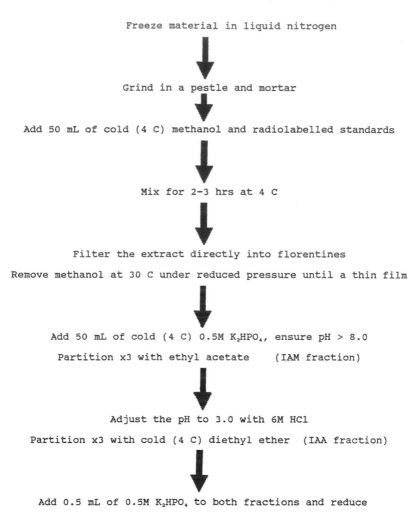

Fig. 1. Extraction and quantitation of IAA using HPLC and fluorimetric detection of derivatized IAA.

three times with deionized-distilled water. Dry the glassware thoroughly and coat with dichlorosilane (Repelcote, a 2% solution of dichlorosilane in 1,1,1,trichloroethane). Air dry the glassware and then bake at about 80°C for 2 h. Rinse with hot water and deionized-distilled water a minimum of three times each. Gloves and appropriate safety clothing should be worn.

2. All solutions should be prepared with freshly deionized-distilled water in glassware prepared as in item 1. They can be stored for 2–3 days at 4°C but fresh solutions must then be prepared. Polyvinylpolypyrrolidone

Apply the aqueous residue to a PVPP column

Rinse with 0.5 mL of 0.5M K₂HPO₄ and appply to the column

Elute with 0.1M K₂HPO₄, pH 9.3 and collect 150 mL

IAM fraction partition against cold (4 C) ethyl acetate

IAA fraction adjust the pH to 3.0 with 6M HCl partition x3

against cold (4 C)

diethyl ether

Freeze at -20 C for 12 hrs

filter the extract through prewashed Whatman no1 into florentines

IAM -Remove the ethyl acetate under reduced pressure at 30 C

take the residue up in a minimum volume of methanol

IAA- reduce to about 1 mL

Transfer both samples to reagent vials and store at -20 C prior

to analysis

Fig. 1. *(continued)*

(PVPP) should be washed thoroughly with deionized-distilled water removing fine particles at the same time or alternatively buy high-quality ready-prepared PVPP.

3. $0.5M$ K_2HPO_4 in deionized-distilled water, pH 9.3.
4. $0.1M$ K_2HPO_4 in deionized-distilled water, pH 9.3, using analytical grade reagents.
5. Solvents: Dried redistilled diethyl ether can be prepared but is most conveniently purchased. Chloroform must be of a very high quality Distol grade or equivalent. Ethyl acetate and methanol are HPLC grade.
6. Pestle and mortar, acid wash and rinse as described in item 1.

7. Liquid nitrogen.
8. Radiolabeled standards: 3-indolyl [2-^{14}C] acetic acid, specific activity 2 GBq/mmol (11.3 MBq/mg) and 3[5(n)-^{3}H] indolyl acetic acid, specific activity 940 GBq/mmol (5.33 GBq/mg) are used.
9. Rotary evaporator.
10. 6M Hydrochloric acid.
11. Separation funnels.
12. Whatman (Maidstone, UK) No. 1 filter paper.
13. Florentines.
14. Sealable reagent vials.
15. HPLC, fluorimetric detector, and reverse phase (ODS) analytical column.
16. Trifluoroacetic acid and acetic anhydride (analytical grade reagents).
17. Magnesium sulfate, analytical grade anhydrous powder.
18. Liquid scintillation counter: Scintillation liquid that tolerates high levels of water (Picoflour 40 or equivalent).
19. 0.01M Tetraethylammonium chloride in 0.001M phosphate buffer, pH 6.6 (analytical grade reagents).
20. Phosphorous pentachloride.
21. Ammonium hydroxide (0.89 g/mL).

3. Methods
3.1. Extraction Protocol

The following procedures should be performed as rapidly as safe practice allows without drying the samples until stipulated to ensure maximal recoveries.

1. Freeze 1 g of fresh material in liquid nitrogen. Grind in a pestle and mortar (clean and dedicated to auxin analysis), repeating the freeze-grind treatment until the sample is a fine powder.
2. Add 50 mL of cold methanol to extract, and radiolabel standards to account for losses during the purification procedure (*see* Note 1). Mix at 4°C for 2–3 h. Longer extraction times may be required and the optimum should be determined. Extraction periods of up to 12 h do not increase the amount of IAA released from sugar beet hairy roots (*see* Note 2). Tritiated standards have a high specific activity allowing the estimation of low levels of IAA, but the purity of the radioactive standard must be monitored weekly as considerable degradation occurs.
3. Filter the extract directly into a 250-mL florentine and reduce under reduced pressure at 30°C to a thin film on the flask. Add 50 mL of cold (0–4°C) 0.5M K$_2$HPO$_4$. This concentration and quantity of buffer are used to ensure that the pH is above 8.0. This should be checked initially and can be reduced, scaling the quantities of other solvents used proportionately.

4. The extract is partitioned three times against equal volumes of ethyl acetate. This removes the IAM and other similar nonacidic precursors/derivatives of IAA.
5. Adjust the pH to 3.0 with $6M$ HCl, ensuring that the acid is added slowly with constant shaking.
6. Partition three times with equal volumes of cold (0–4°C) diethyl ether. IAA partitions into the ether under these conditions. If the phases are very slow to separate include a diethyl ether wash at pH 8.0 prior to partitioning at pH 3.0.
7. For both the ethyl acetate and the diethyl ether fractions, add 0.5 mL of $0.5M$ K_2HPO_4 and reduce to the aqueous phase under reduced pressure at 30°C.
8. Apply the aqueous phase carefully without disturbing the packing to a PVPP column (0.8 × 10 cm) pre-equilibrated with $0.1M$ K_2HPO_4. Rinse the flask with an additional 0.5 mL of $0.5M$ K_2HPO_4 and apply to the column. Allow the extract to enter the PVPP so that the surface is just dry. Carefully fill the rest of the column and elute with $0.1M$ K_2HPO_4, and collect 150 mL.
9. Partition the IAM containing fraction directly three times with one-third volume of ethyl acetate. The pH of the IAA fraction is adjusted as before *(5)*, and partitioned three times against one-third volume of cold (0–4°C) diethyl ether.
10. The samples are then placed in the freezer (no colder than –20°C). This freezes the water, which is then removed by filtration through Whatman No. 1 filter paper that has been prewashed with the appropriate solvent. The filtrate should be collected directly in florentines.
11. The ethyl acetate fraction is removed under reduced pressure as before and the residue taken up in a minimum volume of methanol. Store the extract in a reagent vial sealed with a Teflon-lined cap and store at –20°C until the HPLC analysis is performed.
12. The diethyl ether is reduced to about 1.0 mL and transferred to a reagent vial and stored as in step 11.
13. Blank extractions must be performed to check for interference from any contaminants in the reagents used.

3.2. Derivatization for IAA Estimation: Formation of 2-Methylindolo-2,3:3',4'-pyr-6-one (2-MIP)

1. Further purification prior to derivatization is normally not required. However, extracts from sugar beet hairy roots form a white precipitate when dried and this must be removed. The only method we have found to be successful is to partition the extract in aqueous solution against chloroform. Add 0.5 mL of $0.1M$ K_2HPO_4 to the reagent vial and remove the diethyl ether under a stream of nitrogen. Add 0.5 mL of chloroform and

shake. Assist separation of the phases by centrifuging the vial gently (500 rpm in a benchtop centrifuge). Remove the lower chloroform phase to a fresh vial. Repeat the procedure. *Do not take any of the water phase.*

2. The chloroform from the combined fractions is removed under a stream of nitrogen and derivatized with a 1:1 mixture of trifluoroacetic acid and acetic anhydride (25 µL) for 5 min. (There should not be any significant amount of water at this stage, but if the derivatization fails, add a small quantity of dry $MgSO_4$ to the combined chloroform fractions. Repeat the centrifugation and transfer the dried chloroform to a fresh vial.) This should be sufficient for most samples, but the optimum time should be determined for your particular tissue.

3. HPLC of the derivatized sample is performed on a reverse phase C-18, 5-µm column with a methanol/water (acidified to pH 3.5 with acetic acid) gradient with fluorescence detection, 445 nm excitation, and 480 nm emission. The precise conditions will vary with the type of column used. The following program gives excellent results on a Lichrospher 125 × 4 mm, RP 18.5 µm analytical column using a Merck-Hitachi L-6200 HPLC system.

Time	Methanol, %	Water, %
0	30	70
5	70	30
7	100	0
8	100	0
10	30	70

Flowrate 1.5 mL/min. 2-MIP has a retention time of approx 6 min, 40 s.

4. The peak corresponding to 2-MIP is collected and counted in a liquid scintillation counter. (Unless other acidic indoles 5-OH-IAA or 4-chloro-IAA are present, there is only one major peak.) Recoveries of approx 20% are routinely achieved.

5. Exceptionally, as is the case with extracts from etiolated wheat and maize, ion-pair HPLC purification of the extract is required before derivatization. Tetraethylammonium chloride (0.01*M*) as the counter ion in pH 6.6, 0.001*M* phosphate buffer has been used *(9)* and is recommended as a starting point.

3.3. Measurement of IAM

1. The extract in methanol is reduced to an appropriate volume of approx 50 µL and subjected to HPLC analysis with fluorescence detection, 280 nm excitation, and 350 nm emission.

2. The following profile gives a retention time of approx 14 min, 40 s for IAM on a 125 × 4 mm RP18, 5-µm Lichrospher column. This was considered to be an adequate separation for this material, as further analysis under

different conditions, ion-pair, and ion-suppression HPLC did not alter the specific activity of the recovered sample. This must be verified for each tissue analyzed.

Time	Methanol, %	Water, %
0	20	80
4	45	55
10	100	0
15	100	0
20	20	80

Flowrate 1.0 mL/min.

4. Notes

1. It may be necessary to synthesize IAM. This is readily done using the procedure used by Crosby et al. *(10)*. Briefly, IAA is taken up in dry diethyl ether and kept on ice. Excess phosphorous pentachloride is added and the mixture stirred for 30 min. Ensure that moisture is excluded up to this point. The reaction mixture is then added slowly to a large excess of concentrated ammonium hydroxide with stirring. After 30 min the ether is removed under nitrogen and the IAM extracted into ethyl acetate. Yields of about 70% have been reported, but yields of between 30 and 40% are obtained routinely.
2. If large increases of IAA are found after extended extraction times, the possibility of release from IAA-conjugates should be investigated *(9)*.

References

1. Gartland, K. M. A., McInnes, E., Hall, J. F., Mulligan, B. J., Morgan, A. J., Elliott, M. C., and Davey, A. R. (1991) Effects of Ri plasmid *rol* gene expression on the IAA content of transformed roots of *Solanum dulcamara* L. *Plant Growth Regul.* **10,** 235–241.
2. Stiller, J., Svoboda, S., Nemcova, B., and Machackova, I. (1992) Effects of agrobacterial oncogenes in kidney vetch (*Anthyllis vulneraria* L.). *Plant Cell Rep.* **11,** 363–367.
3. Nilsson, O., Crozier, A., Schmulling, T., Sandberg, G., and Olsson, O. (1993) Indole-3-acetic acid homeostasis in transgenic tobacco plants expressing the *Agrobacterium rhizogenes rolB* gene. *Plant J.* **3,** 681–689.
4. Van Onckelen, H., Rudelsheim, P., Inze, D., Follin, A., Messens, E., Horemans, S., et al. (1985) Tobacco plants transformed with the *Agrobacterium* T-DNA gene 1 contain high amounts of indole-3-acetamide. *FEBS Lett.* **181,** 373–376.
5. Blakesley, D., Hall, J. F., Weston, G. D., and Elliott, M. C. (1983) Simultaneous analysis of indole-3-acetic acid and detection of 4-chloroindole-3-acetic acid and 5-hydroxyindole-3-acetic acid in plant tissues by high-performance liquid chromatography of their 2-methylindolo-α-pyrone derivatives. *J. Chromatogr.* **258,** 155–164.

6. Brown, S. J., Gartland, K. M. A., Slater, A., Hall, J. F., and Elliott, M. C. (1990) Plant growth regulator manipulations in sugar beet, in *Progress in Plant Cellular and Molecular Biology* (Nijkamp, H. J. J., Van Der Plas, L. W. H., and Van Aartrijk, J., eds.), Kluwer, Dordrecht, The Netherlands, pp. 486–492.
7. Morgan, P. W. and Durham, J. I. (1983) Strategies for extracting, purifying and assaying auxins from plant tissues. *Bot. Gaz.* **144,** 20–31.
8. Sandberg, G., Crozier, A., and Ernsten, A. (1987) Indole-3-acetic acid and related compounds, in *Principles and Practice of Plant Hormone Analysis*, vol 2 (Rivier, L. and Crozier, A., eds.), Biological Techniques Series, Academic, London, pp. 169–284.
9. Blakesley, D., Allsopp, A. J. A., Hall, J. F., Weston, G. D., and Elliott, M. C. (1984) Use of reversed-phase ion-pair high performance liquid chromatography for the removal of compounds inhibitory to the formation of the 2-methyl indolo-α-pyrone derivative of indole-3-acetic acid. *J. Chromatogr.* **294,** 480–484.
10. Crosby, D. G., Boyd, J. B., and Johnson, H. E. (1960) Indole-3-alkanamides. *J. Organic Chem.* **25,** 1826,1827.

CHAPTER 23

Quantifying Phytohormones in Transformed Plants

Els Prinsen, Pascale Redig, Miroslav Strnad, Ivan Galís, Walter Van Dongen, and Henri Van Onckelen

1. Introduction

Physicochemical techniques are very specific, sensitive, and accurate techniques widely used for phytohormone analysis (for a review *see* refs. *1–3*). Liquid chromatography mass spectrometry (LC-MS) or gas chromatography-coupled mass spectrometry (GC-MS) in particular, recently have become more important for both qualitative and quantitative analyzes of all phytohormones, except ethylene (for a recent review *see* ref. *3*). Specific purification is, however, necessary prior to high-performance liquid chromatography (HPLC), gas chromatography (GC), LC-MS, or GC-MS. On the other hand, we have immunological techniques available through radioimmunoassay (RIA) and enzyme-linked immunosorbent assay (ELISA). The advantages of immunochemical techniques are their high sensitivity, high specificity, and an extremely short analysis time. There are a lot of reports on poly- and monoclonal antibodies for the analysis of plant hormones including cytokinins *(4–7)*. The mass spectrometric fragmentation pattern from electron impact (EI) GC-MS is a frequently used identification criterion. Using LC-MS for qualitative and quantitative cytokinin analysis, derivatization can be omitted *(8)*. Moreover, cytokinins exhibit strong specific UV absorbance in the 220–300 nm range. Since UV spectroscopy is nondestructive, mass spectro-

From: *Methods in Molecular Biology, Vol. 44:* Agrobacterium *Protocols*
Edited by: K. M. A. Gartland and M. R. Davey Humana Press Inc., Totowa, NJ

metry and UV spectroscopy are complementary for the identification of cyto-kinins *(9)*. Recently, qualitative LC-MS analysis has also been described for abscisic acid (ABA) *(10)*, however, deuterated ABA obtained by active hydrogen exchange *(11)* is not suitable for thermospray conditions (our unpublished results).

In the last decade, procedures described for phytohormone analysis as well as the available hardware have improved substantially, resulting in rapid, efficient, and highly sensitive methods that allow the analysis of different phytohormones starting from a minimal amount of plant tissue often as little as 10–50 mg. The method described here for the purification of indole-3-acetic acid (IAA), ABA, and cytokinins from the same plant extract is based on the previously described solid-phase purification for IAA and ABA *(12)*. The different classes of hormones need specific analytical approaches that are discussed in Section 3.3.

2. Materials

2.1. Preparation of the Diethylaminoethyl (DEAE)-Sephadex A25 (Anion Exchange Column, Formate Conditions)

1. Dissolve 100 g DEAE-Sephadex A25 in 500 mL distilled water for swelling following the manufacturer's guideline.
2. Rinse the swollen DEAE with 3 L, $1M$ NH_4-formate. A separation funnel is very useful for this purpose.
3. Rinse the column material with distilled water until pH 6 is reached.
4. Check if the pH of the column material is between 5 and 7 before use. If not, then rinse with additional water or 50% (v/v) methanol.
5. Always use freshly prepared column material.
6. Pour 2 mL of column material in a 5-mL syringe.

2.2. C18 Column Material

As C18 column material, use one Bond-Elut reversed phase (RP)-C18 (Analitichem Intern., prepacked 1 or 0.5 mL columns) or two consecutive Sep-Pak RP-C18 (Waters Associate). Prerinse the columns with 10 mL technical grade ethanol followed by 20 mL distilled water and 10 mL of the solvent to be used.

2.3. Preparation of Diazomethane (13)

Caution: Work in a fume hood during this procedure. The *N*-nitrosotoluol-4-sulfomethylamide and diazomethane are toxic!

Table 1
Radioactive Tracers Used for Phytohormone Analysis

Abbreviation	Full name	Specific radioactivity	Source
^{3}H-ZR-dialk.	[^{3}H]-(trans)-zeatin-riboside dialcohol	1.4 TBq/mmol	1
^{3}H-Z	[2-^{3}H]-(trans)-zeatin	0.9 TBq/mmol	2
^{3}H-ZR	[2-^{3}H]-(trans)-zeatin-riboside	0.9 TBq/mmol	2
^{3}H-iP	[2-^{3}H]-N^6-2-isopentenyl)-adenine	1.65 TBq/mmol	2
^{3}H-iPA	[^{3}H$_5$]-N^6-(2-isopentenyl)-adenosine	1.65 TBq/mmol	2
^{3}H-(diH)Z	[^{3}H]-DL-dihydrozeatin	1.27 TBq/mmol	1
^{14}C-IAA	3-indolyl-[2-^{14}C]-acetic acid	2.18 GBq/mmol	1
^{14}C-IAA	3-indolyl-[1-^{14}C]-acetic acid	2.18 GBq/mmol	1
^{3}H-IAA	[3-(5[n]-^{3}H]-indole-acetic acid	788 GBq/mmol	1
^{3}H-ABA	DL-(cis,trans)-[G-^{3}H]-abscisic acid	2.26 TBq/mmol	1

Sources: 1. Amersham, Bucks, UK and 2. Institute of Experimental Botany, Isotopic Lab, Prague, Czech Republic.

1. Dissolve 0.5 g *N*-nitrosotoluol-4-sulfomethylamide (Fluka, AG, Buchs, Switzerland) in 20 mL of diethyl ether and add 2.5 mL of a 4% (w/v) KOH in proanalyze grade ethanol solution.
2. Distill the diazomethane that is formed and the diethyl ether at 50°C and collect this yellow diazomethane–ether mixture in a glass recipient at 0°C.
3. Store the diazomethane in a closed vial in the dark at –20°C for no longer than 2 d.

2.4. Other Reagents

All glassware to be used as recipients for samples (i.e., evaporation vials, fraction collector tubes) must be silanized.

1. Radioactive tracers (*see* Table 1).
2. Deuterated tracers (*see* Table 2). In these protocols we use deuterated tracers because they are easy to purchase, however, other "heavy tracers" have also been described *(9,14–16)*.
3. 80% (v/v) Proanalyze grade methanol.
4. 0.04*M* Ammonium acetate pH 6.5. Always adjust pH before use.
5. 6% (v/v) Formic acid.
6. Acidified methanol: 100 mL water-free methanol plus 20 μL 37% (w/v) HCl.
7. Phosphate-buffered saline (PBS): 10 m*M* potassium phosphate pH 7.2, 0.9% (w/v) NaCl.
8. Tris-buffered saline (TBS): 50 m*M* Tris-HCl pH 7.4, 10 m*M* NaCl, 1 m*M* MgCl$_2$.
9. Distilled water.

Table 2
Deuterated Tracers Used for Phytohormone Analysis

Abbreviation	Full name	Mol wt	Source
^{2}H-Z	[^{2}H$_5$]-(trans)-zeatin	224	1
^{2}H-ZR	[^{2}H$_5$]-(trans)-zeatin-riboside	356	1
^{2}H-iP	[^{2}H$_6$]-N^6-(2-isopentenyl)-adenine	209	1
^{2}H-iPA	[^{2}H$_6$]-N^6-(2-isopentenyl)-adenosine	341	1
^{2}H-(diH)Z	[^{2}H$_3$]-DL-dihydrozeatin	224	1
^{2}H-(diH)Z-R	[^{2}H$_3$]-(RS)-dihydrozeatin-riboside	356	1
^{2}H-Z-7-glu	[^{2}H$_5$]-(trans)-zeatin-7B-D-glucoside	386	1
^{2}H-Z-9glu	[^{2}H$_5$](trans)-zeatin-9B-D-glucoside	386	1
^{2}H-IAA	Indole-[2,4,5,6,7-^{2}H$_5$]-3-acetic acid	180	2
^{2}H-ABA	[^{2}H$_6$]-ABA	270	*(11)*

Sources: 1. Apex International, Devon, UK and 2. Aldrich, Brussels, Belgium.

2.5. Preparation of ^{2}H-ABA (11)

1. Add to 1 mg ABA 1000 µL ^{2}H$_2$O (Janssen Chimica, Beerse, Belgium) and 100 µL NaO^{2}H (40% [w/v] in ^{2}H$_2$O, Aldrich, Milwaukee, WI).
2. Remove all air under a nitrogen stream and seal the vial airtight.
3. Keep overnight at room temperature.
4. Acidify to pH 2.0 with 2*N* HCl.
5. Desalt by passing through a C18 column and rinse the column with an additional 10 mL 0.5*N* HCl.
6. Elute the ^{2}H-ABA from the C18 column matrix with 5 mL ether (*see* Section 3.2., steps 9 and 10).
7. Dissolve the ^{2}H-ABA in a minimal amount of 100% methanol and quantify the exact concentration of the ^{2}H-ABA solution on HPLC-UV using an ABA standard solution. This ^{2}H-ABA is stable as long as no alkaline conditions are used.

3. Methods

3.1. Extraction

1. Add 80% methanol to 0.5–1 g tissue (9 mL/g fr wt) and ^{3}H (about 250 Bq) and/or ^{2}H (about 400 ng) tracers for recovery and localization purposes. Note the exact amount of each tracer added!
2. Extract by diffusion (for calli or soft tissue) or by homogenization (for most plant material) at 4°C for 2 h followed by –20°C for 16 h in the dark. A mixer or, for small amounts, a mortar and pestle, are convenient for homogenization.

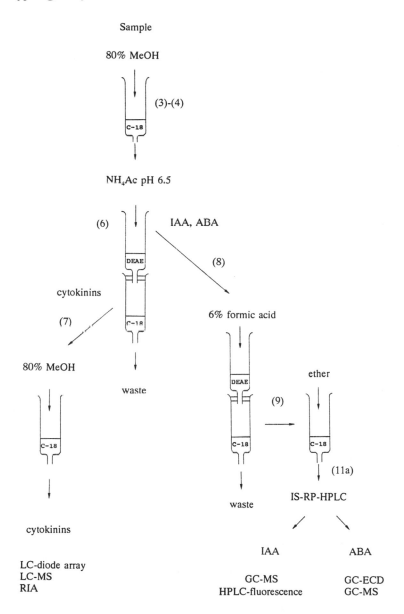

Fig. 1. Purification of IAA, ABA, and cytokinins using solid phase extraction. Numbering used in the protocol is given between brackets.

3.2. Purification

A flow diagram of the purification is shown in Fig. 1.

1. Remove cell debris by centrifugation: 24,000g, 4°C, 15 min, or by filtration.

2. Wash the pellet with an additional 10 mL of 80% methanol and pool, after centrifugation or filtration, with the sample. In order to quantify IAA conjugates, part of the sample should be hydrolyzed at this stage (*see* Note 3).
3. Purify the 80% methanol extract over a RP-C18 column.
4. Rinse the column after application of the sample with an additional 10 mL of 80% methanol and pool this 10 mL with the sample. The effluent contains IAA, ABA, and cytokinins (for cytokinins *see* Note 2).
5. Collect the eluate and evaporate *in vacuo* until all the methanol is removed. Dilute the remaining sample with 0.04M ammonium acetate pH 6.5 (*see* Notes 1 and 4).
6. Purify the sample using DEAE-Sephadex A25 (prerinsed with 100 mL ammonium acetate buffer 0.05M, pH 6.5) and concentrate the eluate on a directly underneath coupled Bond-Elut RP-C18.
   a. Rinse the columns after application of the samples with an additional 20 mL of buffer.
   b. Disconnect the C18-RP from the DEAE-Sephadex column.
      The effluent can be discarded, IAA and ABA are retained on the DEAE column matrix, and all cytokinins (free bases, ribosides, and glucosides) are retained on C18-RP.
7. Rinse the C18 column with 20 mL distilled water to remove the remaining buffer. Elute the cytokinins from the C18 column with 4 mL 80% methanol and evaporate in a speed-vac. Rinse the DEAE column with an additional 20 mL buffer and couple a Bond-Elut RP-C18 column underneath the DEAE-Sephadex.
8. Elute IAA and ABA from the DEAE-Sephadex with 20 mL 6% formic acid. The effluent of the C18 can be discarded. IAA and ABA are retained on the C18 matrix.
9. Elute IAA and ABA from the RP-C18 column with 5 mL diethyl ether. Discard the water fraction underneath the ether phase.
10. Evaporate *in vacuo*, dissolve in 500 μL methanol, and transfer to an Eppendorf tube. Rinse the vial with an additional 500 μL and pool with the sample. Dry in a speed-vac. If GC-MS will be used for the analysis of both IAA and ABA, then methylate the sample at this stage (*see* step 11) prior to GC-MS. If HPLC-fluorescence will be used for IAA analysis or GC-electron capture detection (ECD) is chosen for analysis of ABA, separation of IAA and ABA is necessary (*see* step 12).
11. Dissolve the dry sample in 100 μL acidified methanol. Take care that the vial, which contains the sample, is resistant to ether, i.e., glass.
    a. Methylate IAA and ABA by adding excess diazomethane until a slight yellow color remains for 10 min.

    b. Concentrate the sample under a nitrogen stream to a final volume of ±10 µL. Never dry methylated IAA or ABA *in vacuo*; this will evaporate the methylated compound.

12. Purify and separate IAA and ABA using a preparative HPLC run. Preparative ion suppression reversed phase (IS-RP) HPLC:
    a. Liquid phase: 50/49.5/0.5 (v/v/v; methanol/H_2O/acetic acid).
    b. Stationary phase: Alltech Econosphere C18, 100 × 4.6 mm, 3 µm.
    c. Flow: isocratic, 0.5 mL/min.
    d. Detection: on-line fluorescence detector Shimadzu RF530 or RF10A, $\lambda_{ex:}$ 285 nm, $\lambda_{em:}$ 360 nm.
    e. Collect 1-min fractions during the time interval corresponding with the retention time and the radioactivity of an IAA- and ABA-reference. Count 1/10 or 1/20 of the collected fractions for ^3H of the internal standard.
    f. Dry the remaining in a speed-vac to analyze using ion pairing (IP) HPLC for the IAA fraction(s) and for methylation of the ABA fraction(s).

13. Dissolve the dried ABA-fraction in 100 µL acidified methanol and methylate, as described in step 11, prior to gas chromatography (GC-MS or GC-ECD).

3.3. Analysis

3.3.1. Cytokinins

3.3.1.1. HPLC-DIODE ARRAY

Analysis of UV spectra makes it possible to use spectophotometry for quantification and identification of cytokinins with several advantages. The UV detector has a broad area of linear response to the amount of material, and thus higher precision. The HPLC eluate is sampled many times per second, giving much better resolution of closely eluted peaks. Since the UV light is nondestructive, the cytokinins can be collected for further analysis.

1. Liquid phase: Solvent A: 10% (v/v) methanol, 40 m*M* acetic acid, adjusted to pH 3.35 with triethylamine. Solvent B: 80% methanol in 40 m*M* HAc. Solvent C: 100% MeOH.
2. Gradient: 0–20 min from 90–50% A; 20–30 min 40% A; 30–31 min. 100% C; 31–38 min 100% C; 38–40 min. 90% A *(17)*.
    a. Stationary phase : Alltech Altima C18 150 x 4.6 mm, 5 µm.
    b. Flow: 1.0 mL/min.
    c. Detection: photodiode array detector, sampling time 61 ms × 4, 220–320 nm (1.3 nm resolution).

Figure 2 shows a typical cytokinin-specific spectrum used as an identification criterion.

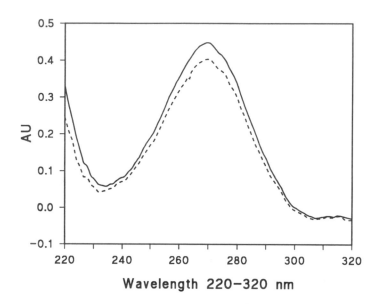

Fig. 2. Cytokinin-specific spectrum obtained with diode array from 220–320 nm. cis (dashed line) and trans (full line) zeatin-riboside are used as examples. AU = absorbance unit.

3.3.1.2. RIA

Analysis by RIA is favored because of its high sensitivity, high specificity, and short analysis time. A high amount of rather crude samples (*see* Note 2) can be handled in parallel. Different crossreactivities of the cytokinin antibodies available toward the individual different cytokinin-conjugates, present in the plant extract, are, however, a severe disadvantage. Therefore, data are expressed as cytokinin-equivalents. The use of a preparative off line HPLC run (*see* Section 3.3.1.1.) provides separation of the different cytokinins prior to RIA.

RIAs are performed as described by Weiler *(4)* after slight modifications. Standards and samples were tested in triplicate. Plant samples were analyzed at 10 different dilutions using zeatin-riboside (ZR)-, isopentenyl-adenosine (iPA)-, or dihydrozeatin-riboside (diH)ZR-specific antibodies in a RIA. The immunoreaction takes place in Eppendorf tubes.

1. Pipet in each Eppendorf tube:
    a. 50 µL ^{3}H-cytokinin-dilution (±1 kBq = 0.7 pmol/50 µL); and
    b. 100 µL Blanco-dilution (10 nmol cytokinin/100 µL) (blanco); or
    c. 100 µL TBS (B_0); or

    d. 100 μL Sample; or

    e. 100 μL Known antigen concentration (calibration curve 0.05–500 pmol).

2. Add to each tube 100 μL antibody solution and shake for 1 min.
3. Incubate for 1 h at room temperature to obtain binding equilibrium.
4. Add to each tube 100 μL bovine serum albumin solution (0.5 mg/100 μL distilled water) and 1 mL saturated ammonium sulfate solution (pH 7) to precipitate all proteins.
5. Incubate for 1 h on ice.
6. Pellet all proteins by centrifugation at 13,000g for 5 min and aspirate the supernatants off.
7. Dissolve the pellet in 100 μL distilled water.
8. Add 1.2 mL scintillation fluid and insert the Eppendorf tubes in scintillation vials.
9. Count each vial using a liquid scintillation counter.

Standard curves are plotted after a logit transformation *(18):*

$$\log\left(\left[(B - Bl)/(B_0 - Bl)\right]/\left\{1 - \left[(B - Bl)/(B_0 - Bl)\right]\right\}\right) = a + b \log c \quad (1)$$

where B_0 = maximal binding capacity (dpm) of the radioactive ligand at these experimental conditions; Bl = aspecific binding (dpm) of the radioactive ligand in presence of excess (10 nmol) antigen; B = binding capacity (dpm) of the radioactive ligand in presence of a known standard concentration or sample; and c = antigen concentration.

The detection limit is 0.05 pmol for cytokinin-ribosides or 1 pmol for free cytokinins. When free bases, cytokinin-ribosides, or cytokinin-glucosides are not separated on HPLC prior to RIA (*see* Section 3.3.1.1.), the results are expressed as cytokinin-riboside equivalents.

3.3.1.3. HPLC-MS

Quantification can be performed with the mass spectrometer in scan— or in selected ion monitoring (SIM) mode. Scan mode is essential if mass spectral information is required. In case the amount of cytokinins to be analyzed in the sample is too low for scan mode, SIM mode, using specific diagnostic ions for the appropriate cytokinins, is a less informative but more sensitive alternative.

Since a careful tuning with the appropriate solvent composition is necessary to obtain an acceptable detection limit for all cytokinin-riboside conjugates, gradient runs for MS should be avoided. This implies that zeatin- and dihydrozeatin-type cytokinins on the one hand and isopentenyladenine-type cytokinins on the other hand are analyzed on separate HPLC runs.

1. Zeatin- and dihydrozeatin-type cytokinins: The following HPLC conditions provide separation of all zeatin-type (zeatin, zeatin-riboside, zeatin-O-glucoside, zeatin-N_7-glucoside, zeatin-N_9-glucoside) and dihydrozeatin-type (dihydrozeatin, dihydrozeatin-riboside, dihydrozeatin-O-glucoside, dihydrozeatin-N_7-glucoside, dihydrozeatin-N_9-glucoside) cytokinins.

   a. Liquid phase: 35/65 (v/v; methanol/0.1M ammonium acetate).
   b. Stationary phase: Superspher RpSelect B C18 (Merck), 250×4 mm, 5 μm.
   c. Flow: Isocratic, 0.8 mL/min.
   d. Injection volume : 5–35 μL.
   e. Detection: On-line UV (260 nm) and MS using thermospray conditions (TS$^+$).
   f. Source temperature: 250°C; capillary temperature: 220–225°C; Repeller 200 V; selective ion response (SIR) dwell time 0.1 s; Span 0.5 Amu.
   g. The selective diagnostic ion mass for Z ([M + H]$^+$), ZR ([M − ribose + H]$^+$), Z-O-glu ([M − glucose + H]$^+$), Z-N$_7$-glu ([M − glucose + H]$^+$), and Z-N$_9$-glu ([M − glucose + H]$^+$) is 220 (*see* Fig. 3A, B).
   h. The selective diagnostic ion mass for ZR ([M + H]$^+$) is 352 (*see* Fig. 3B).
   i. The selective diagnostic ion mass for (diH)Z ([M + H]$^+$), (diH)ZR ([M − ribose + H]$^+$), (diH)Z-O-glu ([M − glucose + H]$^+$), (diH)Z-N$_7$-glu ([M − glucose + H]$^+$), and (diH)Z-N$_9$-glu ([M − glucose + H]$^+$) is 222 (*see* Fig. 4A, B).
   j. The selective diagnostic ion mass for (diH)ZR ([M + H]$^+$) is 354 (*see* Fig. 4B).
   k. The selective diagnostic ion for all ^{2}H-tracers is 225 (*see* Table 2).
   l. Typical retention times (min): ZR 7.76, (diH)ZR 8.82, Z 9.90, (diH)Z 11.8, Z-N$_9$-glu 3.4, Z-O-glu 4.2, (diH)Z-N$_9$-glu 5.4, and (diH)Z-O-glu 3.7.
   Specific thermo spray mass spectra of zeatin (*see* Fig. 3A), zeatin-riboside (*see* Fig. 3B), dihydrozeatin (*see* Fig. 4A), and dihydrozeatin-riboside (*see* Fig. 4B) are shown in Figs. 3 and 4.

2. Isopentenyladenine-type cytokinins: The following HPLC conditions provides separation of all isopentenyl-type cytokinins (isopentenyl-adenine [iP] and isopentenyl-adenosine [iPA]).

   a. Liquid phase: 65/35 (v/v; methanol/0.1M ammonium acetate).
   b. Other conditions are as described for Z and (diH)Z type cytokinins (*see* Section 3.3.1.3., step 1).
   c. The selective diagnostic ion mass for iP ([M + H]$^+$) and iPA ([M − ribose + H]$^+$) is 204 (*see* Fig. 5A,B).
   d. The selective diagnostic ion mass for iPA ([M + H]$^+$) is 336 (*see* Fig. 5B).
   e. The selective diagnostic ion mass for all ^{2}H-tracers is 210 (*see* Table 2).
   f. Typical retention times (min): iPA 4.9 and iP 6.3.

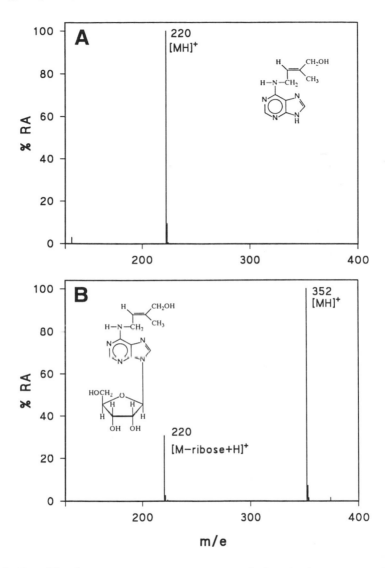

Fig. 3. Specific thermospray mass spectra and chemical structure of zeatin (**A**) and zeatin-riboside (**B**).

Specific thermospray mass spectra of isopentenyl adenine and isopentenyl-adenosine are shown in Figs. 5A and 5B, respectively.

Endogenous concentrations of the different cytokinins are calculated from the data obtained using the following equation:

$$(area_X/area_Y) = (x/y) \qquad (2)$$

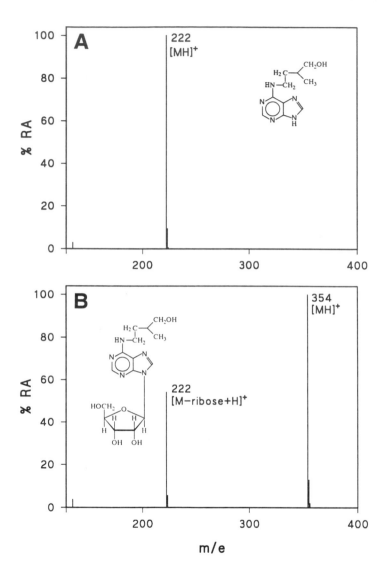

Fig. 4. Specific thermospray mass spectra and chemical structure of dihydrozeatin (**A**) and dihydrozeatin-riboside (**B**).

where $area_X$ = peak integration value for the specific diagnostic ion for the nonlabeled compound; $area_Y$ = peak integration value for the specific diagnostic ion for the ^{2}H-labeled compound; x = endogenous content of the compound initially present in the tissue analyzed; and y = amount of ^{2}H-labeled compound initially added to the extract before purification.

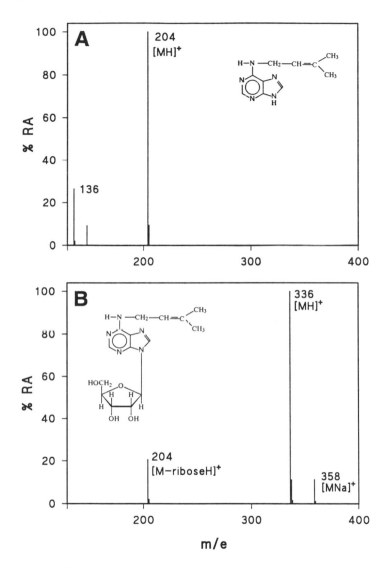

Fig. 5. Specific thermospray mass spectra and chemical structure of isopentenyl adenine (**A**) and isopentenyl-adenosine (**B**).

3.3.2. Indole-3-Acetic Acid

3.3.2.1. HPLC-FLUORESCENCE (12)

Depending on the type of fluorometer used, HPLC-fluorescence is an extremely sensitive method for IAA analysis. In case only a little plant material is available (<100 mg) or in case alkaline hydrolysis is used (*see* Note 3), HPLC-fluorescence is preferred for analysis.

1. Ion pairing reversed phase (IP-RP) HPLC:
   a. Liquid phase: 40/60 (v/v; methanol/1 mM KH$_2$PO$_4$) K$_2$HPO$_4$, 10 mM tetra butyl ammonium hydroxide, pH 6.6.
   b. Stationary phase: Alltech Adsorbosphere C18, 100 × 4.6 mm, 3 μm or Merck Lichrospher 100 RP-18, 100 × 4.6 mm, 5 μm.
   c. Flow: isocratic, 0.5 mL/min.
   d. Detection: on-line fluorescence detector, Shimadzu RF551 λ_{ex}: 285 nm, λ_{em}: 360 nm.
   e. Use ^{14}C-IAA (10–100 Bq) for calibration.
2. Integration and calculation: The specific fluorescence of the IAA is calculated as follows:

$$SF = (dpm_{peak}/SR_{14C\text{-}IAA})/H_{peak} \tag{3}$$

where dpm_{peak} = amount of dpm present in the ^{14}C-IAA peak; SR ^{SR14}C-IAA = specific radioactivity of the ^{14}C-IAA (cf. batch technical data); and H_{peak} = height of the IAA peak.

3. The amount of endogenous IAA extracted from the tissue (y) is calculated following the principles of Isotope Dilution:

$$y = H_{peak} \times (dpm_i/dpm_{peak}) \times SF - (dpm_i/SR_{3H\text{-}IAA}) \tag{4}$$

where H_{peak} = height of the IAA peak in the sample; dpm_i = initial dpm applied to the sample as internal standard; dpm_{peak} = amount of dpm present in the IAA peak of the sample; and $_{3H\text{-}IAA}$ = specific radioactivity of the ^{3}H-IAA (cf. batch technical data) initially applied to the sample as internal standard.

The detection limit of this method is 10 fmol IAA injected.

3.3.2.2. GC-MS

The common derivatization procedure for IAA and ABA followed by a good separation of the methylated IAA and methylated ABA on GC allows us to analyze both phytohormones using one single GC run omitting the more time-consuming preparative HPLC run (*see* Section 3.2., step 12).

1. Gas phase: He.
2. Stationary phase: 15 m DB1, 0.25 mm ID, Df = 0.25 μm.
3. Flow: 1 mL/min.
4. Temperatures: Temperature gradient from 120–225°C, 15°C/min, $T_{inj.}$= 250°C.
5. Detection: Electron impact MS using SIM or scan mode.

6. Injection: Splitless, 1–2 μL.
7. The selective diagnostic ion masses for IAA and ^{2}H-IAA are 130 and 135, respectively.
8. Tuning: Inject a methylated ^{2}H-IAA standard solution to determine the 135/130 ratio.

If the sample is analyzed by GC-MS at point 10, both IAA- and ABA-methyl esters can be monitored in the same run at the methylated IAA- and methylated ABA-specific retention times (4.6 and 6.5 min, respectively).

Endogenous IAA concentrations are calculated using Eq. [2]. Correct for the 135/130 ratio of the ^{2}H-IAA tracer if necessary. The detection limit of this method is 1 pmol IAA injected.

3.3.3. Abscisic Acid

3.3.3.1. GC-ECD *(12)*

ECD is more sensitive, however, less specific than mass spectrometric detection using selective ion monitoring. This implicates, in comparison to GC-MS, the necessity of two extra preparative HPLC runs (*see* Section 3.2., step 12, and step 1 below). GC coupled to ECD is therefore preferred only in case less than 10 mg starting plant material is available.

1. The methylated ABA is separated from the nonmethylated ABA using an RP-HPLC run (50/50 [v/v; methanol/H$_2$O], same column and flow as for ion suppression [IS]-HPLC) (*see* Section 3.2., step 12).
2. The ^{3}H-methyl-ABA-containing fractions are collected and evaporated under a nitrogen stream.
3. GC:
   a. Gas phase: Argon-methane Ar/CH$_4$; 90/10.
   b. Stationary phase: Capillary OV 101 WCOT, 25 m, 0.25 mm ID.
   c. Flow: Through column 1 mL/min; through detector 30 mL/min.
   d. Temperatures: Column 220°C, Injection 280°C, Detection 280°C.
   e. Injection: Solid phase injector.
   f. Detection: ECD.
4. Use a methylated ^{3}H-ABA standard solution (about 300 dpm) for calibration. Keep the solution on ice and concentrate the solution by evaporation of the solvent.
5. Count a known amount of the standard solution and samples to know the exact amount injected. cis- and trans-ABA retention times are identified on GC before and after UV illumination resulting in a cis–trans conversion.
6. Concentrations are calculated using Eqs. (3) and (4).

The detection limit of this method is 2 fmol ABA injected.

3.3.3.2. GC-MS *(19)*

1. Use the same conditions as described in Section 3.3.2.2.
2. The selective diagnostic ion masses for ABA and $^2H_6$-ABA are 190 and 194, respectively *(20)*. Tuning: Inject a methylated ^2H-ABA standard solution to determine the 194/190 ratio.
3. Endogenous ABA concentrations are calculated using Eq. (2).

The detection limit of this method is 1 pmol ABA injected.

4. Notes

1. Before purification on DEAE, the pH should be controlled for each individual sample, and adjusted to pH 6.5 if necessary.
2. In case only cytokinins will be measured and RIA is used for analysis, only a crude purification is required. In this case, all purification steps after step 4 in Section 3.2. can be omitted.
3. To quantify IAA conjugates, a known fraction of the sample can be hydrolyzed prior to purification (Section 3.2., step 3).
   a. IAA amino acid conjugates are hydrolyzed in 7N NaOH at 100°C for 3 h under a water-saturated nitrogen stream *(21)*.
   b. IAA ester conjugates are hydrolyzed in milder conditions; 1 h 1N NaOH at room temperature *(22)* under a water-saturated nitrogen stream.
   c. Cool down in ice.
   d. Dilute fourfold with distilled water.
   e. Acidify the sample with 2N HCl to pH 2.
   f. Desalt using a C18-column in acid conditions (*see* Section 2., step 7).
   g. Continue purification from step 6 in Section 3.2. ^2H-IAA is partially exchanged for ^1H-IAA during alkaline hydrolysis. Therefore in case IAA-conjugates are to be measured, HPLC-fluorescence detection or MS using other "heavy" tracers, has to be preferred. The amount of free IAA, analyzed in half of the sample, should be subtracted from the amount of total IAA (after hydrolysis), to obtain the total amount of IAA conjugates present in the sample.
4. If cytokinin ribotides will be measured, a known fraction of the sample can be treated after step 5 in Section 3.2. with acid phosphatase (1 mg/sample) for 30 min in the dark at 25°C. Otherwise they will be coupled to the DEAE. Continue the procedure from step 6 in section 3.2.

References

1. Rivier, L. and Crozier, A. (1987) *Principles and Practice of Plant Hormone Analysis*, vols. 1 and 2, Academic, London.

2. Takahashi, N. (1986) *Chemistry of Plant Hormones*, CRC, Boca Raton, FL.
3. Hedden, P. (1993) Modern methods for the quantitative analysis of plant hormones. *Annu. Rev. Plant Physiol. Mol. Biol.* **44**, 107–129.
4. Weiler, E. W. (1980) Radioimmunoassay for *trans*-zeatin and related cytokinins. *Planta* **149**, 155–162.
5. Weiler, E. W. (1984) Immunoassay of plant growth regulators. *Annu. Rev. Plant Physiol.* **35**, 85–95.
6. Eberle, J., Arnscheidt, A., Klix, D., and Weiler, E. W. (1986) Monoclonal antibodies to plant growth regulators III. Zeatinriboside and dihydrozeatin-riboside. *Plant Physiol.* **81**, 516–521.
7. Khan, S. A., Humayun, M. Z., and Jacob, T. M. (1977) A sensitive radioimmunoassay for isopentenyl-adenosine. *Anal. Biochem.* **83**, 632–635.
8. Imbault, N., Moritz, T., Nilsson, O., Chen, H.-J., Bollmark, M., and Sandberg, G. (1993) Separation and identification of cytokinins using combined capillary liquid chromatography/mass spectrometry. *Biol. Mass Spec.* **22**, 201–210.
9. Horgan, R. and Scott, I. M. (1987) Cytokinins, in *Principles and Practice of Plant Hormone Analysis*, vol 2, Academic, London, pp. 303–365.
10. Hogge, L. R., Abrams, G. D., Abrams, S. R., Thibault, P., and Pleasance, S. (1992) Characterization of abscisic acid and metabolites by combined liquid chromatography-mass spectrometry with ion-spray and plasma-spray ionization techniques. *J. Chromatogr.* **623**, 255–263.
11. Milborrow, B. V. (1971) Abscisic acid, in *Aspects of Terpenoid Chemistry and Biochemistry* (Goodwin, T. W., ed.), Academic, London, pp. 137–151.
12. Prinsen, E., Rüdelsheim, P., and Van Onckelen, H. (1991) Extraction, purification and analysis of endogenous indole-3-acetic acid and abscisic acid, in *A Laboratory Guide for Cellular and Molecular Plant Biology* (Negrutiu, I. and Gharti-Chhetri, G., eds.), Birkhäuser Verlag, Basel, Switzerland, pp. 175–185, 323–324.
13. Schlenk, H. and Gellerman, J. L. (1960) Esterification of fatty acids with diazomethane on a small scale. *Anal. Chem.* **32**, 1412–1414.
14. Bialek, K. and Cohen, J. D. (1986) $^{13}C_6$-[Benzene-ring]indole-3-acetic acid. *Plant Physiol.* **80**, 14–19.
15. Zeevaart, J. A. D., Heath, T. G., and Gage, D. A. (1989) Evidence for a universal pathway of abscisic acid biosynthesis in higher plants from ^{18}O incorporation patterns. *Plant Physiol.* **91**, 1594–1601.
16. Koshimizu, K. and Iwamura, H. (1986) Cytokinins, in *Chemistry of Plant Hormones* (Takahashi, N., ed.), CRC, Boca Raton, FL, pp. 153–199.
17. Strnad, M., Vaněk, T., Binarová, P., Kamínek, M., and Hanuš, J. (1990) Enzyme immunoassays for cytokinins and their use for immunodetection of cytokinins in alfalfa cell cultures, in *Molecular Aspects of Hormonal Regulation of Plant Development* (Kutáček, M., Elliott. M. C., and Macháčková, I., eds.), SPB Academic Publishers, The Hague, The Netherlands, pp. 41–54.
18. Rodbard, D. (1974) Statistical quality control and routine data processing for radioimmunoassays and immunoradiometric assays. *Clin. Chem.* **20(10)**, 1255–1270.

19. Rivier, L., Milon, H., and Pilet, P.-E. (1977) Gas chromatography-mass spectrometric determinations of abscisic acid levels in the cap and the apex of maize roots. *Planta* **134,** 23–27.
20. Gray, R. T., Mallaby, R., Ryback, G., and Williams, V. P. (1974) Mass spectra of methyl abscisate and isotopically labelled analogues. *J. C. S. Perkin II,* 919–924.
21. Bialek, K. and Cohen, J. D. (1989) Quantization of indoleacetic acid conjugates in bean seeds by direct tissue hydrolysis. *Plant Physiol.* **90,** 398–400.
22. Bialek, K. and Cohen, J. D. (1986) Isolation and partial characterization of the major amide-linked conjugate of indole-3-acetic acid from *Phaseolus vulgaris* L. *Plant Physiol.* **80,** 99–104.

CHAPTER 24

Manipulating Photosynthesis in Transgenic Plants

Jacqueline S. Knight, Francisco Madueño, Simon A. Barnes, and John C. Gray

1. Introduction

Agrobacterium tumefaciens-mediated transformation of plants provides a relatively straightforward means of altering the amounts of individual proteins involved in photosynthetic processes. By the introduction of additional copies of genes or of chimeric cDNA constructs, it is possible to produce transgenic plants with increased or decreased amounts of specific photosynthetic components. These plants can then be analyzed to examine the effects of the altered protein levels on a wide range of physiological or biochemical processes.

The introduction of additional copies of genes normally will result in consequential increases in the levels of specific mRNA and protein. This has been demonstrated for photosynthetic components by the transformation of tobacco with genes encoding plastocyanin and ferredoxin-NADP+ oxidoreductase, resulting in transgenic plants containing increased amounts of these electron transfer proteins *(1,2)*. Occasionally, however, the introduction of additional copies of genes may result in decreased amounts of the protein being produced *(3,4)*. This phenomenon, which has not yet been observed with genes for photosynthetic components, is termed cosuppression and may provide a means of altering protein levels in the future. The mechanism of cosuppression is not known.

From: *Methods in Molecular Biology, Vol. 44:* Agrobacterium *Protocols*
Edited by: K. M. A. Gartland and M. R. Davey Humana Press Inc., Totowa, NJ

The amounts of individual proteins in transgenic plants can be decreased more consistently using antisense methodology *(5,6)*. Antisense RNA forms part of a naturally occurring mechanism for gene regulation in bacteria and this methodology has been applied to photosynthetic genes in plant systems with considerable success *(7–9)*. The amounts of ribulose 1,5-bisphosphate carboxylase *(7,8)* and a 10-kDa photosystem II protein *(9)* have been decreased markedly by the introduction of antisense constructs into tobacco or potato plants. In this method, a chimeric construct consisting of the cDNA of the target protein placed in the reverse orientation behind a strong promoter is introduced into the plant genome. Transcription of this construct in the plant produces an antisense RNA complementary to the mRNA for the target protein. The synthesis of the antisense RNA usually results in a decrease in the amount of mRNA for the target protein and a consequent decrease in the synthesis of the target protein. The mechanism of action of antisense RNA in plants is not fully understood, but it provides an excellent system for decreasing the amount of specific individual proteins.

This chapter gives detailed protocols for the preparation of transgenic tobacco plants containing antisense constructs, although these procedures may also be used to produce plants containing cDNA in the sense orientation to increase amounts of specific proteins. Procedures for the introduction of a binary vector containing the antisense construct into *A. tumefaciens* by electroporation, subsequent transformation of tobacco leaf discs, and the screening and analysis of transgenic plants are given in this chapter. Transgenic tobacco plants containing antisense constructs for phosphoribulokinase and the chloroplast Rieske FeS protein are used to illustrate analysis of the plants by Northern and Western blots.

2. Materials

2.1. Transformation of **Agrobacterium** by Electroporation

1. Construction of antisense RNA gene: A cDNA homologous to the mRNA of the target protein is required for the production of antisense constructs. The cDNA is cloned in the reverse orientation into derivatives (pROK2, pROK8, *see* Fig. 1) of the binary vector pBIN19 *(10)*. pROK2 contains the strong, nominally constitutive, 35S promoter from cauliflower mosaic virus *(11)* and pROK8 contains the light-regulated promoter from a tobacco *rbc*S gene encoding the small subunit of ribulose-1, 5-bisphosphate car-

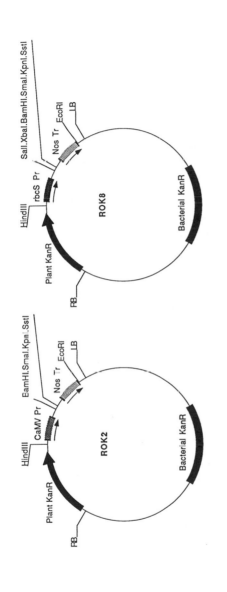

Fig. 1. A diagramatic representation of the binary vectors pROK2 and pROK8, showing the restriction enzyme sites available for construction of chimeric genes. The vectors are based on pBIN19 (*10*) and contain the 35S promoter from cauliflower mosaic virus (CaMV Pr) or the tobacco *rbcS* promoter (*rbcS* Pr) separated from the transcription terminator of the nopaline synthase gene (*nos* Tr) by a multiple cloning site. The T-DNA borders (RB and LB) enclose the chimeric gene and a neomycin phosphotransferase gene (Plant KanR), which provides a selectable marker in plants.

boxylase *(12)*. These pBIN19 derivatives contain the T-DNA borders required for *Agrobacterium*-mediated transfer of the antisense construct into tobacco, as well as the gene for neomycin phosphotransferase, which confers kanamycin resistance and enables selection of transformed plants.

2. Preparation of DNA for electroporation: Each electroporation experiment requires 10–100 ng of plasmid DNA prepared by the alkaline lysis method *(13)*.

3. Luria-Bertani (LB) medium: 10% (w/v) bactotryptone (Difco Laboratories, Detroit, MI), 5% (w/v) yeast extract (Oxoid, Basingstoke, UK), 10% (w/v) NaCl adjusted to pH 7.5 with 10*M* KOH and sterilized by autoclaving; LB agar consists of LB medium plus 1.5% (w/v) bacto agar (Difco).

4. Antibiotics: Filter-sterilized stock solutions of 10 mg/mL streptomycin sulfate and 25 mg/mL kanamycin sulfate.

5. Competent *Agrobacterium*: Cultures of *A. tumefaciens* LBA4404 are grown in 4 × 10 mL LB medium supplemented with 100 µg/mL streptomycin with shaking (200 rpm) at 30°C for 24–30 h. The cells are cooled on ice and pelleted by centrifugation in a benchtop centrifuge (700*g*, 5 min). The cells are washed successively with 40, 20, 0.8, and 0.4 mL of sterile 10% (v/v) glycerol and then resuspended in 0.4 mL 10% (v/v) glycerol. The cells are quick frozen (liquid nitrogen) in 40-µL aliquots and can be stored at –140°C for up to 6 wk.

6. Gene Pulser™ with pulse controller unit manufactured by Bio-Rad Laboratories Ltd. (Hemel Hempstead, UK) and sterile 0.2 cm Bio-Rad cuvets.

7. Incubate at 30°C for static and orbital incubations.

8. Benchtop centrifuge (MSE Scientific Instruments supplied by Fisons Scientific Equipment, Loughborough, UK).

2.2. Agrobacterium-*Mediated* Tobacco Leaf Disc Transformation

1. Stock solutions: α-naphthaleneacetic acid (NAA; 5 mg/mL); 6-benzylamino purine (BAP; 1 mg/mL in 0.1*M* HCl); 100X B$_5$ vitamins (1% [w/v] meso-inositol [inactive], 0.1% [w/v] thiamine-HCl, 0.01% [w/v] nicotinic acid, 0.01% [w/v] pyridoxine-HCl); 2,4-dichlorophenoxyacetic acid (2,4-D; 1 mg/mL in ethanol); kanamycin sulfate (25 mg/mL); carbenicillin (5 mg/mL).

2. Murashige and Skoog (MS) medium: 0.46% (w/v) MS salts (ICN Biomedicals Ltd., High Wycombe, UK) and 3% (w/v) sucrose adjusted to pH 5.9 with 10*M* KOH; MS agar, MS medium plus 1% (w/v) bacto agar.

3. NBM plates: MS agar with the addition of 20 µL/L NAA stock, 1 mL/L BAP stock, 10 mL/L 100X B$_5$ vitamins.

4. Cork borer (No. 6, 1-cm diameter); sterile filter discs 8.5-cm diameter (Whatman [Maidstone, UK] No. 1 filter paper). Both are sterilized by wrapping in aluminum foil, followed by autoclaving.

5. Tobacco leaf discs: Discs are prepared from 4-wk-old, healthy tobacco plants (*Nicotiana tabacum* var. Samsun) grown under sterile conditions in the plant growth room. To subculture the plants, take tobacco stem segments containing a node and internode from plants established in tissue culture and transfer to sterile screwcapped jars (60 mL) each containing 15 mL of fresh MS agar (1 plant/jar). Subculture plants every 1–2 mo.

6. Feeder plate suspension culture: This comprises tobacco *(Nicotiana benthamiana)* leaf callus dispersed in liquid culture by shaking at 120 rpm at 25°C. The culture should be subcultured weekly by placing a volume (*see* Note 6) of old suspension into 70 mL medium in a 250 mL conical flask. A 70 mL volume of medium contains 0.54 g MS salts, 2.1 g sucrose, 0.7 mL 100X B_5 vitamin stock, 70 μL 2,4-D stock, 14 μL BAP stock, and is adjusted to pH 5.9 with $10M$ KOH. The medium can be stored for up to 3 mo after autoclaving.

7. Transformed *Agrobacterium* suspension culture: Inoculate 10 mL of LB medium containing 50 μg/mL kanamycin and 100 μg/mL streptomycin with the transformed *Agrobacterium* strain and grow at 30°C with shaking (200 rpm) for 24–30 h. Pellet the cells in a benchtop centrifuge ($700g$, 5 min) and resuspend in 10 mL of MS medium.

8. Storage of transformed *Agrobacterium*: Transformed and nontransformed *Agrobacterium* can be stored either on minimal-T plates at 4°C for 1–2 mo, or as glycerol stocks at –20°C for up to 1 yr. For minimal-T plates, autoclave 3.2. g agar in 350 mL of water and add 20 mL 20X T salts, 20 mL 20X T buffer and 10 mL 20% (w/v) sucrose, plus the appropriate antibiotic (50 μg/mL kanamycin and/or 100 μg/mL streptomycin). The additives are sterilized separately. 20X T salts consist of 20 g NH_4Cl, 4 g $MgSO_4 \cdot 7H_2O$, 0.04 g $MnCl_2$, 0.2 g $CaCl_2$, 0.1 g $FeSO_4 \cdot 7H_2O$ in 1 L. 20X T buffer consists of 210 g K_2HPO_4 and 90 g KH_2PO_4 in 1 L. Glycerol stocks are made by the addition of 2 mL of sterile 60% (v/v) glycerol to 1 mL of culture.

9. Plant growth room at 25°C with a photosynthetically active irradiance of 80 μmol photons/m²/s on a 16-h photoperiod.

10. Laminar flow hood (supplied by Merck Ltd., Poole, UK).

11. Incubators: Orbital at 25, 30, and 37°C and static at 30 and 37°C.

2.3. Screening for Transgenic Plants with Altered mRNA and Protein Levels

2.3.1. Northern Blotting

1. All glassware, electrophoresis equipment, and solutions should be prepared to ensure that they are RNase free. All solutions are prepared by adding diethylpyrocarbonate (DEPC) to a final concentration of 0.1% (v/v) and

incubating at 37°C for 12 h. Solutions are then autoclaved for 15 min at 103 kPa. All glassware (including mortars and pestles) is treated by baking at 180°C for 8 h. Electrophoresis equipment is treated with a 10M NaOH for 60 min, followed by extensive rinsing with sterile distilled water.

2. RNA extraction buffer: 100 mM Tris-HCl, pH 8.5, 100 mM NaCl, 20 mM Na$_2$EDTA, 1% (w/v) N-lauroyl sarcosine (Sarcosyl), 1.7% (w/v) diethyl-dithiocarbamate (add just prior to use).

3. Phenol/chloroform: Equal volumes of phenol and chloroform equilibrated with 50 mM Tris-HCl, pH 7.5.

4. Absolute ethanol and 70% (v/v) ethanol.

5. 3M Sodium acetate, pH 6.0.

6. DEPC-treated distilled water.

7. Pestles and mortars.

8. Microcentrifuge (MSE supplied by Fisons).

9. Spectrophotometer capable of measurements at 260 and 280 nm.

10. Stock 5X running buffer for electrophoresis: 0.2M 4-morpholine pro-panesulfonic acid (MOPS), 50 mM sodium acetate, 5 mM Na$_2$EDTA, pH 8.3.

11. Gel for electrophoresis: 1.2 g agarose (Sigma, Poole, UK, Molecular Biology Grade A 9539), 20 mL 5X running buffer, 74 mL H$_2$O. Heat to solubilize the agarose, cool to 50°C and add 5.4 mL 37% (v/v) formaldehyde. Make up to 100 mL with H$_2$O and pour into a 10 × 15 cm gel former.

12. Sample buffer for electrophoresis: 48% (v/v) formamide, 1X running buffer, 6% (v/v) formaldehyde, 5% (v/v) glycerol, 0.001% (w/v) bromophenol blue.

13. 25 mM Sodium phosphate buffer, pH 6.4.

14. Equipment for Northern blot as described by Sambrook et al. *(13)*.

15. Nylon membrane (Hybond N™, Amersham International Ltd., Amersham, UK).

16. Hybridization reagents: 20X SSPE (3M NaCl, 0.18M NaH$_2$PO$_4$, 20 mM Na$_2$EDTA adjusted to pH 7.4 with 10M NaOH); 20X SSC (3M NaCl, 0.3M trisodium citrate); 10% (w/v) SDS; 100X Denhardt's solution (2% [w/v] Ficoll [M_r 400,000], 2% [w/v] polyvinylpyrrolidone [M_r 300,000], 2% [w/v] bovine serum albumin [BSA] [Fraction V]); dextran sulfate; fish milt DNA (10 mg/mL), sheared by vortexing for 10 min.

17. Radiolabeled probe synthesized and purified using the Prime-it™ II Random Primer Labeling Kit and Nuctrap™ Push Columns as outlined by the supplier (Stratagene Ltd., Cambridge, UK). The probe is synthesized using 10 ng DNA/5 µCi [α-^{32}P]dATP/mL hybridization mix.

18. Hybridization oven and hybridization vessels (Hybaid, Teddington, UK).

19. Light source: UV short wavelength (302 nm).

20. X-ray film (NIF RX, Fuji Photo Film Co. Ltd. supplied by KP Professional Ltd., Cambridge, UK).

2.3.2. Western Blotting

1. Sand (Sigma).
2. 50 mM Tris-HCl, pH 7.5.
3. Sodium dodecyl sulfate (SDS) protein loading buffer: 100 mM Tris-HCl, pH 7.5, 10% (w/v) glycerol, 10% (w/v) SDS, 0.002% (w/v) bromophenol blue, 5% (v/v) 2-mercaptoethanol. If to be stored, 2-mercaptoethanol is added just prior to use.
4. Protein electrophoresis: Proteins are electrophoresed using the Laemmli discontinuous gel system *(14)* and the Bio-Rad Mini-protean II™ electrophoresis apparatus.
5. Nitrocellulose filters (Schleicher and Schuell, BA83, pore size 0.2 μm, supplied by Orme Scientific Ltd., Manchester, UK).
6. Blotting of proteins: Use a semidry blotter model AE-6670 Horizblot (Atto Co., supplied by Genetic Research, Instrumentation Ltd., Dunmow, UK).
7. Transfer buffer for Western blotting: 25 mM Tris, 192 mM glycine, 20% (v/v) methanol, 0.1% (w/v) SDS.
8. Tris-buffered saline (TBS): 20 mM Tris-HCl, pH 7.5, 0.5M NaCl.
9. 2% BSA-TBS: 2% (w/v) BSA (Sigma) in TBS.
10. 0.1% NP40-TBS: 0.1% (v/v) Nonidet P40 (NP40, BDH Chemicals, Poole, UK) in TBS.
11. ^{125}I-labeled protein A (Amersham).
12. X-ray film (NIF RX, Fuji Photo Film Co. Ltd.).

3. Methods

3.1. Transformation of Agrobacterium by Electroporation

Antisense RNA constructs are introduced into *A. tumefaciens* by electroporation using the method of Shen and Forde *(15)*. Transformants are selected by their resistance to kanamycin.

1. Thaw the competent *Agrobacterium* suspension on ice and add the binary vector containing the antisense RNA construct (10–100 ng DNA).
2. Pipet the *Agrobacterium*-DNA mixture into an ice-cooled 0.2-cm cuvet using a cutoff Gilson tip. Ensure that the liquid is at the bottom of the cuvet by tapping it gently.
3. Pulse the sample with the Gene-Pulser™ set to give a potential difference of 2.5 kV, a resistance of 400 Ω, and a capacitance of 25 μF.
4. After the pulse has been applied wash the sample out of the cuvet and into an Eppendorf tube with 1 mL of LB medium and incubate at 30°C with shaking (200 rpm) for 3 h.

5. Pellet the *Agrobacterium* in a microfuge at 6500*g* for 5 min and resuspend in 100 µL of LB medium.
6. Spread this preparation on LB agar plates containing 50 µg/mL kanamycin and 100 µg/mL streptomycin and incubate at 30°C for 3 d.

To ensure the correct construct has been introduced into the *Agrobacterium*, a crude plasmid preparation is performed. Due to the low yield and difficulty of restriction enzyme digests with *Agrobacterium* plasmid DNA, this preparation is then used to transform *Esherichia coli*.

3.2. Transformation of E. coli

1. Grow individual *Agrobacterium* transformants in 10 mL of LB medium containing 50 µg/mL kanamycin and 100 µg/mL streptomycin at 30°C for 24–30 h with shaking (200 rpm).
2. Remove clumps of *Agrobacterium* cells in 1.5 mL of culture to an Eppendorf tube and pellet the cells at 6500*g* for 5 min. Discard the supernatant, add an additional 1.5 mL of culture, and centrifuge once again, discarding the supernatant.
3. Use the minipreparation procedure outlined by Sambrook et al. *(13)* to prepare crude *Agrobacterium* plasmid DNA from these cells. After the first ethanol precipitation, resuspend the pellet in 20 µL of 10 m*M* Tris-Cl, pH 8.0, and 1 m*M* Na₂EDTA and use this to transform competent *E. coli* cells *(13)*. Select transformants on LB agar plates containing 25 µg/mL kanamycin.
4. Plasmid DNA can be prepared from these transformants by the standard procedure *(13)* and analyzed by restriction enzyme digests.

3.3. Agrobacterium-*Mediated* Tobacco Leaf Disc Transformation

The antisense gene can be transferred to the tobacco genome using the *Agrobacterium*-mediated leaf disc transformation method *(16)*. The following manipulations are all carried out under sterile conditions.

3.3.1. Day 1

1. Prepare feeder plates by spreading 1 mL of the tobacco cell suspension culture (Section 2.2.6.) evenly onto 9-cm NBM agar plates (minus antibiotics) and cover with a sterile filter disc.
2. Prepare leaf discs by dissecting healthy, fully expanded leaves from stock tobacco plants. Place these in a Petri dish and cut leaf discs using a flame-sterilized cork borer.
3. Transfer the leaf discs to the *Agrobacterium* suspension (Section 2.2.7.) in a 9-cm Petri dish. Shake gently to ensure the cut edges are wetted by the suspension. Incubate for 30 min.

4. Transfer the leaf discs to the feeder plates (5 discs/9-cm plate) with the lower epidermis uppermost. Seal the plate with Nescofilm (Nippon Shogi Kaisha Ltd. supplied by Orme) and place in the growth room for 2 d.

3.3.2. Day 3

5. Transfer the leaf discs to NBM plates containing 100 µg/mL kanamycin and 500 µg/mL carbenicillin and return to the growth room.

3.3.3. Week 3

6. To induce roots, dissect shoots from the discs avoiding callus material and insert in MS agar containing 100 µg/mL kanamycin and 200 µg/mL carbenicillin. Place all the shoots from one leaf disc on one 9 cm Petri dish.

3.3.4. Week 5

7. After roots have appeared, transfer individual plants to screwcapped jars (60 mL) containing 15 mL MS agar and 100 µg/mL kanamycin and 200 µg/mL carbenicillin. Plants can be maintained on MS agar containing antibiotics by subculturing as described in Section 2.2.5.
8. When plants are approx 5 cm high they can be transferred to the glass house. Plants are removed from the tissue-culture vessels, the agar is washed off their roots, and the plants potted in damp compost (Levington M3, Fisons, Ipswich, UK). The potted plants are placed in a plant propagator and acclimatized to glasshouse conditions over a 2-wk period.

3.4. Screening for Transgenic Plants with Altered mRNA and Protein Levels

3.4.1. Northern Blotting

Analysis of transgenic plants by Northern blotting provides a rapid screening procedure for plants with altered expression of the target gene. Those plants that show changes in mRNA levels may then be selected for further analysis by Western blotting or enzyme assay. A rapid, small-scale method for the extraction of total nucleic acid from young tobacco leaves is described in the following. This procedure is modified from that of Apel and Kloppstech *(17)*. The majority of nucleic acids extracted by this method is RNA *(see* Fig. 2A).

1. Freeze approx 0.3 g of leaf tissue in liquid nitrogen and homogenize in a mortar (precooled with liquid nitrogen) with 1 mL of extraction buffer.
2. Incubate the homogenate at room temperature for 15 min.
3. Transfer the homogenate to a 2-mL Eppendorf tube and extract three times with 0.75 mL phenol/chloroform. Shake for 15 min per extraction and centrifuge at 10,000g for 5 min in a microfuge after each extraction.

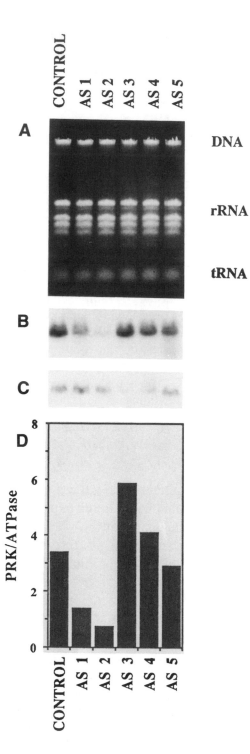

4. Add 2 vol of absolute ethanol (–20°C) and 50 μL 3*M* sodium acetate to the upper aqueous phase; place at –20°C for 3 h.
5. Centrifuge at 10,000*g* for 10 min in a microfuge, remove the supernatant, and resuspend the pellet in 300 μL H_2O.
6. Add 750 μL ethanol (–20°C) followed by 30 μL 3*M* sodium acetate, and place at –20°C overnight.
7. Centrifuge at 10,000*g* for 10 min in a microfuge and remove the supernatant.
8. Wash the pellet with 70% ethanol and resuspend in 50 μL H_2O.

The yield of nucleic acids prepared from each tissue sample may now be estimated spectrophotometrically by the absorbance at 260 nm (RNA at a concentration of 40 μg/mL has an absorbance of 1). Contamination of the sample with protein or phenol may be estimated by assessing the ratio of absorbance at wavelengths of 260 and 280 nm. This should be between 1.8 and 2.0 *(13)*. Sample concentrations should be adjusted to 1 mg RNA/mL. The quantity and quality (amount of degradation) of RNA can also be assessed by electrophoresing a sample in a TBE-agarose gel *(see* Fig. 2A). The samples may now be analyzed by Northern blotting.

1. Add 5-μL sample containing 5 μg RNA to 15-μL sample buffer and incubate at 65°C for 15 min.
2. Load the samples onto a denaturing (formaldehyde) 1.2% agarose gel, and electrophorese the samples at 40 mV until the dye-front has moved approx 75% of the way along the gel.
3. Soak the gel in 200 mL 25 m*M* sodium phosphate buffer, pH 6.4, for 20 min with agitation. Repeat this procedure.

Fig. 2. *(previous page)* Preliminary analysis of tobacco plants transformed with an antisense tobacco phosphoribulokinase (PRK) gene by Northern blotting. Nucleic acids were extracted from control and 5 kanamycin-resistant PRK antisense plants (AS 1–5). A 1-μg sample was analyzed by TBE-1% agarose gel electrophoresis *(13)*. Ethidium bromide staining of this gel **(A)** shows a high molecular weight genomic DNA band, ribosomal RNA bands (rRNA), and a transfer RNA band (tRNA). The same extract (5 μg) was electrophoresed in MOPS-formaldehyde 1.2% agarose, transferred to nylon membrane, and probed with a tobacco PRK cDNA **(B)**. This membrane was then stripped at 65°C for 2 h in 5 m*M* Tris-HCI, pH 8.0, 2 m*M* Na_2EDTA, and 0.1X Denhart's solution and reprobed for the β subunit of the tobacco mitochondrial ATPase mRNA **(C)**. PRK and ATPase mRNA levels were quantified by scanning densitometry and the ratio of PRK to ATPase plotted **(D)**. Subsequent analysis of AS 2 by assaying PRK activity showed the activity to be 10% of control levels (from J. S. Knight, A. M. Loynes, and J. C. Gray, unpublished data).

4. Transfer the electrophoresed RNA from the gel to the nylon membrane by blotting overnight from 25 mM sodium phosphate buffer, pH 6.4 as described by Sambrook et al. *(13)*.
5. Fix the RNA to the membrane by exposing the membrane, RNA-side down to UV light (302 nm) for 3 min.
6. Rinse the membrane in 2X SSC and place in the hybridization vessel.
7. Prepare the prehybridization mix (5 mL formamide, 2 mL 20X SSPE, 1 mL 10% [w/v] SDS, 0.75 mL 100X Denhardt's solution, and 1.25 mL H$_2$O) and add this to the hybridization vessel.
8. Boil 0.5 mL fish-milt DNA (10 mg/mL) for 1 min and add this to the hybridization vessel immediately.
9. Prehybridize at 42°C for 1–2 h.
10. Prepare the radiolabeled probe (*see* Section 2.3.1., item 17).
11. Prepare the hybridization mix (4.7 mL formamide, 1.5 mL 20X SSPE, 1 mL 10% [w/v] SDS, 0.65 mL 100X Denhardt's solution and 2.45 mL H$_2$O). Dissolve 1 g of dextran sulfate in the hybridization mix at 42°C.
12. Add the radiolabeled probe to 0.5 mL fish-milt DNA (10 mg/mL), boil for 1 min, and add this to the hybridization vessel immediately.
13. Hybridize at 42°C for 16 h.
14. Remove the membrane from the hybridization vessel and rinse in 2X SSC, wiping the dextran sulfate off the membrane using *gloved* fingers.
15. Wash the membrane in 500 mL 0.1X SSC/0.5% (w/v) SDS at 42°C. Repeat this procedure three times.
16. Wrap the membrane in clingfilm and expose to X-ray film for 4 h to 3 d.

To ensure that any apparent differences in mRNA levels between samples are not due to differences in sample loading, the filter should be stripped (*see* Fig. 2) and reprobed for a different mRNA, which should not have been affected by the genetic manipulation. We and others *(18)* have used a probe for the β-subunit of the tobacco mitochondrial ATPase *(19)*. Differences in mRNA between transformed plants may then be quantified with respect to the unchanged control mRNA in each sample (Fig. 2). This approach is crucial when screening by Northern blot analysis for differences in the expression of one particular gene.

3.4.2. Western Blotting

The amount by which protein levels are decreased in plants transformed with an antisense gene can be assessed by Western blotting. Protein extracts from control and transformed plants grown under constant conditions (*see* Note 8) are separated by electrophoresis in polyacrylamide gels, blotted to nitrocellulose filters, and inmunodetected with antibodies against the protein under study.

Depending on the protein to be analyzed, we use either total leaf extracts or thylakoid-enriched extracts. The latter are prepared as follows:

1. Grind 0.2 g of leaf material using a mortar and pestle with a small amount of sand (about 100 mg). If frozen leaves are used, the tissue is allowed to thaw before grinding.
2. Add 1 mL of ice-cold 50 mM Tris-HCI, pH 7.5 (lysis buffer), homogenize with the pestle, and transfer the suspension to an Eppendorf tube.
3. Centrifuge for 30–60 s at 1000g to pellet the sand, transfer the supernatant to a fresh Eppendorf tube, and centrifuge again (same conditions) to pellet the remaining cell debris. Transfer the supernatant to a fresh Eppendorf tube and incubate for 5 min on ice.
4. Centrifuge for 3 min at 13,000g to pellet membranes, discard the supernatant, resuspend the pellet in 1 mL of lysis buffer, and pellet the membranes (3 min, 13,000g).
5. Resuspend the pellet in a volume of protein loading buffer equivalent to the volume of supernatant in step 3.
6. Punch a hole in the top of the tube and boil for 2 min. Centrifuge for 3 min at 13,000g and remove the supernatant (extract, ready to load on the gel) to a fresh Eppendorf tube.

Total leaf extracts are prepared by grinding 0.1–0.2 g of leaf tissue in liquid nitrogen using a mortar and pestle. Protein loading buffer (1 mL) is then added and the mixture homogenized with the pestle. Once thawed, the homogenate is transferred to an Eppendorf tube and treated as in step 6 above.

Proteins in leaf extracts are separated by electrophoresis in SDS-polyacrylamide gels *(14)* and then transferred to nitrocellulose with a semidry blotter, following the manufacturer's recommendations (*see* Section 2.3.2., item 6). We electrophorese proteins on 12–15% polyacrylamide minigels (see Section 2.3.2., item 4), loading up to 1 µg chlorophyll (*see* Note 10) from total extracts or 3 µg chlorophyll from thylakoid-enriched extracts, into each 5 mm × 0.75 mm well.

3.4.3. Immunodetection

To immunodetect these blots, use antibodies against the protein under study and ^{125}I-labeled protein A. The method, described below, produces images (autoradiographs or fluorographs) that, when quantified by laser-scanning densitometry, give a linear response over a broad range of protein loadings (Fig. 3).

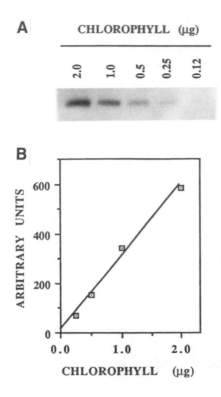

Fig. 3. Detection of the chloroplast Rieske FeS protein in tobacco thylakoids by Western blotting. The proteins in tobacco thylakoid-enriched extracts (0.12–2.0 µg chlorophyll) were resolved in a 15% SDS-polyacrylamide gel and blotted to nitrocellulose. The membrane was immunodecorated with antibodies against the chloroplast Rieske protein and ^{125}I-labeled protein A, and then exposed overnight to X-ray film with an intensifying screen (**A**). The amount of Rieske protein at each chlorophyll loading was quantified by scanning densitometry and these data plotted against the amount of chlorophyll loaded on the gel (**B**) (from F. Madueño, J. S. Knight, and J. C. Gray, unpublished data).

All incubation and wash steps are carried out with gentle agitation.

1. Wash the filter containing the blotted proteins for 5 min in TBS.
2. Incubate the filter in 2% BSA-TBS for 2 h at 25°C (blocking).
3. Incubate the filter in a small volume (10–15 mL) of 2% BSA-TBS containing the antibody (a dilution of between 1:100 and 1:1000 is usually suitable) for 2 h.
4. Wash the filter once with TBS for 5 min, once with 0.1% NP40 TBS for 5 min, and once again with TBS for 5 min.

5. Incubate the filter in a solution of 2% BSA-TBS containing ^{125}I-labeled protein A (0.03 mCi/mL) for 2 h at 25°C.
6. Wash the filter three times with TBS for 5 min and once with 0.1% NP40-TBS for 5 min.
7. Dry the filter completely at room temperature and expose to X-ray film overnight.

4. Notes

4.1. Transformation of Agrobacterium by Electroporation

1. Antisense RNA studies suggest that a full-length cDNA is not essential and that antisense genes encoding one-half or even a quarter-length cDNA originating from either the 5' or 3' end of the cDNA are effective *(20)*.
2. The DNA preparation to be used to transform *Agrobacterium* by electroporation should have a low salt concentration to avoid "arcing" when the current is applied.

4.2. Agrobacterium-Mediated Tobacco Leaf Disc Transformation

3. Unless otherwise stated, chemicals are obtained from Sigma or are reagent grade.
4. To prepare the BAP stock, first dissolve 10 mg in 1 mL of 1*M* HCl and, once dissolved, dilute to 10 mL.
5. Storage of stock solutions: NAA, 3AP, kanamycin at 4°C; 2,4-D at 4°C away from light; 100X B5 vitamins at 4°C for 2–3 wk or at –20°C indefinitely; carbenicillin, streptomycin at –20°C.
6. It is essential to have a healthy log-phase feeder plate suspension culture. However, the growth rate of these cultures is variable. It is usual, therefore, to subculture several volumes (30, 20, 10, and 1 mL) to ensure that a healthy culture is available for the transformation procedure.
7. The times stated for shoot (wk 3) and root (wk 5) appearance are variable.

4.3. Screening for Transgenic Plants with Altered mRNA and Protein Levels

8. It is important to grow the plant material to be analyzed under conditions that are as constant as possible. The authors use Fi-totron 600H growth cabinets supplied by Sanyo Gallenkamp (Loughborough, UK) or growth rooms where conditions such as light and temperature can be controlled, as in some cases a comparison of plants grown under different light regimes (intensity or photoperiod) may be of importance. As many of the polypeptides of the photosynthetic machinery have been shown to follow circadian rhythms *(21)*, it is preferable to collect the tissue samples for RNA or pro-

tein extractions at a fixed time of the day. The samples can be frozen in liquid nitrogen and stored at –80°C prior to use. It is important to note that material may be analyzed from plants that have been transferred from tissue culture to compost (primary transformants), or plants that have been grown from seed produced by the primary transformants (F_1 generation). Equivalent material must be chosen for analysis from each plant.

9. Leaf protein extracts can be stored at –20°C for several weeks, allowed to thaw, and boiled again for 1 min prior to use.

10. The chlorophyll concentration of the leaf extracts in SDS sample buffer containing marker dye can be determined as described by Arnon *(22)*. An aliquot of the extract (10–20 µL) is placed in 1 mL of 80% (v/v) acetone, membranes are removed by centrifuging for 1 min at 13,000g, and the absorbance at 652 nm is measured using a blank consisting of an extract of the same volume of loading buffer (chlorophyll at a concentration of 29 µg/mL in the acetone extract has an absorbance of 1).

11. The optimal dilution for the antibodies, as well as the loading range in which a linear response is obtained, have to be determined for each protein and antibody.

12. When Western blots are exposed to X-ray film, the use of an intensifying screen reduces the required exposure time without affecting the linearity of the response.

13. Although further analysis of transgenic plants beyond the characterization of altered protein levels is beyond the scope of this chapter, it should be noted that usually the next step in this characterization is the measurement of changes in the activity of the enzyme or photosynthetic component that is being manipulated. The change in protein level or enzyme activity can be correlated with resultant changes in flux (if any) through the biochemical pathways of which the protein forms part.

Measurements of the overall rate of photosynthesis can be made in two ways. The maximal rate O_2 evolution in saturating CO_2 and varying light intensities can be measured using the Hansatech leaf disc electrode *(23,24)*, and the altered protein level or enzyme activity can be measured in the same leaf disc *(25)*. This eliminates the problem of sampling equivalent tissues to make measurements of photosynthetic rates and enzyme activities. The rate of carbon assimilation can be measured using the infrared gas analysis (IRGA) LCA-2 system fitted with a Parkinson leaf chamber (Analytical Development Co., Hoddesdon, UK). This is appropriate for measurements of photosynthetic rates in atmospheric air as opposed to the saturating CO_2 used by the leaf disc electrode, and is hence a measure of actual rates of photosynthesis in physiological conditions. Both systems may be used to produce complementary sets of data *(25,26)*.

References

1. Last, D. I. and Gray, J. C. (1990) Synthesis and accumulation of pea plastocyanin in transgenic tobacco plants. *Plant Mol. Biol.* **14,** 229–238.
2. Gray, J. C., Last, D. I., Dupree, P., Newman, B. J., and Slatter, R. E. (1990) Expression of genes for photosynthetic electron transfer components in transgenic plants, in *Genetic Engineering of Crop Plants* (Lycett, G. W. and Grierson, D., eds.), Butterworths, London, pp. 191–205.
3. Napoli, C., Lemieux, C., and Jorgensen, R. (1990) Introduction of a chimeric chalcone synthase gene into petunia results in reversible co-suppression of homologous genes in trans. *Plant Cell* **2,** 279–289.
4. van der Krol, A. R., Mur, L. A., Beld, M., Mol, J. N. M., and Stuitje, A. R. (1990) Flavonoid genes in petunia: addition of a limited number of gene copies may lead to suppression of gene expression. *Plant Cell* **2,** 291–299.
5. van der Krol, A. R., Lenting, P. E., Veenstra, J., van der Meer, I. M., Koes, R. E., Gerats, A. G. M., et al. (1988) An anti-sense chalcone synthase gene in transgenic plants inhibits flower pigmentation. *Nature* **333,** 866–869.
6. Smith, C. J. S., Watson, C. F., Ray, J., Bird, C. R., Morris, P. C., Schuch, W., and Grierson, D. (1988) Antisense RNA inhibition of polygalacturonase gene expression in transgenic tomatoes. *Nature* **334,** 724–726.
7. Rodermel, S. R., Abbott, M. S., and Bogorad, L. (1988) Nuclear–organelle interactions: nuclear antisense gene inhibits ribulose bisphosphate carboxylase enzyme levels in transformed tobacco plants. *Cell* **55,** 673–681.
8. Hudson, G. S., Evans, J. R., von Caemmerer, S., Arvidsson, Y. B. C., and Andrews, T. J. (1992) Reduction of ribulose-1, 5-bisphosphate carboxylase/oxygenase content by antisense RNA reduces photosynthesis in transgenic tobacco plants. *Plant Physiol.* **98,** 294–302.
9. Stockhaus, J., Hofer, M., Renger, G., Westerhoff, P., Wydrzynski, T., and Willmitzer, L. (1990) Anti-sense RNA efficiently inhibits formation of the 10 kd polypeptide of photosystem II in transgenic potato plants: analysis of the role of the 10 kd protein. *EMBO J.* **9,** 3013–3021.
10. Bevan, M. (1984) Binary *Agrobacterium* vectors for plant transformation. *Nucleic Acids Res.* **12,** 8711–8721.
11. Odell, J. T., Nagy, F., and Chua, N.-H. (1985) Identification of DNA sequences required for the activity of cauliflower mosaic virus 35S promoter. *Nature* **313,** 810–812.
12. Mazur, B. J. and Chui, C.-F. (1985) Sequence of a genomic DNA clone for the small subunit of ribulose bis-phosphate carboxylase-oxygenase from tobacco. *Nucleic Acids Res.* **13,** 2373–2386.
13. Sambrook, J., Fritsch, E. F., and Maniatis, T. (1989) *Molecular Cloning: A Laboratory Manual.* Cold Spring Harbor Laboratory, Cold Spring Harbor, NY.
14. Laemmli, U. K. (1970) Cleavage of structural proteins during the assembly of the head of the bacteriophage T4. *Nature* **227,** 680–685.
15. Shen, W. and Forde, B. G. (1989) Efficient transformation of *Agrobacterium* spp. by high voltage electroporation. *Nucleic Acids Res.* **17,** 8385.

16. Horsch, R. B., Fry, J. E., Hoffmann, N., Eichhloltz, D., Rogers, S. G., and Fraley, R. T. (1985) A simple and general method for transferring genes to plants. *Science* **227,** 1229–1231.

17. Apel, K. and Kloppstech, K. (1978) The plastid membrane of barley *(Hordeum vulgare).* Light-induced appearance of mRNA coding for the apoprotein of the light harveting chlorophyll a/b protein. *Eur. J. Biochem.* **85,** 581–588.

18. Vaucheret, H., Kronenberger, J., Lepingle, A., Vilaine, F., Boutin, J.-P., and Caboche, M. (1992) Inhibition of tobacco nitrite reductase activity by expression of antisense RNA. *Plant. J.* **2,** 559–569.

19. Boutry, M. and Chua, N.-H. (1985) A nuclear gene encoding the beta subunit of the mitochondrial ATP synthase in *Nicotiana plumbaginifolia. EMBO J.* **4,** 2159–2165.

20. van der Krol, A. R., Mur, L. A., de Lange, P., Mol, J. N. M., and Stuitje, A. R. (1990) Inhibition of flower pimentation by antisense CHS genes: promoter and minimal sequence requirements for antisense effect. *Plant Mol. Biol.* **14,** 457–466.

21. Adamska, I., Scheel, B., and Kloppstech, K. (1991) Circadian oscillations of nuclear-encoded chloroplast proteins in pea *(Pisum sativum). Plant Mol. Biol.* **17,** 1055–1065.

22. Arnon, D. I. (1949) Copper enzymes in isolated chloroplasts. Polyphenol oxidase in *Beta vulgaris. Plant Physiol.* **24,** 1–15.

23. Walker, D. A. (1987). *The Use of the Oxygen Electrode and Fluorescence Probes in Simple Measurements of Photosynthesis.* Oxygraphics Ltd., Sheffield, UK.

24. Walker, D. A. (1989) Automated measurement of leaf photosynthetic O_2 evolution as a function of photon flux density. *Phil. Trans. R. Soc. Lond. B* **323,** 313–326.

25. Quick, W. P., Schurr, U., Scheibe, R., Schulze, E.-D., Rodermel, S. R., Bogorad, L., and Stitt, M. (1991) Decreased ribulose-1,5-bisphosphate carboxylase-oxygenase in transgenic tobacco transformed with "antisense" rbcS. I. Impact on photosynthesis in ambient growth conditions. *Planta* **183,** 542–554.

26. Stitt, M., Quick, W. P., Schurr, U., Schulze, E.-D., Rodermel, S. R., and Bogorad, L. (1991) Decreased ribulose-1,5-bisphosphate carboxylase-oxygenase in transgenic tobacco transformed with "antisense" rbcS. II. Flux-control coefficients for photosynthesis in varying light, CO_2, and air humidity. *Planta* **183,** 555–566.

CHAPTER 25

Gene Activation by T-DNA Tagging

Klaus Fritze and Richard Walden

1. Introduction

Agrobacterium-mediated transformation of plants has not only become a useful tool for the introduction of chimeric genes into plants, but it has also recently been utilized in gene tagging (for reviews, *see* refs. *1–3*). T-DNA tagging has proven a powerful genetic approach for the isolation of novel plant genes. Compared with other approaches to gene isolation (for a review, *see* ref. *4*), mutagenesis by insertion of defined DNA sequences allows the rapid cloning and characterization of the mutated locus. Furthermore, T-DNA is an ideal gene tag, as it has the potential to insert preferentially into potentially transcribed regions of the genome *(5)*.

The versatility of T-DNA-based vectors is such that they can be engineered to contain a variety of sequences allowing a diversity of different tagging strategies. The initial use of T-DNA as a tag was to generate mutations resulting from gene inactivation by disruption of coding, or regulatory, regions. Generally, such insertions cause recessive mutations in diploid species, which cannot be visualized in a pool of primary transformants and require selfing and screening of the resulting T2 generations to identify specific phenotypes (for a review, *see* ref. *6*). An alternative strategy involves the use of T-DNA containing promoter-less reporter genes in order to identify active promoter sequences within the plant genome *(7–10)*. In this case, T-DNAs used are constructed in a way that reporter genes lacking promoters are cloned as a passive element next to one of the T-DNA border sequences. The expression of the reporter genes becomes induced should translational or transcriptional

From: *Methods in Molecular Biology, Vol. 44:* Agrobacterium *Protocols*
Edited by: K. M. A. Gartland and M. R. Davey Humana Press Inc., Totowa, NJ

gene fusions occur with plant genomic sequences located near the T-DNA following insertion into the plant genome.

When using T-DNA as an insertional mutagen to create recessive mutations, the appearance of a specific phenotype is a matter of chance. This limitation can be overcome if the insertion of T-DNA produces a dominant mutation, allowing direct selection for a specific phenotype. We have developed a strategy to carry out T-DNA tagging in such a way that, following insertion into the plant genome, flanking plant genes are transcriptionally activated or overexpressed, producing dominant mutations. To do this, we engineered T-DNA so that it contains a set of multiple transcriptional enhancer sequences derived from the 35S RNA promoter of Cauliflower mosaic virus cloned in tandem next to the T-DNA right border sequence. The T-DNA-induced activation of expression of tagged plant genes will result in a dominant mutation allowing selection for a specific phenotype to be carried out from a population of primarily transformed hemizygous individuals. In addition to the transcriptional enhancer elements, the T-DNA contains a plant-specific marker gene based on antibiotic resistance allowing identification of transgenic clones. Moreover, the T-DNA contains a bacterial origin of replication and an ampicillin resistance gene that enables us to re-isolate the T-DNA physically linked to tagged plant DNA sequences directly from plant tissue by plasmid rescue.

To ensure isolation of a specific mutation from the plant genome, selection needs to be carried out on an extremely large number of primary transformants. Such numbers can be readily obtained by protoplast *Agrobacterium* cocultivation. Selection can be carried out by scoring the ability of transgenic cells to grow under specific selective conditions. This allows biochemical selection to be applied. The advantage of this is that selection is not dependent on a detailed biochemical understanding of the process under investigation. Hence, studies can be initiated on processes about which currently little or nothing is known. We are interested in the molecular basis of plant growth substance action. In order to isolate genes that may play a role in the perception or action of such substances, we have utilized T-DNA tagging using the tag containing multiple transcriptional enhancers. We have produced mutant cell lines able to grow either in the absence of auxin *(11)* or cytokinin (Miklashevichs and Walden, unpublished) in the culture media. Similarly, cell lines have been generated able to grow in toxic levels of Methylglyoxal bis (guanyl)hydrazone

(MGBG), an inhibitor of polyamine biosynthesis (Fritze and Walden, in preparation).

Here we describe a typical experimental procedure for a gene-tagging experiment using the enhancer tag vector with tobacco protoplast transformation. The experiments involve:

1. The cocultivation of single cells derived from tobacco protoplasts with *Agrobacterium* containing gene-activating T-DNA vectors followed by suitable culture and selection procedures to identify specific tagging events.
2. The culture of the transgenic lines and the re-isolation of T-DNA and its flanking plant DNA sequences.
3. The re-introduction of isolated sequences in wildtype protoplasts in order to reconstitute the (selected) mutant phenotype. The complete process is summarized in Figure 1.

2. Materials

1. All material used for plant tissue culture must be thoroughly cleaned and autoclaved. Solutions which cannot be autoclaved should be filter-sterilized by commercially available 45-µm filter units (Millipore, Bedford, MA).
2. Axenically grown tobacco plants for the isolation of mesophyll protoplasts (*see* Note 1; *12*).
3. Enzyme solutions for the preparation of protoplasts (*see* Note 2).
4. For the isolation of protoplasts:
    a. Sterile sieves of 100-µm mesh size (nylon or high grade steel).
    b. Peristaltic pump.
    c. Rubber tubes with an inner diameter of 1 mm.
    d. Glass capillaries of 200-mm length and an inner diameter of approx 1 mm.
    e. Screwcapped 12-mL polystyrene or glass tubes (Nunc, Wiesbaden, Germany).
    f. Benchtop centrifuges (Hettich, Tuttlingen, Germany) with swingout rotors (speed: 500–3000 rpm).
    g. Sterile 1- and 10-mL pipets.
    h. Inverted phase contrast microscope.
    i. Hemocytometer (Fuchs-Rosenthal type) for estimation of cell density.
    j. K3 Media: 0.4*M* sucrose, *see* Table 1.
    k. W5 Media: *see* Table 1.
5. For protoplast culture and maintainance of developing callus tissue:
    a. Disposable Petri dishes of 90- and 150-mm diameter (Greiner, Solingen, Germany).
    b. Suitable dark and illuminated phytocabinets.

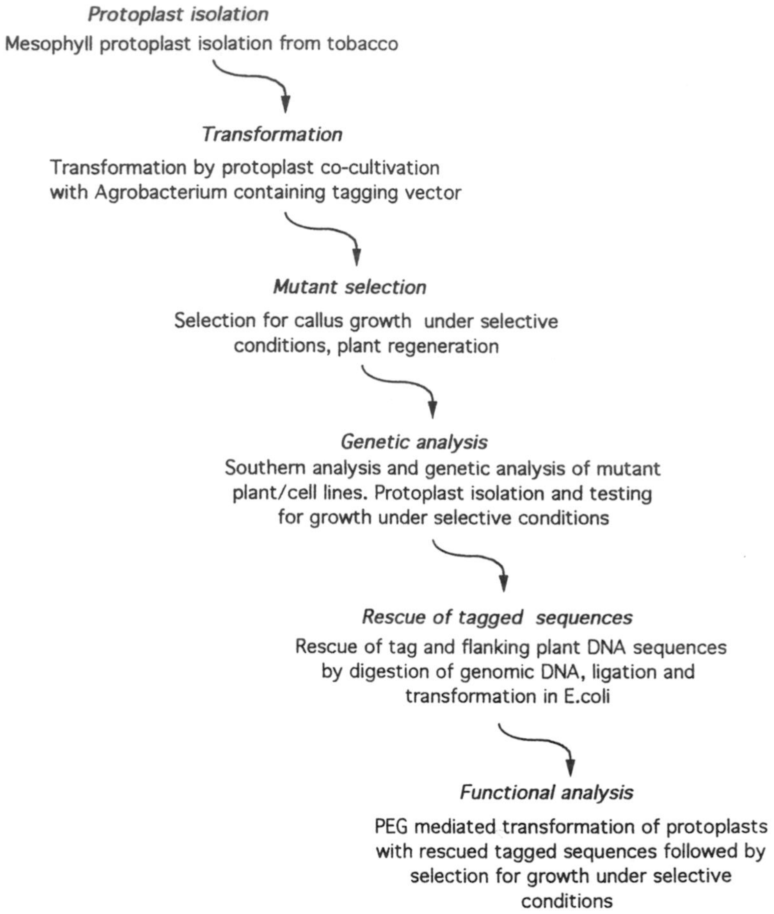

Protoplast isolation
Mesophyll protoplast isolation from tobacco

Transformation
Transformation by protoplast co-cultivation
with Agrobacterium containing tagging vector

Mutant selection
Selection for callus growth under selective
conditions, plant regeneration

Genetic analysis
Southern analysis and genetic analysis of mutant
plant/cell lines. Protoplast isolation and testing
for growth under selective conditions

Rescue of tagged sequences
Rescue of tag and flanking plant DNA sequences
by digestion of genomic DNA, ligation and
transformation in E.coli

Functional analysis
PEG mediated transformation of protoplasts
with rescued tagged sequences followed by
selection for growth under selective
conditions

Fig. 1. Scheme for the isolation of plant genes by the use of T-DNA as gene activator in transgenic tobacco.

c. K3 Media: 0.1–0.4M sucrose, *see* Table 1.
d. Stocks of phytohormones: 1 mg/L naphthalene acetic acid (NAA) and 0.1 mg/L kinetin (Sigma, St. Louis MO) dissolved in 100% ethanol.
e. Stocks of antibiotics: 15 mg/L hygromycin (Sigma), 100 mg/L cefotaxime (Claforan, Hoechst, Frankfurt, Germany).
f. Low-gelling temperature agarose (autoclaved in glass bottles) to be dissolved in K3 media.
g. MS3 and MS1 Media: *see* Table 1.
h. Sterile jars of 200- and 500-mL total volume.
i. Parafilm.

6. For the cocultivation of *Agrobacterium* with protoplasts: The *Agrobacterium* T-DNA vector pPCVIC En4 Hpt (*see* Fig. 2) as described *(11)* is maintained in *Agrobacterium tumefaciens* strain GV 3101 (*see* Note 3; *14,15*).
7. Molecular analysis of mutants: Materials and methods as described in refs. *16* and *17*.
8. For the rescue of T-DNA from mutant plants:
   a. Restriction enzymes.
   b. Plant DNA prepared as described in ref. *18*.
   c. Ligase (Biolabs, Beverly, MA).
   d. Electroporation apparatus (Bio-Rad, Richmond, CA).
   e. Commercially available competent *E. coli* DH10 cells (Gibco-BRL, Gaithersburg, MD).
9. For DNA uptake experiments:
   a. Freshly prepared tobacco protoplasts: *see* Section 3.1.
   b. Sterile plasmid DNA preparations.
   c. W5-Media: *see* Table 1.
   d. MaMg Transformation media: *see* Table 1.
   e. PEG Solution: *see* Table 1.
   f. K3 Media: 0.4*M* sucrose, *see* Table 1.
   g. 45°C Water bath.
   h. Sterile polystyrene tubes (Nunc).
   i. Equipment for cultivation of protoplasts.

3. Methods

3.1. Isolation of Tobacco Protoplasts

1. Cut fully expanded leaves from 8–10-wk-old axenically grown tobacco plants.
2. Dissect leaves from midribs, cut into 10 mm^2 small pieces, and submerge them in prepared enzyme solutions.
3. Incubate preparations overnight (14–16 h) in a dark cabinet at 26°C.
4. Release protoplasts by gentle agitation of the digest mixture for 10 min and subsequent passage through a sterile sieve unit placed in sterile beakers or jars.
5. To separate intact protoplasts from undigested material, cell walls, and the enzyme, carefully transfer the sieved suspension to screwcapped tubes and centrifuge for 6 min at approx 3000 rpm in the swing-out rotor of a benchtop centrifuge. Under these conditions intact mesophyll protoplast float in the upper part of the tubes.
6. Remove enzyme solution and sedimenting material from below the floating protoplasts by suction through glass capillaries.
7. For washing, carefully resuspend the remaining protoplasts in K3 media (0.4*M* sucrose) and centrifuge as described in step 5.

Table 1
Plant Cell and Tissue-Culture Media

MS-medium (13)		
NH_4NO_3	1650	mg/L
KNO_3	1900	mg/L
$CaCl_2 \cdot 2H_2O$	440	mg/L
$MgSO_4 \cdot 7H_2O$	370	mg/L
KH_2PO_4	170	mg/L
H_3BO_3	6.2	mg/L
$MnSO_4 \cdot 1H_2O$	22.3	mg/L
$ZnSO_4 \cdot 4H_2O$	8.6	mg/L
KI	0.83	mg/L
$Na_2MoO_4 \cdot 2H_2O$	0.25	mg/L
$CuSO_4 \cdot 5H_2O$	0.025	mg/L
$CoCl_2 \cdot 6H_2O$	0.025	mg/L
Na_2 EDTA	37.2	mg/L
$FeSO_4 \cdot 7H_2O$	27.8	mg/L
Glycine	2	mg/L
Nicotinic acid	0.5	mg/L
Pyridoxine	0.5	mg/L
Myo-inositol	100	mg/L
Thiamine	0.4	mg/L
Sucrose	1% w/v (MS1)/3% w/v (MS3)	
pH 5.7 before autoclaving the media, with KOH		
K3 medium (21)		
NH_4NO_3	250	mg/L
KNO_3	2500	mg/L
$CaCl_2 \cdot 2H_2O$	900	mg/L
$MgSO_4 \cdot 2H_2O$	250	mg/L
$NaH_2PO_4 \cdot 1H_2O$	150	mg/L
$(NH_4)_2SO_4$	134	mg/L
$CaHPO_4 \cdot 1H_2O$	50	mg/L
H_3BO_3	3	mg/L
$MnSO_4 \cdot 1H_2O$	10	mg/L
$ZnSO_4 \cdot 4H_2O$	2	mg/L
KI	0.75	mg/L
$Na_2MoO_4 \cdot 2H_2O$	0.25	mg/L
$CuSO_4 \cdot 5H_2O$	0.25	mg/L
$CoCl_2 \cdot 6H_2O$	0.025	mg/L
Na_2EDTA	37.2	mg/L
$FeSO_4 \cdot 7H_2O$	27.8	mg/L
Inositol	100	mg/L
Sucrose	137	g/L (= 0.4M)

Table 1 *(continued)*

Xylose	250	mg/L
Nicotinic acid	1	mg/L
Pyridoxine	1	mg/L
Thiamine	10	mg/L
Adjust to pH 5.8 with KOH		
W5 medium		
KCl	80	mg/L
CaCl$_2$	18.375	g/L
NaCl	9	g/L
Glucose	9	g/L
Adjust to pH 6 with NaOH		
MaMg medium		
Mannitol	0.45M	
MgCl$_2$	15 mM	
MES	0.1%	(w/v)
Adjust to pH 5.6 with KOH		
PEG solution		
Ca(NO$_2$)$_2$ · 4H$_2$O	0.1M	
Mannitol	0.4M	
PEG 4000 (Merck)	40%	(w/v)
Adjust to pH 7–9 with NaOH		

8. Repeat centrifugation and resuspension in fresh media once.
9. After the final washing, estimate the cell density of protoplasts using a hemocytometer, and adjust suspension to a final concentration of 10^6 cells/mL.

3.2. Cocultivation of Protoplasts with A. tumefaciens

1. Resuspend 10^6 isolated protoplasts in 10 mL K3 media (0.4M sucrose, 1 mg/L NAA and 0.2 mg/L kinetin), transfer to 90-mm diameter Petri dishes.
2. Incubate for 48 h in a dark phytocabinet at 26°C.
3. Transfer the protoplasts to an illuminated phytochamber for an additional 3–5 d. Protoplasts should be monitored daily so that cocultivation with *A. tumefaciens* is initiated as soon as protoplasts begin to synthesize cell walls and carry out the first cell divisions. At this point the protoplasts appear oval shaped.
4. For *Agrobacterium* cocultivation, an appropriate inoculum of bacterial supension is calculated. Estimate the density of an overnight bacterial culture by measurement of optical density at 600 nm and calculate the volume of the inoculum by $100/(1.8 \times OD_{688}) =$ microliter bacteria/10^6 protoplasts.

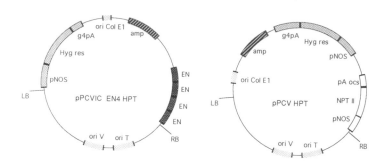

Fig. 2. The plant transformation vectors pPCVIC EN4 and pPCV HPT are derivatives from the series of pPCV vectors described in ref. *14*. pPCVIC En4 Hpt contains: (a) A direct tandem repeat of four transcriptional enhancer elements (EN) derived from the 35S RNA promoter of the *Cauliflower mosaic virus* (position −90 to −427 as described by ref. *19* approx 20 bp from the right border [RB] sequence). (b) A hygromycin resistance selectable marker (Hyg res), driven by the *nos* promoter (pNOS) and the polyadenylation signal from gene 4 of T-DNA (g4pA) next to the left border (LB) of T-DNA. (c) An *E. coli* colEl plasmid origin of replication (ori col El) and an ampicillin resistance gene (amp), derived from pIC 19H *(20)*. (d) Origins of replication (ori V) and plasmid conjugational transfer (ori T) of the wide host range plasmid RK2 that are functional in *A. tumefaciens* and are derived from pPCV002 *(14)*. pPCV HPT contains no transcriptional enhancer elements, but a hygromycin selectable marker and in addition to the functions described above, a selectable kanamycin resistance marker (*npt*II) driven by the *nos* promoter and the polyadenylation sequence derived from the *Agrobacterium* octopine synthase gene.

5. Centrifuge this volume in a sterile 1.5-mL Eppendorf tube for 3 min at room temperature in a benchtop centrifuge, resuspend bacteria in K3 media and add to protoplast cultures.
6. Cocultivate in the dark at 26°C for 48 h. During cocultivation, successful attachment of bacteria to the walls of plant cells can be monitored by viewing through an inverted phase contrast microscope.
7. Following cocultivation, wash the cells twice with W5 media by mixing 1 vol of cell suspension with 1 vol of W5 media in screwcapped polystyrene tubes and centrifuge for 3 min at 700 rpm in a swingout rotor of a benchtop centrifuge.

8. Gently pour off the supernatant and resuspend the sedimented cells in K3 media (0.4M sucrose, 1 mg/L NAA, and 0.2 mg/L kinetin) with additional 500 mg/L of cefotaxime. Cefotaxime stops the further growth of *Agrobacterium*.
9. Transfer the cells to 90-mm diameter Petri dishes.
10. Grow cells for 2–3 d in an illuminated phytochamber to recover from cocultivation.

3.3. Culture and Selection of Transformed Protoplasts

Depending on the mode of selection, selective pressure can be initiated directly after cocultivation, or can be started in any step after embedding of protoplasts (*see* the following and Notes 4 and 5).

1. Embed the cells by the gentle addition of 1 vol of hand warm low gelling temperature agarose (1.6% [w/v] to produce a final concentration of 0.8%) dissolved in K3 media (0.2M sucrose, 1 mg/L NAA, 0.2 mg/L kinetin, and 500 mg/L cefotaxime) to the cell cultures in 90-mm diameter Petri dishes.
2. Cut solidified agarose into four segments with a broad blade spatula and transfer these to a 150-mm diameter Petri dish.
3. Add 20 mL of liquid K3 media (0.3M sucrose, 1 mg/L NAA, 0.2 mg/L kinetin, and 500 mg/L cefotaxime).
4. Maintain cultures under continuous illumination at 26°C and change the liquid media at weekly intervals. Reduce the osmolarity of the culture media stepwise by an 0.05M decrease of the sucrose concentration to a final concentration of 0.1M.
5. Depending on the mode of selection, viable calli become visible 2–6 wk after embedding and are removed to solidified MS3 containing 1 mg/L NAA, 0.2 mg/L kinetin, and 250 mg/L cefotaxime in 90-mm diameter Petri dishes once they have reached a size of approx 5 mm.
6. Transfer these calli to fresh media at 3-wk intervals.
7. For the regeneration of plantlets, culture callus on MS3 media containing a shoot inducing concentration of plant hormones (0.1 mg/L NAA, 0.5 mg/L benzaminopurin, 250 mg/L cefotaxime).
8. Excise shoots and place them on hormone-free MS1 media to induce root formation. Rooting plantlets may be grown further in the greenhouse to flowering.

3.4. Analysis of Mutants and Rescue of the T-DNA from Mutants

For genetic analysis, regenerated plants are selfed and/or cross-pollinated with wild-type pollen. Segregation of the hygromycin resistance marker and any new phenotypic traits can be monitored at this stage. Ideally, a novel phenotype should be discriminated from the wild type at the seedling level.

1. To demonstrate the inheritance of the selected phenotype, prepare protoplasts from three to five leaves (digested by 6–8 mL of enzyme solution in a 90-mm Petri dish) of offspring plants.
2. Score for growth under selective conditions following the initial isolation protocol.
3. Number and organization of T-DNA insertions within the mutants can be assessed by genomic Southern blotting as described by ref. *17* using hybridization probes derived from sequences of both ends of the T-DNA (*see* Note 6).
4. For plasmid rescue, 5 µg of isolated plant genomic DNA *(18)* is digested with 40 U suitable restriction enzyme in a vol of 200 µL. Perform digests for 4–6 h and monitor digestion on agarose gels.
5. Extract completely digested preparations using phenol/chloroform and precipitate with ethanol.
6. Ligate digested and precipitated plant DNA in a vol of 200 µL containing 2 Weiss units of Ligase (Biolabs) at 15°C for 16 h.
7. Treat the ligation with phenol/chloroform and precipitate the DNA with ethanol.
8. Wash the DNA pellet extensively with 70% (v/v) ethanol to remove all traces of salt and redissolve in 40 µL sterile water.
9. Ligated plant DNA preparations are electroporated into bacteria according to the manufacturer's protocols (*see* Note 7).
10. Five microliters of ligated DNA were routinely transformed into 20 µL of electrocompetent cells and plated out on Luria-Bertani (LB) plates containing 100 mg/L ampicillin to identify clones containing a single copy of T-DNA.

3.5. Transfection of Protoplasts by DNA Uptake

The physical linkage of T-DNA to a tagged gene can be demonstrated by DNA uptake by wild-type protoplasts followed by the reconstitution of a given phenotype under appropriate selective conditions. DNA uptake experiments can be easily performed by PEG-mediated transformation.

1. Prepare supercoiled plasmid DNA representing rescued sequences by a standard isolation protocol (for details *see* ref. *17*), which includes two CsCl gradients, treatment with *n*-butanol, dialysis, and ethyl alcohol (ETOH) precipitation.
2. Resuspend precipitated DNA in sterile water and keep under sterile conditions.

3. Isolate SR1 wild-type protoplasts as described in Section 3.1., but resuspend them in MaMg media (*see* Table 1) following the final washing step.
4. Transfer aliquots of 3.3×10^5 protoplasts in a vol of 330 μL MaMg solution to a single screwcapped plastic tube.
5. Heat shock for 5 min in a 45°C water bath.
6. Cool the tubes to room temperature.
7. Add 10 μg of sterile DNA in a final vol of 10 μL.
8. Incubate the cells for 10 min at room temperature.
9. Add an equal volume of freshly prepared and filter-sterilized PEG solution to the protoplasts and mix gently.
10. Incubate for 20 min at room temperature.
11. Carefully resuspend the protoplasts in 5 mL of K3 medium ($0.4M$ sucrose, 1 mg/L NAA, 0.2 mg/L kinetin).
12. Culture further as described in Section 3.3.
13. To demonstrate the reconstitution of the mutant phenotype, culture transfected protoplasts under selective conditions. The efficiency of transformation should be monitored by parallel transfection of plasmid containing selectable marker genes based on antibiotic resistance.

4. Notes

1. Culture of axenically grown tobacco plants: Tobacco seeds (*Nicotiana tabacum*, Petit Havanna SR1 *[12]*) are surface sterilized by short rinsing with 70% (v/v) ethanol, a short wash in 7% (w/v) $NaClO_4$, 0.1% SDS (w/v) for 10 min, washed three times with sterile H_2O and air-dried. Seeds are germinated on MS1 media *(13)*, solidified with 0.8% agar in Petri dishes. Two weeks post-germination, two to three individual seedlings are transferred to MSl media in sterile Weck®jars of 500-mL total volume. Cultivation of seedlings and plantlets should be performed under symmetrical illumination (≈2000 lux/≈150–160 μ Einstein/m²/s), 22–24°C, and light/dark photoperiod (16/8 h) in suitable phytocabinets or growth chambers.
2. Enzyme solutions for the preparation of protoplasts: Enzyme solutions are prepared by dissolving 1.5% (w/v) cellulase and 0.5% (w/v) macerozyme (Serva, Heidelberg, Germany) in K3 media containing $0.4M$ sucrose (*see* Table 1), gentle mixing for 6–8 h at room temperature, and adjusting the pH value to 5.8. The enzyme solution is filter sterilized and can be stored at –20°C for some weeks. Prior to protoplast isolation, 20-mL aliquots are pipeted into small sterile polystyrene pots (Greiner 100 mL total volume). Five grams of leaf material harvested from two 8–10-wk-old tobacco plants is digested in 20 mL of enzyme solution and typically yields $2–3 \times 10^6$ protoplasts.
3. Culture of *A. tumefaciens*: *Agrobacterium* strains containing the activating T-DNA tagging vector are maintained as glycerol stocks. For the initiation

of cultures, bacteria are streaked from the glycerol stock on YEB culture media *(15)* solidified with 1% (w/v) agar. Single colonies of overnight-grown bacteria are streaked out on solid YEB media with selective concentrations of rifampicin (100 mg/L), carbenicilin (100 mg/L), gentamycin (40 mg/L), and kanamycin (25 mg/L). *Agrobacterium* are cultivated at 28°C. The selection of bacteria should be performed on YEB plates containing combinations of two of the described antibiotics in each case. Liquid cultures of YEB media, supplemented with carbenicilin and kanamycin, are inoculated with single bacterial colonies and incubated with vigorous shaking approx 48 h prior to cocultivation experiments so that the culture reaches an OD_{688} of 1.0.

4. Selection of protoplast cultures: The selective conditions for the isolation of specific mutants (i.e., kill curves) need to be established using culture conditions similar to those following cocultivation (i.e., pH value, cell density, phytohormones, antibiotics). Growth of transgenic cells is selected by the presence of hygromycin (15 mg/L). This can be used to estimate transformation frequency.

5. Design of tagging experiments: Tagging experiments can be initiated with $20–50 \times 10^6$ protoplasts. In addition to transformation with *Agrobacterium* containing an activating T-DNA vector, controls, such as the growth of untransformed protoplasts under selection and of protoplasts transformed with a vector lacking transcriptional enhancers (e.g., pPCV HPT; *see* Fig. 2), are advisable. These provide a means of testing the selection conditions and establishing the transformation frequency.

6. Rescue of T-DNA and flanking sequences: Using pPCVIC En4 HPT as the tagging vector and following the cocultivation protocol described in Section 4.1., more than 2/3 of the tagged populations of plants that we have studied contain single T-DNA inserts. Rescue of T-DNA with flanking plant DNA from mutant lines is feasible because the T-DNA contains a bacterial origin of replication together with a selectable marker gene. Depending on the organization of the T-DNA insert, genomic DNA from the tagged mutant line is digested with a restriction enzyme so that fragments of DNA are obtained that contain a single bacterial origin of replication and the ampicillin resistance gene, and are as large a segment of flanking plant DNA as possible. Digested DNA is religated and electroporated into competent *E. coli* strains. T-DNA, together with flanking plant DNA, creates a replicative unit that can be identified by its bacterial ampicilin resistance.

7. Electroporation of *E. coli:* Typically a Gene Pulser (Bio-Rad) may be used in connection to a pulse controller (Biorad). For electroporation of "ElectroMax" DH10 β-cells (Gibco-BRL) pulse conditions are set at 25 uF, 1.6 kV, and 200 Ohm.

Acknowledgments

The authors thank Jeff Schell for support and encouragement during the process of developing the concept of T-DNA activation tagging, Inge Czaja and Elke Bongartz for excellent technical assistance, and Hinrich Harling and Dirk Prüfer for interest and comments on the manuscript. K. Fritze is supported by the Max Planck Society.

References

1. Walden, R., Hayashi, H., and Schell, J. (1991) T-DNA as a gene tag. *Plant J.* **3,** 281–288.
2. Koncz, C., Nemeth, K., Redei, G., and Schell, J. (1993) T-DNA insertional mutagenesis in *Arabidopsis. Plant Mol. Biol.* **20,** 963–976.
3. Coomber, S. and Feldmann, K. A. (1993) Gene tagging in transgenic plants, in *Transgenic Plants*, vol. 1 (Kung, S. and Wu, R., eds.), Academic, New York, pp. 225–240.
4. Gibson, S. and Somerville, C. (1993) Isolating plant genes. *TibTech* **11,** 306–313.
5. Koncz, C., Martini, N., Mayerhofer, R., Koncz-Kalmann, Zs., Körber, H., Redei, G. P., and Schell, J. (1989) High frequency T-DNA mediated gene tagging in plants. *Proc. Natl. Acad. Sci. USA* **86,** 8467–8471.
6. Feldmann, K. A. (1991) T-DNA insertion mutagenesis in *Arabidopsis*: mutational spectrum. *Plant J.* **1,** 71–82.
7. Teeri, T. H., Herrera-Estrella, L., DePicker, A., Van Montagu, M., and Palva, E. T. (1986) Identification of plant promoters *in situ* by T-DNA mediated transcriptional fusions of the NPT II gene. *EMBO J.* **5,** 1755–1760.
8. Kerbundit, S., De Greve, H., Deboeck, F., Van Montagu, M., and Hernalsteens, J.-P. (1991) *In vivo* random β-glucuronidase gene fusions in *Arabidopsis thaliana. Proc. Natl. Acad. Sci. USA* **88,** 5212–5216.
9. Goldsborough, A. and Bevan, M. (1990) New patterns of gene activity in plants detected using an *Agobacterium* vector. *Plant Mol. Biol.* **16,** 263–269.
10. Topping J. F., Wei, W., and Lindsay, K. (1991) Functional tagging of regulatory elements in the plant genome. *Development* **112,** 1009–1019.
11. Hayashi, H., Czaja, I., Lubenow, H., Schell, J., and Walden, R. (1992) Activation of a plant gene by T-DNA tagging: auxin independent growth *in vitro. Science* **258,** 1350–1353.
12. Maliga, P., Sz.-Breznovitis, A., and Marton, L. (1973) Streptomycin resistant plants from callus culture of haploid tobacco. *Nature* **244,** 623–625.
13. Murashige, T. and Skoog, F. (1962) A revised medium for rapid growth and bioassays with tobacco tissue cultures. *Physiol. Plant* **15,** 473–497.
14. Koncz, C. and Schell, J. (1986) The promoter of the T_L-DNA gene 5 controls the tissue-specific expression of chimeric genes carried by a novel type of *Agrobacterium* binary vector. *Mol. Gen. Genet.* **213,** 285–290.
15. Van Larebeke, N., Genetello, C., Hernalsteens, J-P., DePicker, A., Zaenen, I., Messens, E., et al. (1977) Transfer of Ti-plasmids between *Agrobacterium* strains by mobilisation with the conjugative plasmid RP4. *Mol. Gen. Genet.* **152,** 119–124.

16. Draper, J., Scott, R., Armitage, P., and Walden, R. (1988) *Plant Genetic Transformation and Gene Expression: A Laboratory Manual*. Blackwell, Oxford, UK.
17. Sambrook, J., Fritsch, E. F., and Maniatis, T. (1989) *Molecular Cloning: A Laboratory Manual*, 2nd ed. Cold Spring Harbor Laboratory Press, Cold Spring Harbor, NY.
18. Dellaporta, S. L., Wood, J., and Hicks, J. B. (1983) A plant DNA minipreparation: version II. *Plant Mol. Biol. Rep.* **1,** 19–21.
19. Odell, J. T., Nagy, F., and Chua, N. H. (1985) Identification of sequences required for activity of the Cauliflower mosaic virus 35 S promoter. *Nature* **313,** 810–812.
20. Marsh, J. L., Erfle, M., and Wykes, E. J. (1984) The pIC plasmid and phage vectors with versatile cloning sites for recombinant selection by insertional inactivation. *Gene* **32,** 481–485.
21. Nagy, J. I. and Maliga, P. (1976) Callus induction and plant regeneration from mesophyll protoplasts of *Nicotiana sylvestris*. *Z. Pflanzenphysiol.* **78,** 453–455.

CHAPTER 26

Assessing Cadmium Partitioning in Transgenic Plants

George J. Wagner

1. Introduction

Several research groups have undertaken gene transfer studies to test the feasibility of using animal metallothionein (MT; a Cd, Zn, Cu, Au, Ag, etc. binding peptide) to partition and sequester pollutant metals in plant tissues. To date, several mammalian MT genes have been expressed constitutively in *Nicotiana* and *Brassica* species and the effects of their presence on Cd accumulation and Cd tolerance have been assessed. General objectives of these studies have been:

1. To sequester metal in unconsumed tissues of food crops in order to reduce the transfer of undesirable metals from crops to humans, and
2. To make plants more tolerant to the presence of accumulated metal so as to develop super-accumulators that might be useful for bioremediation of contaminated environments or for biomining of valuable metals.

In the present chapter, the author has focused on objective 1 and has utilized animal MT in what is referred to here as a MT-Cd sequestration strategy. To date, the author has tested the hypothesis that constitutive expression of an animal MT gene will result in sequestration of Cd by MT as it enters the root, thus reducing Cd translocation to the shoot.

Mechanisms underlying the partitioning of ions in plants (including so-called "heavy" metals [1] such as Cd) are not well understood (2–4). Ions may accumulate in the root free space, in tissue cell walls, and in vacuoles of mature cells throughout the plant (2,3,5). They may be per-

From: *Methods in Molecular Biology, Vol. 44:* Agrobacterium *Protocols*
Edited by: K. M. A. Gartland and M. R. Davey Humana Press Inc., Totowa, NJ

manently accumulated in tissues or become remobilized during growth and development *(6,7)*. In some cases, they may be secreted, as by salt glands or trichomes *(8)*. In the absence of well-defined targets for manipulating endogenous metal accumulation and sequestration mechanisms, the well-characterized animal MT gene and its protein have been the focus of initial attempts to manipulate metal accumulation in plants. Cd, a pollutant metal that is an undesirable contaminant in agricultural crops and in general environments, has been a focus in many of these studies. MT has a high affinity for class IB and IIB transition elements (i.e., stability constant of 10^{17} to 10^{15} for Cd and coordinates metals through mercaptide bonds with cysteine), is about 7 kDa, and binds seven atoms of metal/peptide. The properties of naturally occurring and recombinant MTs were recently reviewed *(9,10)*.

There is considerable evidence that higher plants contain MT-like genes *(11–15)*. However, to date, no report in the refereed literature describes the isolation of the protein product of these genes from a plant extract. An MT-like gene isolated from pea has been expressed in *Escherichia coli* by two groups *(16,17)*. The work of Kille et al. *(16)* shows that the metal-binding domains at the molecules termini are animal-fungal MT-like, but the linker region is much larger and is sensitive to proteolysis. The work of Tommey et al. *(17)* indicates that a glutathione-S-transferase-pea MT fusion protein expressed in *E. coli* has metal binding characteristics like that of animal-fungal MT. Attempts to isolate mouse MT-I protein and the endogenous plant-MT gene product from plants expressing the mouse MT-I gene is discussed in Section 3.6.

It may soon be possible to test in plants an alternative to the MT-Cd sequestration strategy. Ow and coworkers isolated a gene designated as *hmt*I from an *S. pombe* mutant deficient in formation of a sulfide–phytochelatin–Cd complex that is formed during growth of this yeast (and higher plants) in the presence of very high levels of Cd *(18)*. The gene appears to encode for a tonoplast-membrane transport protein having homology to ABC-type membrane transport proteins found in various eukaryotic and prokaryotic cells. Members of this family of membrane transport proteins facilitate movement of ions, peptides, and other small molecules in mammalian and bacterial cells. Yeasts overexpressing this gene were shown to have enhanced Cd accumulation and Cd tolerance under high-level Cd exposure conditions *(18)*. It will be interesting to test the effects of *hmt*I expression on Cd accumulation and partitioning in plants.

The methods used in transgenic experiments to manipulate Cd accumulation and partitioning in tobaccos expressing animal MT are described here. Studies to date, though limited, have yielded some promising results. But they have also exposed possible difficulties associated with the MT-Cd sequestration strategy. These obstacles are discussed in Note 10. Other methods not yet applied to transgenic plants but directly applicable to the study of metal tissue-partitioning in transgenic plants are briefly described in Notes 5, 6, and 15.

To minimize the Cd content of crops it will undoubtedly be desirable to express a Cd-sequestration mechanism in specific-tissues-only using tissue-specific promoters. For most crops, root-specific sequestration would be desirable *(5)*. As our understanding of gene elements conferring tissue specificity increases, this refinement of current experiments will undoubtedly be possible. Some preliminary experiments have been conducted to this end *(19)*.

2. Materials

1. Shoot-forming selective medium: Murashige and Skoog (MS) medium 1 consists of 4.3 g MS salts (Gibco-DRL, Gaithersburg, MD), 1 mL B5 vitamins (2.5 g thiamine-HCl, 0.25 g nicotinic acid, 0.25 g pyridoxine-HCl, 25 g myo-inositol in 500 mL H_2O), 30 g sucrose, 1 mg 6-benzylaminopurine, 0.1 mg napthyleneacetic acid, in 1 L H_2O. Adjust pH to 5.75–5.8, add 8 g phytoagar, autoclave for 35 min, and cool to 50°C. Add, via a sterile filter (0.2 μM), 400 mg mefoxin (Merck, Rahway, NJ) and 200 mg kanamycin monosulfate in 10 mL H_2O. Stir and pour into Petri plates.
2. Gel filtration medium: 25 mM KH_2PO_4/K_2HPO_4, pH 7.2, 10 mM β-mercaptoethanol. For exchange binding, about 100,000 dpm of ^{109}Cd was added prior to chromatography on Sephadex G-25 *(20)*.
3. Root-forming selective medium: Same as shoot-forming selective medium but without 6-benzylaminopurine and napthyleneacetic acid.
4. 1.5-mL Conical Eppendorf tubes with tip, 0.5-cm portion, and cap removed.
5. Pyrex baking dishes (38 × 25 × 5 cm).
6. One-half strength Hoagland's solution *(21)*.
7. Nitric acid:perchloric acid, 9:1 (v/v).
8. Protoplasting solution: 2% (w/v)-final cellulysin (Calbiochem, La Jolla, CA), 0.5% final pectinase (72 U/mL, Sigma [St. Louis, MO], P-5146) desalted on G-25 coarse then made 0.3M with mannitol.
9. Protoplast purification step gradient: 4.0 mL 15% (w/v) Ficoll in 0.3M mannitol.

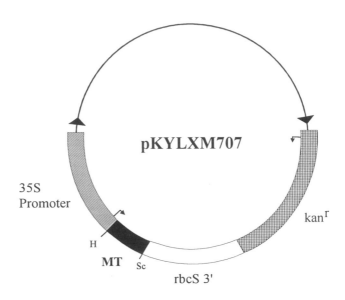

Fig. 1. Expression vector pKYLXM707 *(23)* derived from the expression cassette pKYLX7 *(22)* containing the mouse MT-I cDNA and selectable markers.

10. Protoplast lysis medium: $0.18M$ mannitol, 1.0 mM EGTA, 0.5 mM CHAPS (Sigma), 20 mM HEPES/Tris, pH 8.0.
11. Vacuole purification step gradient 20% (w/v) Ficoll in 0.24 mM mannitol, 1.0 mM EGTA, 0.5 mM CHAPS, 20 mM HEPES/Tris, pH 8.0.
12. Anti-MT antibody (Alpha Gamma Laboratory, Sierra Madre, CA).

3. Methods

3.1. Transformation of Tobacco with a Chimeric Mouse MT-I Gene and Characterization of Transformants

In our studies, transformation of *Nicotiana tabacum* cv KY 14 and cv Petit Havana is achieved using the binary vector pKYLX7 *(22)* containing the constitutive 35S promoter from the cauliflower mosaic virus, the poly(A) signal from the pea *rbc-S*-E9 gene, several selectable markers, and an inserted mouse *MT-I* cDNA (Fig. 1; *21*). The plasmid is transferred to *Agrobacterium tumefaciens* C58C1:GV3850 and the leaf disk method is used for tissue transformation *(23)*.

1. Regenerate shoots from callus on modified MS medium (shoot-forming selective medium) containing 200 µg/mL kanamycin and 400 µg/mL mefoxin.

2. Obtain rooted plantlets using the medium of Nitsch *(22)* containing antibiotics (root-forming, selective medium).
3. Confirm integration and expression of the *MT-I* gene using Southern, Northern, and Western blot analysis *(23)*.
4. Estimate the amount of MT-I protein in leaves of transgenic KY 14 plants using Western blots. Typically, this is ≈0.1% of leaf-soluble protein *(23)*.
5. Gel filtration-[109]Cd binding assay should show that MT extracted from leaves is capable of binding Cd *(20)*. Prepare tissue extracts using gel filtration medium.
6. Add ≈100,000 dpm [109]Cd to 1 mL of extract and hold 5–10 min at room temperature prior to loading on a Sephadex G-25 column (90 × 15 cm).
7. Elute using gel filtration medium and analyze fractions directly for [109]Cd. Several alternate approaches and genes have been used to introduce MT into plants (*see* Note 2).

3.2. Bare-Root Solution Culture of Plants to Assess Tissue Partitioning of Cd

1. Transfer size-matched, control (nontransformed), and kanamycin-resistant transformed seedlings from R1 plants after 2 wk growth on root-forming, selective medium to vermiculite-filled 1.5-mL plastic microfuge tubes from which the bottom 1 cm and cap has been removed (*see* Fig. 2; *21*).
2. Fill the bottom 1/3 of tubes with unmilled vermiculite and the top 2/3 with vermiculite milled to pass a 20-mesh screen.
3. About 16 transformed and 16 control seedlings in tubes are held by a sheet of aluminum foil (5 cm between tubes) over a shallow (38 × 25 × 5 cm) Pyrex dish containing 2 L of 1/2 Hoagland's nutrient solution (see Fig. 2; *20,21*).
4. After 1 wk of growth under constant cool-white light at 23–26°C, 10 µCi of [109]Cd (2.5 Ci/gm, Research Products International, Mount Prospect, IL) is added and plants are grown for an additional 15 d. Generally, six independent experiments are performed for each R1 line or variety tested *(20)*.
5. Bare-root solution culture may also be used to grow mature plants (*see* Note 5) and grafting may be used to assess the dominance of roots and shoots in metal partitioning (*see* Note 6).

3.3. Tissue Preparation and Quantitation of [109]Cd and [112]Cd Content

Individual tops and roots of seedlings reach a size of about 0.1 and 0.04 gm dry wt, respectively, after growth for 15 d with [109]Cd.

1. Wash roots free of vermiculite using distilled H_2O.
2. Freeze tissues (–80°C), lyophylize, then digest with nitric:perchloric acid (9:1, v/v).

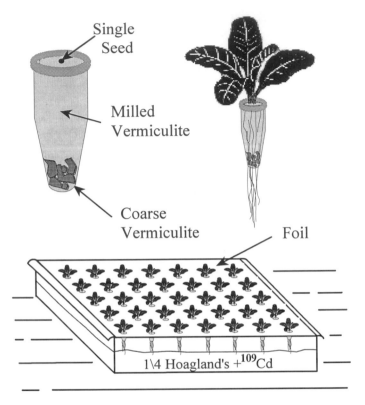

Fig. 2. Bare-root solution culture system.

3. Evaporate digests and dissolve the ash with $1N$ HCl.
4. Count an aliquot by gamma spectroscopy.
5. Analyze variance using the SAS procedure and significance using the Student's t-test *(20)*.

3.4. Field Studies with Transgenic Plants

1. Homozygous seedlings are selected with kanamycin from self-fertilized seed of one primary transformant segregating 3:1 *(24)*.
2. Seedlings are grown in peat pots and transferred to the field after 4 wk.
3. A randomized complete block field design is used with six or eight replications. Each experiment contains 12 plants per replication with 46 cm between plants and 107 cm between rows. Two border rows of nontransformed plants flank each experiment.
4. Irrigation is applied as necessary and insects may be controlled with acephate.
5. Adventitious shoots are removed by hand and plants are decapitated to remove flower initials as is the common cultural practice with burley tobacco.

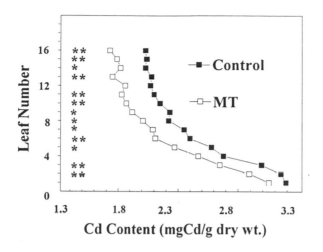

Fig. 3. Cd accumulation in leaves of field-grown *N. tabacum* cv KY 14, MT-expressing and nontransformed control.

6. At harvest, leaves of a given stalk position in a replication are pooled. Leaves are washed to dislodge soil and midribs are removed.
7. Lamina are forced, hot-air-dried, weighed, and ground to pass a 40 mesh.
8. An aliquot is digested as described in Section 3.3. and analyzed by flame atomic absorption spectroscopy.
9. Soil samples are removed at the beginning and end of the experiment (or at weekly intervals) from the center of each replication at a depth of 15 cm. For soil solution pH measurement, a 1:1 (v/v) soil/freshly boiled-H_2O slurry is maintained at room temperature for 30 min, centrifuged at 10,000g for 10 min, and the pH of the supernatant is measured at 25°C.
10. Soil Cd is determined by wet oxidation of 0.25-g aliquots of dry soil followed by atomic absorption analysis.
11. Statistical analysis is as for seedling data. Typical results are shown in Fig. 3 (*also see* Notes 10–12).

3.5. Subcellular and Cell-Type Localization of Accumulated Cd and MT

The subcellular location of accumulated Cd or of MT in transgenic tobaccos or *Brassica* plants has not been determined. However, Cd has been shown to be predominantly vacuolar in leaves of nontransformed tobacco seedlings grown in 20-μM Cd *(25)* and in *Datura inoxia* cultured cells grown in 30-μM Cd *(26)* using protoplasts and isolated vacuoles to determine vacuole/extravacuole distribution. Using the same approach, intracellular Cd has been found to be principally vacuolar in tobacco seedlings grown with trace levels of Cd (Wagner, unpublished).

1. To assess vacuole/extravacuole distribution, isolate protoplasts from leaves of seedlings or tissue-cultured cells that were grown in the presence of Cd *(25,26)*.
2. Prepare protoplasts by digestion of plasmolyzed tissues with wall-degrading enzymes (protoplasting solution).
3. Purify protoplasts using a protoplast purification step gradient containing Ficoll.
4. Release vacuoles by osmotic shock in protoplast lysis medium.
5. Purify vacuoles on a vacuole purification step gradient.
6. Compare vacuole and protoplast Cd content on a per-structure basis or using the enzyme α-mannosidase as a vacuolar marker enzyme *(25,26)*.
7. Observe MT protein by Western blotting and gel filtration-[109]Cd binding assay of soluble extracts of seedlings and mature leaves of transgenic plants *(23)*. In preliminary experiments, it is possible to assess the distribution of Cd in cell types of tissues using tissue printing (*see* Note 15).

3.6. Attempts to Monitor MT and Bound-Cd in Plant Extracts

Soluble-bound Cd has been observed in plant extracts, primarily using gel-filtration, ion exchange, and reversed phase high-performance liquid chromatography (RP HPLC) methods. Most studied of the bound-Cd forms are the so-called phytochelatins that are prominent Cd-binding peptides of plant tissues and cultured cells grown with moderate-to-high Cd levels *(27)*. The extent of their occurrence in plants exposed to Cd levels generally encountered in natural and agricultural environments is not yet known *(5)*.

MT (≈9 kDa-apparent) has been observed by gel filtration of transgenic (but not control) leaf extracts incubated with [109]Cd to affect exchange-binding of [109]Cd *(19,20)*. Plants used in these experiments were grown without added Cd and little or no phytochelatin was produced. A second [109]Cd-binding (4.3 kDa-apparent) component was found in transgenic plants but not controls. This component does not appear to be a phytochelatin, but its identity is not known.

4. Notes

1. Genetic analysis of self-pollinated R1 plants should indicate if inheritance of the marker is as a dominant Mendelian trait *(22)*.

2. Several other laboratories have reported on Ti-plasmid mediated transformation using various animal MT-genes and animal MT-reporter gene fusions *(28–30)*. In one case, an Ri-plasmid was utilized *(31)*.
3. In the system described in Fig. 2, roots grow through the vermiculite into nutrient solution and recovery of all roots—free of vermiculite—is facilitated. Since vermiculite binds Cd, its complete removal from roots without loss of roots is essential.
4. A typical final concentration of Cd added to the system described by Fig. 2 is 0.018 μM; however, effective concentration may be lower owing to complexation with nutrient medium salts. In any case, uptake is sufficient to allow analysis of Cd partitioning and one may estimate that a concentration of about 0.01 μM is similar to that expected in a soil solution of an uncontaminated crop soil *(5)*.
5. It is possible to grow plants to maturity using a gravel-solution culture system *(32)*, even mature tobacco plants that may reach a height of >3 m. Roots of young solution-cultured plants are placed under acid-washed and neutralized, nonporous stones contained in a 20-L plastic pail. Stones are irrigated every 20 min (with free draining) with 1/2-strength Hoagland's solution. This system has been utilized to characterize the tissue partitioning of Cd in tobacco plants exposed to nongrowth inhibiting concentrations (3 μM) of ^{112}Cd. Tissues can be analyzed for Cd content by atomic absorption spectroscopy. This system allows recovery of most roots by simply pouring out the root-stone contents of containers and carefully washing roots free, then collecting dislodged fine roots by filtration. The method has not yet been applied to the study of transgenic plants, but this is feasible. It has been used to make detailed analysis of tissue partitioning of Cd in untransformed, mature *Nicotiana tabacum* cv KY 14 plants *(21)*. The order of decreasing Cd concentration (dry wt basis) in tissues of flowering plants was found to be: oldest leaves > young leaves = floral structures = seed = midribs = roots > stems. The profile obtained with *N. rustica* was similar except that smallest roots had Cd concentration similar to that of oldest leaves. The desirable low leaf-very high root Cd accumulation characteristic of this species was simply correlated with its formation of a large mass of high Cd accumulating fine roots *(21)*. Cd was about evenly distributed within lamina in both *N. tabacum* and *N. rustica (21)*.
6. Grafting experiments can also be used to assess the dominance of roots or shoots in partitioning. This method has not yet been applied to transgenic tobacco, but it was useful in demonstrating that the root controls Cd translocation to the shoot in *N. tabacum* and *N. rustica (33)*. When high leaf-high root accumulating *N. tabacum* tops were grafted onto low leaf-very high root Cd accumulating *N. rustica* roots and established grafts were

exposed to Cd in gravel-solution culture, the level of accumulated Cd in tops was *N. rustica*-like. The inverse graft resulted in *N. tabacum*-like Cd accumulation and results with self-graft controls were as expected. This method is directly applicable to transgenic plants and therefore is noted here.

7. When mature Cd-exposed tissues are analyzed, materials are forced hot-air dried, ground to pass a 40 mesh, and a weighed aliquot is digested and analyzed by flame atomic absorption spectroscopy *(21,24)*.

8. Transportation, field growth, and post-harvest disposal of transgenic plants were as prescribed under permits from the USDA Animal and Plant Health Inspection Service (APHIS).

9. Since tobacco soils generally contain elevated Cd owing to repeated use of phosphatic fertilizers which contain Cd *(5)*, endogenous Cd in such soils ($\approx$0.4 µg/g dry wt) is sufficient to allow Cd monitoring of tobacco tissues using flame atomic absorption spectroscopy.

10. Results of field trials of MT expressing and control *N. tabacum* cv KY 14 (Fig. 3) show that transformed *N. tabacum* cv KY 14 (burley-type tobacco) lamina contained about 14% lower leaf Cd than did control lamina. Differences were significant at the $p \leq 0.1$ level in 13 of 16 leaf positions. Dry leaf weights did not differ significantly but transformed plants had 12% fewer leaves and 9% lower height *(24)*. Zn levels of control and transformed plants were similar but Cu content was about 10% higher in the lower leaf positions of transformed plants, suggesting that perhaps reduced leaf number and plant height were owing to Cu deficiency or toxicity resulting from MT binding of Cu. Alternatively, somaclonal variation (tissue culture derived) or gene position effects may have been involved. These results perhaps indicate the need to engineer the MT gene to remove the metal binding domain that principally binds Cu. Since it is not possible to recover the smallest roots (high Cd accumulating, *21*) from field-grown plants, it was not possible to compare root/shoot accumulation of solution-cultured and field-grown plants. However, comparison of leaf data (seedling vs mature field plants) showed that although tops (above root) of transformed seedlings had 24% lower Cd than controls, the difference was 14% in field-grown plants. Roots of transformed seedlings had about 5% higher Cd than controls *(20)*.

11. Results obtained with another MT expressing tobacco (cv Petit Havana-cigar wrapper-type tobacco), which had a substantially different growth habit than cv KY 14, differed from that with KY 14. In seedling experiments, tops contained no less Cd than controls, but roots had 48% higher Cd content *(24)*. However, transformed seedling weights were 25% greater than controls. In the field, transformed and controls were similar in leaf Cd content, plant height, and leaf number. Again, roots could not be adequately recovered from soil.

12. Experimental results with KY 14 support the hypothesis that the MT-Cd sequestration strategy may result in lower translocation of Cd to above root tissues. However, reduced leaf number and plant height and modified Cu content suggest secondary effects of the presence of MT. Results obtained with cv Petit Havana indicate the need to test the MT-sequestration strategy using other primary transformants of KY 14, other varieties of tobacco and other plant species. In a related study, Brandle et al. *(30)* studied the impact of a GUS-Chinese hamster-*MT-II* fusion gene in mature field-grown plants using *N. tabacum* cv Delgold (flue-cured-type tobacco). No significant differences in leaf Cd content were found between transformed and control plants. The fusion protein was not isolated and its Cd binding capacity was not established. However, metal binding ability of MT-fusion proteins have been shown in at least two studies. A streptavidin-mouse *MT-I* chimera capable of binding both 1 mol of biotin and the normal MT-compliment of seven metal ions/mol MT was recently shown to be efficiently expressed in *E. coli (34)*. Similarly, a glutathione-S-transferase pea MT chimera capable of binding metals was expressed in *E. coli (17)*.

13. As no targeting signals were employed in the transformation vector used *(23)*, it is expected that MT occurs in the cytoplasmic fraction. This could be verified in future experiments by vacuole/extravacuolar analysis and separation of organelles on sucrose gradients.

14. The use of immunocytochemistry to localize MT as applied in studies of MT animal cells will be facilitated by the commercial availability of anti-rat MT antibody from Alpha Gamma Laboratory. Anti-rat MT-I antibody obtained from this source does not bind <60 kDa peptides on Western blots of desalted tobacco leaf extracts (unpublished).

15. The distribution of Cd and MT in cell types of tissues (i.e., stem, leaf, and root cross sections) may be possible. It has been observed that [109]Cd is efficiently absorbed to nitrocellulose filters *(35)* under conditions used in tissue printing. Filters were immersed in $0.2M$ $CaCl_2$ for 30 min, then air-dried. Fresh cut, blotted tissue sections were pressed on to the filters for 15–20 s. Prints prepared from adjacent tissue sections were dried. One was stained with 0.1% toluidine blue in H_2O to reveal cell outlines and the other autoradiographed using X-ray film to visualize [109]Cd. Outlines of stained and radioactivity images were compared directly (Wang, Hunt, and Wagner, unpublished). The isotope [113]m Cd may be more suited to this purpose than [109]Cd. MT may be similarly visualized on tissue prints using anti-MT antibody.

16. As noted earlier, MT-like genes have been observed in several plant species, but their protein products have not been isolated. The high cysteine content of MT and that predicted for plant MT-like products (30 residue

percent) may contribute to problems encountered in recovering these from plant extracts. Polyphenols common in plants (abundant in tobacco) are known to bind sulfhydryl groups of proteins, making recovery of these proteins difficult. MT from transgenic tobacco has been observed by Western blotting of leaf extracts prepared in the presence of 50 mM Tris-HCl, pH 7.4, 50 mM 2-mercaptoethanol, 0.1 mM KCN, and 55 mM ascorbate *(23)*. Recovery of animal MT protein added to such extracts is not quantitative. Other efforts to separate MT from transgenic tissues—to which standard MT had been added—using RP HPLC *(36)* or ion-exchange HPLC *(37)* have not been successful. Methods for observing carboxymethylated MT on silver-stained SDS-PAGE gels *(38)* may offer some promise to a problem that has plagued efforts to isolate products of plant MT genes and our efforts to quantitatively compare MT partitioning and Cd partitioning in plants expressing animal MT.

References

1. Nieboer, E. and Richardson, D. H. S. (1980) The replacement of the nondescript term "heavy metals" by a biologically and chemically significant classification of metal ions. *Environ. Pollution (Series B)* 1, 3–26.
2. Clarkson, D. T. (1983) Movement of ions across roots, in *Solute Transport in Plant Cells and Tissues* (Baker, D. A. and Hall, J. L., ed.), Wiley, New York, pp. 251–305.
3. Marschner, H. (1983) General introduction to the mineral nutrition of plants, in *Encyclopedia of Plant Physiology* (Laüchli, A. and Bieleski, R., eds.), Springer-Verlag, Berlin, pp. 5–49.
4. Verkleij, J. A. C. and Schat, H. (1990) Mechanisms of metal tolerance in higher plants, in *Evolutionary Aspects of Heavy Metal Tolerance in Plants* (Shaw, J., ed.), CRC Press, Boca Raton, FL, pp. 179–193.
5. Wagner, G. J. (1993) Accumulation of Cd in crop plants and its consequences to human health *Adv. Agronomy* 51, 173–211.
6. Cataldo, D. A., Garland, T. R., and Wildung, R. E. (1981) Cadmium distribution and chemical fate in soybean plants. *Plant Physiol.* 68, 835–839.
7. Hill, J. (1980) The remobilization of nutrients from leaves. *J. Plant Nutr.* 2, 407–444.
8. Fahn, A. (1988) Secretory tissues in plants. *New Phytol.* 108, 229–257.
9. Kagi, J. H. R. (1991) Overview of metallothionein. *Meth. Enzymol.* 205, 613–626.
10. Kay, J., Cryer, A., Darke, B. M., Kille, P., Lees, W. E., Norey, C. G., and Stark, J. M. (1991) Naturally occurring and recombinant metallothioneins: structure, immunoreactivity and metal-binding functions. *Int. J. Biochem.* 23, 1–5.
11. Robinson, N. J., Gupta, A., Fordham-Skelton, A. P., Coy, R. R. D., Whitton, B. A., and Huckle, J. W. (1990) Procaryotic metallothionein gene characterization and expression: chromosome crawling by ligation-mediated PCR. *Proc. R. Soc. Lond. B* 242, 241–247.
12. Evans, I. M., Gatehouse, L. N., Gatehouse, J. A., Robinson, N. J., and Croy, R. R. D. (1990) A gene from pea (*Pisum sativum* L.) with homology to metallothionein genes. *FEBS Lett.* 262, 29–32.

13. DeMiranda, J. R., Thomas, J. R., Thurman, M. A., and Tomset, A. B. (1990) Metallothionein genes from the flowering plant Mimulus gullatus. *FEBS Lett.* **260,** 277–280.

14. de Framond, A. J. (1991) A metallothionein-like gene from maize. *FEBS Lett.* **290,** 103–106.

15. Kawashima, I., Inokuchi, Y., Chino, M., Kimura, M., and Shimizv, N. (1991) Isolation of a gene for a metallothionein-like protein from soybean. *Plant Cell Physiol.* **32,** 913–916.

16. Kille, P., Winge, D. P., Harwood, J. L., and Kay, J. (1991) A plant metallothionein produced in *E. coli. FEBS Lett.* **295,** 171–175.

17. Tommey, A. M., Shi, J., Lindsay, W. P., Urwin, P. E., and Robinson, N. J. (1991) Expression of the pea gene PsMT$_A$ in *E. coli. FEBS Lett.* **292,** 48–52.

18. Ortiz, D. F., Kreppel, L., Speiser, D. M., Scheel, G., McDonald, G., and Ow, D. W. (1992) Heavy metal tolerance in fission yeast requires an ATP-binding cassette-type vacuolar membrane transporter. *EMBO J.* **11,** 3491–3499.

19. Maiti, I. B., Wagner, G. J., and Hunt, A. G. (1991) Light inducible and tissue-specific expression of a chimeric mouse metallothionein cDNA gene in tobacco. *Plant Sci.* **76,** 99–107.

20. Maiti, I. B., Wagner, G. J., Yeargan, R., and Hunt, A. G. (1989) Inheritance and expression of the mouse metallothionein gene in tobacco. *Plant Physiol.* **91,** 1020–1024.

21. Wagner, G. J. and Yeargan, R. (1986) Variation in Cd accumulation potential and tissue distribution of Cd in tobacco. *Plant Physiol.* **82,** 274–279.

22. Schardl, C. L., Byrd, A. D., Benzion, G., Altschuler, M. A., Hildebrand, D. F., and Hunt, A. G. (1987) Design and construction of a versitile system for the expression of foreign genes in plants. *Gene* **61,** 1–11.

23. Maiti, I. B., Hunt, A. G., and Wagner, G. J. (1988) Seed-transmissible expression of mammalian metallothionein in transgenic tobacco. *Biochem. Biophys. Res. Comm.* **150,** 640–647.

24. Yeargan, R., Maiti, I. B., Nielsen, M. T., Hunt, A. G., and Wagner, G. J. (1992) Tissue partitioning of Cd in transgenic tobacco seedlings and field grown plants expressing the mouse metallothionein I gene. *Transgenic Res.* **1,** 261–267.

25. Vogeli-Lange, R. and Wagner, G. J. (1990) Subcellular localization of cadmium and cadmium-binding peptides in tobacco leaves—implication of a transport function for cadmium-binding peptides. *Plant Physiol.* **92,** 1086–1093.

26. Krotz, R. M., Evangelou, B. P., and Wagner, G. J. (1989) Relationships between Cd, Zn, Cd-peptide and organic acid in tobacco suspension cells. *Plant Physiol.* **91,** 780–787.

27. Rauser, W. E. (1990) Phytochelatins. *Annu. Rev. Plant Physiol.* **59,** 61–86.

28. Lefebvre, D. D., Miki, B. L., and Laliberte, J. (1987) Mammalian metallothionein functions in plants. *Biotechnol.* **5,** 1053–1056.

29. Misra, S. and Gedamu, L. (1989) Heavy metal tolerant transgenic *Brassica* and *Nicotiana* plants. *Theor. Appl. Genet.* **78,** 161–168.

30. Brandle, J. E., Labbe, H., Hattori, J., and Miki, B. L. (1993) Field performance and heavy metal concentrations of transgenic flue-cured tobacco expressing a mammalian metallothionein-β-glucuronidase gene fusion. *Genome* **36,** 255–260.

31. Pautot, V., Ryszard, B., and Tepfer, M. (1989) Expression of a mouse metallothionein gene in transgenic plant tissues. *Gene* **77,** 133–140.
32. Hatfield, J. L., Egli, D. B., Leggett, J. E., and Peaslee, D. E. (1974) Effect of applied nitrogen on the nodulation and early growth of soybeans. *Agron. J.* **66,** 112–114.
33. Wagner, G. J., Sutton, T. G., and Yeargan, R. (1988) Root control of leaf cadmium accumulation in tobacco. *Tobacco Sci.* **190(20),** 64–68.
34. Sano, T., Glazer, A. N., and Cantor, C. R. (1992) A streptavidin-metallothionein chimera that allows specific labeling of biological materials with many different heavy metal ions. *Proc. Natl. Acad. Sci. USA* **89,** 1534–1538.
35. Ye, Z. and Varner, J. E. (1991) Tissue-specific expression of cell-wall proteins in developing soybean tissues. *Plant Cell* **3,** 23–37.
36. Richards, M. (1991) Purification and quantification of metallothioneins by reversed-phase high performance liquid chromatography. *Methods Enzymol.* **205,** 217–238.
37. Klaassen, C. D. and Lehman-McKeeman, L. D. (1991) Separation and quantification of isometallothioneins by high performance liquid chromatography-atomic absorption spectrometry. *Methods Enzymol.* **205,** 190–198.
38. Otsuka, F., Koizumi, S., Kimura, M., and Ohsawa, M. (1988) Silver staining for carboxymethylated metallothioneins in polyacrylamide gels. *Anal. Biochem.* **168,** 184–192.

CHAPTER 27

Overexpression of Chloroplastic Cu/Zn Superoxide Dismutase in Plants

Randy D. Allen

1. Introduction

Molecular oxygen, although required for aerobic life, is toxic. The cellular damage caused by oxygen derivatives is known as oxidative stress. To survive, all aerobic organisms must posses the ability to protect themselves from oxidative stress. These protective mechanisms include both enzymatic and nonenzymatic systems. Superoxide dismutases (SODs) catalyze the vital first step in the enzymatic oxidative stress protective system.

1.1. Characteristics of SODs

SODs are a group of metalloenzymes found in nearly all aerobic organisms. Although the function of SOD was not identified until 1968 *(1)*, it is among the most extensively studied enzymes. The superoxide radical (O_2^-) is formed by the addition of a single electron to molecular oxygen. This can occur through a number of pathways in vivo, the most important of which are probably the membrane-bound electron transport chains of mitochondria, endoplasmic reticulum, and chloroplasts *(2,3)*. In plant cells, chloroplasts are especially susceptible to oxidative stress because they have a high internal O_2 concentration and the transfer of electrons from electron receptors, such as ferredoxin, can directly reduce O_2. SOD catalyzes the dismutation of O_2^- to H_2O_2, as shown in Fig. 1. Although O_2^- is unstable in aqueous media and dismutates spontaneously to H_2O_2, this reaction occurs relatively slowly under typical biological

From: *Methods in Molecular Biology, Vol. 44:* Agrobacterium *Protocols*
Edited by: K. M. A. Gartland and M. R. Davey Humana Press Inc., Totowa, NJ

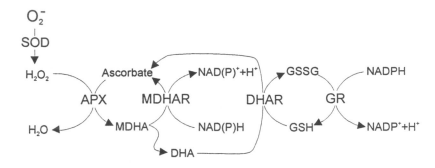

Fig. 1. Proposed active oxygen scavenging pathway of plant chloroplasts. SOD and APX scavenge superoxide radicals (O_2^-) and hydrogen peroxide (H_2O_2), respectively. Monodehydroascorbate reductase (MDHAR), dehyro-ascorbate reductase (DHAR), and glutathione reductase (GR) are necessary for the rapid regeneration of ascorbate.

conditions ($k_2 = 5 \times 10^5/M/s$ at pH 7.0). The V_{max} of this reaction catalyzed by Cu/Zn SOD is about $1.6 \times 10^9/M/s$ *(3)*. The H_2O_2 produced in this reaction can be metabolized through a number of pathways, the most damaging of which is its reaction with O_2^- in the presence of Fe^{3+} to form the highly reactive and toxic hydroxyl radical ($\cdot OH$). It is believed that most of the biological damage associated with oxidative stress is attributable to $\cdot OH$. Thus, cellular systems that scavenge both H_2O_2 and O_2^- appear to be critical for effective protection from oxidative stress. Enzymes involved in scavenging of H_2O_2 in plant cells include catalase and ascorbate peroxidase (APX) (Fig. 1). Catalase is limited primarily to peroxisomes in plant cells where it scavenges H_2O_2 produced during photorespiration and β-oxidation of lipids. Although catalase has a high catalytic capacity, its affinity for H_2O_2 is relatively low (measured K_m approaches $1M$). Since chloroplast H_2O_2 concentrations as low as 10 μM can inhibit photosynthesis by 50% *(4)*, catalase would probably not be as effective as APX for protection of chloroplasts.

SODs contain either copper and zinc, manganese, or iron prosthetic groups. Although Mn SODs are found in both bacteria and eukaryotes, Cu/Zn SODs appear to be restricted to eukaryotes. Fe SODs are present in prokaryotes and in plants but they do not appear in animals. In plant cells, Cu/Zn SODs are found both in the cytosol and in chloroplasts. Chloroplasts of some plants, including tobacco, contain abundant Fe SOD in addition to Cu/Zn SOD. The majority of the Mn SOD in plants is located in mitochondria, although there is evidence that, at least in pea, a fraction of the Mn SOD exists in peroxisomes *(5)*. A high degree of sequence similarity exists between SODs of each type, and Fe and Mn SOD are highly similar to each other. However, there is no apparent similarity between Cu/Zn SODs and Mn or Fe SODs. These observations have led some to suggest that Cu/Zn SOD is eukaryotic in origin and Mn and Fe SODs are prokaryotic enzymes that have been introduced into eukaryotes by endosymbiosis *(3)*.

1.2. Role of SOD in Plant Stress Protection

Active oxygen species are considered to be important damaging factors in plants exposed to stressful environmental conditions *(6)*. Enhanced levels of SOD have been correlated with increased resistance to drought, air pollutants (e.g., ozone and sulfur dioxide), hypoxia, and high light intensity. Exposure to these and other stresses can also induce the expression of SODs. Paraquat-resistant biotypes of *Conyza bonariensis* (hairy-fleabane) have been identified that also have increased resistance to other oxidative stresses including drought, air pollutants, and photoinhibition *(7)*. The oxidative stress resistance in these plants correlates with constitutively elevated levels of SOD, APX, and glutathione reductase. Interestingly, this pleiotropic trait is inherited in *C. bonariensis* as a dominant allele of a single locus *(8)*.

1.3. Overexpression of SODs Provides Increased Oxidative Stress Tolerance

Attempts to analyze the active oxygen scavenging pathway by modification of the expression of SOD and other individual enzymes using plant transformation techniques has been undertaken in a number of laboratories. Initial reports indicated that expression of high levels (30- to 50-fold above endogenous SOD activity levels) of *Petunia* chloroplastic Cu/Zn SOD in tobacco plants did not lead to a significant increase in resistance to paraquat, photoinhibition, or ozone *(9,10)*. However, subsequent reports from several research groups have clearly shown that increased expression of SOD in plants can lead to significant improvements in oxidative stress tolerance *(11,12)*. Sen Gupta et al. *(12)* reported that transgenic tobacco plants, which express moderate levels ($\approx$threefold increase over endogenous SODs) of chloroplast-localized Cu/Zn SOD from pea, have significantly increased resistance to light-mediated paraquat damage and to photoinhibition caused by high light intensity and low temperature. Genetic analysis of progeny of self-pollinated primary regenerated transgenic plants showed perfect cosegregation of SOD overexpression and resistance to photoinhibition *(13)*.

1.3.1. Overexpression of Cu / Zn SOD Affects Other Enzymes

If the H_2O_2 scavenging capacity in plant cells is limiting, overexpression of SOD could lead to deleterious effects owing to the accumulation of this toxic product. Interestingly, overexpression of chloroplastic Cu/Zn SOD in tobacco plants leads to a compensatory increase in the expression of endogenous APX genes *(13)*. However, specific activities of subsequent enzymes of the Halliwell-Asada pathway including monodehydroascorbate reductase, dehydroascorbate reductase, and glutathione reductase are not elevated in transgenic SOD overexpressing plants. Induction of APX, the enzyme primarily responsible for the removal of H_2O_2 in chloroplasts and cytosol of plant cells, could provide the increased scavenging capacity necessary to cope with the likely increase in H_2O_2 production caused by elevated SOD levels.

2. Materials
2.1. Vector Development and Plant Transformation
2.1.1. Nucleic Acids

1. Fully characterized, full length Cu/Zn SOD cDNA inserted into an appropriate plasmid vector such as pUC19 *(see* Note 1).
2. Expression vector containing appropriate promoter and terminator sequences such as pRTL2.
3. Binary shuttle vector such as pBIN19.
4. Mobilization plasmid pRK2013 for triparental mating.

2.1.2. Enzymes and Other Materials for DNA Manipulations

1. Appropriate restriction enzymes and DNA modifying enzymes.
2. Polymerase chain reaction kit (Amplitaq, Perkin-Elmer Cetus, Norwalk, CT).
3. Powdered glass DNA purification kit (e.g., US Bioclean, US Biochemicals, Cleveland, OH).

2.1.3. Bacterial Strains

1. *Escherichia coli* strain DH5α for growth of pRTL2 and pBIN19 constructs.
2. *Agrobacterium tumefaciens* strain LBA 4404 for plant transformation using pBIN19 constructs.

2.1.4. Plant Materials

1. *Nicotiana tabacum* cv Xanthi.

2.1.5. Culture Media

1. Luria broth media (Gibco-BRL, Gaithersburg, MD) solidified with 1.5% Bacto-Agar and supplemented with appropriate antibiotics for growth of *E. coli* and *A. tumefaciens.*
2. Plant tissue-culture media for transformation and regeneration of transgenic tobacco plants (all components from Sigma [St. Louis, MO] except where noted).
3. MSA: 4.33 g/L M&S salts, 1 ml/L Gamborg's B-5 vitamin mixture, 3% sucrose, 0.1 mg/L naphthalene acetic acid (NAA), 1 mg/L benzyladenine (BA). Solidified with 0.8% agar or 0.2% Gel-Grow (ICN).
4. MSB: Same components as MSA plus 350 mg/L kanamycin and 100 mg/L cefotaxime.
5. MSC: Same as MSB but without NAA or BA.
6. Culture plates and magenta boxes.

2.2. Enzyme Assays

2.2.1. Polyacrylamide Gel SOD Assay

This assay allows identification of SOD isoforms by separation on a nondenaturing polyacrylamide gel and negative staining for SOD activity (*see* Note 2).

1. Resolving gel: 10% polyacrylamide (29:1 acrylamide:bisacrylamide), 0.37M Tris-HCl pH 8.8.
2. Stacking gel: 5% polyacrylamide 0.126M Tris-HCl pH 6.8.
3. Tank buffer: 25 mM Tris, 250 mM glycine, pH 8.3.
4. Sample buffer: 50 mM Tris-HCl pH 6.8, 10% glycerol (v/v), and 0.1% (w/v) bromophenol blue.
5. SOD gel staining solutions: Solution A: 50 mM phosphate buffer pH 7.5, 2.45 mM nitro blue tetrazolium (NBT, Sigma). Solution B: 50 mM phosphate buffer pH 7.5, 28 mM tetramethylethylenediamine (TEMED), and 28 µM riboflavin (*see* Note 3).

2.2.2. Spectrophotometric Enzyme Assays

1. Homogenization solution for spectrophotometric SOD assay: 50 mM potassium phosphate buffer pH 7, 0.1 mM EDTA, and 1% (w/v) polyvinylpolypyrrolidone.
2. Spectrophotometric SOD assay is performed in a 1-mL reaction volume containing 50 mM phosphate buffer pH 7.8, 56 µM NBT, 10 mM methionine, and 0.17 mM riboflavin (*see* Note 4).

3. Homogenization solution for APX consists of 50 mM HEPES, pH 7.0, 1 mM ascorbate, 1 mM EDTA, and 1% (v/v) Triton X-100.
4. Spectrophotometric APX is assayed in a 1-mL reaction volume containing 50 mM HEPES, pH 7.0, 1 mM EDTA, 1 mM H$_2$O$_2$, 0.5 mM ascorbate, and 25 µL of enzyme extract.
5. UV/Vis spectrophotometer.

2.3. Oxidative Stress Analysis

2.3.1. Paraquat Treatments

1. Paraquat solutions of 0.3, 0.6, 1.2, and 2.4 µM prepared in distilled H$_2$O. Triton X-100 can also be included if necessary to facilitate infiltration.
2. Orion model 120 conductivity meter or equivalent and an immersion conductivity cell with 3-mL capacity.

2.3.2. Photoinhibition Analysis

1. Hansatech Oxygen electrode set up or equivalent.
2. Apparatus for high light intensity and low temperature exposure of leaf disks (Fig. 2).

3. Methods

3.1. Development of a Chimeric Cu/Zn SOD Gene

3.1.1. PCR-Based Mutagenesis

If necessary, appropriate restriction enzyme sites should be added to the SOD cDNA to allow convenient ligation into an expression vector. For example, Sen Gupta et al. *(12)* used polymerase chain reaction (PCR) mutagenesis to introduced a *Nco*I site at the start codon and a *Xba*I site in the 3' untranslated sequence of the pea Cu/Zn SOD cDNA. This allowed for the insertion of the cDNA fragment into the pRTL2 expression vector into the existing start codon, resulting in a translation product that consists only of Cu/Zn SOD coding sequences.

1. Amplification of plasmid DNA is carried out using primers complementary to target sequences with the 1- or 2-base alterations necessary to create the restriction sites.
2. Purify the amplification products using powdered glass. Digest with appropriate restriction enzymes and purify again with glass powder.

3.1.2. SOD Expression Vector Construction

1. Ligate digested SOD cDNA fragment or PCR products into the appropriate restriction sites of the expression vector (Fig. 3).

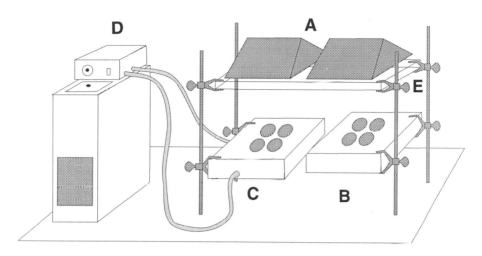

Fig. 2. Diagram of apparatus for exposure of leaf disks to high light intensity and low temperature. Illumination is provided by high intensity quartz-halogen lamps (**A**). Light intensity is adjusted by varying the distance between the light sources and the plexiglass sample stages (**B** and **C**). Leaf disks are placed on wet filter paper on the sample stages. Sample stage B is maintained at room temperature for pre-equilibration and nonstress control treatments. Stage C is temperature controlled by internal circulation with 50% ethylene glycol from a refrigerated circulator bath (**D**). Excess heating of leaf disks from the lamps is prevented with a plexiglass diffuser and water filter (**E**).

2. The completed chimeric gene construct can be excised from the expression vector and subcloned into the binary shuttle vector.
3. The shuttle vector can be mobilized to and appropriate *A. tumefaciens* strain by triparental mating using pRK2013 as a mobilization plasmid or by direct transformation using freeze thaw or electroporation techniques (*see* Note 5 and Chapter 33).

3.2. Plant Transformation and Regeneration

1. Surface sterilize young tobacco leaves with 10% Clorox solution, rinse with sterile water, and cut leaf pieces with a sterile scalpel or cork borer.
2. Grow an overnight culture of LBA 4404 that contains the Cu/Zn SOD::pBIN19 plasmid in minimal media containing appropriate antibiotics.
3. Dilute the overnight culture 1:10 with liquid MSA media and dip the leaf disks into the diluted culture for approx 5 min.
4. Inoculated leaf pieces are blotted dry and placed upside down on MSA plates for 2–3 d.

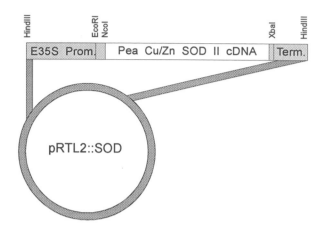

Fig. 3. Example chimeric gene construct for overexpression of chloroplast-localized Cu/Zn SOD in plant cells. The chimeric gene includes a CaMV 35S promoter with a duplicated enhancer element and a CaMV terminator-poly A addition sequence. Pea Cu/Zn SOD II cDNA encodes chloroplast localized Cu/Zn SOD. The Cu/Zn SOD cDNA is fused at the at the star codon of the 5' untranslated sequence of TEV (*see* Note 1).

5. Transfer the leaf pieces to MSB media (MSB contains 300 mg/L kanamycin) for callus growth and shoot formation.
6. After shoots are large enough to handle (4–6 wk) cut them from the callus with a sterile scalpel and place onto MSC media until roots appear (usually within 2–3 wk).
7. Transfer the rooted shoots to sterile soil in magenta boxes and grow for several weeks until an extensive root system is observed.
8. Plantlets are then hardened to ambient humidity by opening the magenta box slowly over a 1-wk period. Hardened plants can be transplanted into pots and grown in a greenhouse.
9. Generally, more than 20 independent transgenic plants (T_0) should be regenerated and analyzed for expression of additional SOD isoforms by the SOD gel assay (*see* Section 3.3.1.). Plants that express detectable levels of the unique SOD isoforms should be grown to maturity, self-pollinated, and seeds (T_1) collected (*see* Note 6).

3.3. Analysis of Enzyme Activity

3.3.1 Polyacrylamide Gel SOD Assay

Electrophoretic separation of plant extracts on nondenaturing polyacrylamide gels provides a semiquantitative assay that allows for visualization of individual SOD isoforms. This assay is a convenient way to screen large numbers of individual transgenic plants for expression of unique SOD isoforms.

1. Grind approx 1 g of leaf tissue on ice in a glass homogenizer in 100 μL of extraction buffer.
2. Centrifuge the resulting slurry for 2 min in a microcentrifuge and load approx 60 μL of the supernatant onto a nondenaturing polyacrylamide gel.
3. Run the gel at 80 V until the bromophenol blue in the sample solution has reached the bottom of the gel.
4. Remove gel from the glass plates and soak in solution A in total darkness for 30 min.
5. Rinse gel with distilled water and soak in solution B, in the dark for 30 min.
6. Develop the gel by exposure to light until maximum contrast is achieved. Color development is stopped by soaking the gel in 10% acetic acid (*see* Note 7).

3.3.2. Spectrophotometric SOD Assay

Quantitative measurement of SOD can be carried our using a number of spectrophotometic assays. The author prefers the NBT reduction assay described by Beyer and Fridovich *(14)* A commercially available kit (SOD-525, Bioxytech, Lyon, France), which is simple to perform and seems to give reliable results, has also been used. However, the kit is quite expensive and the components are proprietary.

1. Leaf tissue ($\approx$0.5 g) is ground in liquid N_2, suspended in SOD homogenization solution, and rapidly homogenized on ice in a glass tissue grinder.
2. Centrifuge the resulting slurry in a microcentrifuge for 2 min and remove the supernatant (*see* Note 8).
3. Prepare the assay solution using various concentrations of enzyme extract in clean glass culture tubes. Add the methionine last to start the reaction.
4. Expose tubes to a light source such as a fluorescent light box and watch carefully as the solution changes from yellow to blue (*see* Note 9).
5. Place the tubes in the dark to stop the reaction.
6. Read absorbance in spectrophotometer at 560 nm (*see* Note 10).

3.3.3. Ascorbate Peroxidase Assay

1. Grind 0.5 g of leaf tissue in liquid N_2, suspend in APX homogenization solution, and homogenize in a glass homogenizer on ice.
2. Centrifuge in a microcentrifuge for 2 min and remove supernatant
3. Prepare APX assay solution adding the H_2O_2 last to start the reaction.
4. Immediately read the change in absorbance (ΔOD) at 290 nm using the kinetic channel of the spectrophotometer (*see* Note 11).

3.4. Analysis of Oxidative Stress Tolerance

3.4.1. Paraquat Treatment

Paraquat (methyl viologen), a contact herbicide that causes massive, light-mediated production of superoxide radicals in photosynthetic tissues, is useful for determining the protective effects of Cu/Zn SOD overexpression in transgenic plants. Although it is possible to assay paraquat resistance using visible damage or regrowth of treated tissues as criteria, these methods are either highly subjective or time consuming. Therefore, an electrolyte leakage assay has been used to measure the extent of membrane damage in paraquat-treated tissues.

1. Collect leaf disks (1.5 cm^2) of each transgenic SOD overexpressing plant and control plants to be tested (*see* Note 12).
2. Infiltrate leaf disks in paraquat by soaking them in 3 mL of paraquat solutions of various concentrations in 3.5-cm diameter Petri dishes for 16 h at 23°C in total darkness (*see* Note 13).
3. Expose leaf disks to light (500 µmol quanta/m/s^2) for 2 h.
4. Incubate in the dark at 30°C for 16 h to allow oxidative damage to occur.
5. Remove the paraquat solutions and measure the conductance using a conductivity meter and record the value (*see* Note 14).
6. Pour the solutions from the conductivity cell into clean glass culture tubes and place the treated leaf disks into the solutions.
7. Autoclave the tubes to destroy all membrane integrity in the leaf disk, releasing all cellular solutes.
8. After the solutions have cooled to room temperature, determine and record the conductivity of each sample again (*see* Note 15).

3.4.2. Analysis of Photoinhibition

Exposure of leaves to high light intensity, especially under conditions that inhibit carbon fixation, such as low temperature, can cause long-term damage to the photosystems. This damage is known as photoinhibition and occurs, at least in part, because of the increased photoreduction of O_2 under these conditions. Analysis of plants that overexpress chloroplast-localized Cu/Zn SOD can therefore provide information about the role of oxidative stress in causing photoinhibition (*see* Note 16).

1. Collect 10 cm^2 leaf disks from appropriate leaves of SOD overexpressing and control plants (*see* Note 17).
2. Pre-equilibrate the leaf disk by placing it on moist filter paper on the high intensity light apparatus adjusted to an appropriate light intensity (for tobacco use 1500 µmol quanta/m^2/s) at room temperature for 1 h (*see* Note 18).

3. Remove leaf disks one at a time to measure photosynthesis at room temperature. This measurement provides a prestress photosynthetic rate for comparison with rates measured after stress treatment (*see* Note 19).
4. Move leaf disks to cold block adjusted to the stress temperature under high-intensity light for an appropriate time (*see* Note 20).
5. Remove leaf disks from cold block one at a time to measure photosynthesis at room temperature (*see* Note 21).

4. Notes
4.1. Vector Development and Plant Transformation

1. The choice of coding sequences, vectors, and host cell lines is up to the individual investigator. For example, the author routinely uses the expression vector pRTL2, which includes a Cauliflower mosaic virus (CaMV) 35S promoter with a duplicated enhancer region and a CaMV 35S 3' terminator and poly A addition signal sequence. In addition, the 5' untranslated sequence from tobacco etch virus (TEV) is provided with a *Nco*I restriction site located at the TEV start codon. Chimeric gene constructs developed in this vector can be easily excised and subcloned into the binary shuttle vector pBIN 19.

4.2. Enzyme Assays

2. The polyacrylamide gel for SOD analysis is similar to a typical sodium dodecyl sulfate (SDS)-polyacrylamide gel with the exception that SDS is omitted. We generally run SOD gels in a Bio-Rad (Richmond, CA) Mini Protean gel system. The 10 mL of resolving gel solution is sufficient for two gels. The gel is polymerized by the addition of 30 µL of 15% ammonium persulfate and 8 µL of TEMED just before pouring. The 5-mL stacking gel solution is polymerized by addition of 20 µL of ammonium persulfate and 5 µL of TEMED just before pouring.
3. MTT (thiazol blue, Sigma) can be used in place of NBT and is somewhat less expensive.
4. The amount of enzyme extract is adjusted to give a 50% inhibition of NBT reduction.

4.3. Development of a Chimeric Cu/Zn SOD Gene

5. Since it is often difficult to obtain significant amounts of pBIN19-based constructs presumably owing to the relative instability of this vector, triparental mating is routinely used to transfer from the plasmid to *Agrobacterium*. However, direct transformation methods may be preferable in many cases.
6. Since primary transgenic plants containing single functional Cu/Zn SOD transgene insertions will produce offspring that segregate 3 to 1 for SOD

overexpression when self-pollinated, seeds collected from these plants should be grown and analyzed for expression of the SOD transgene. T_1 plants that express the introduced SOD transgene can be used for subsequent experimental analyses, whereas nonexpressing plants can be used as controls. Although untransformed plants and plants transformed with an irrelevant construct, such as β-glucuronidase, can be used as controls, we consider T_1 plants that do not overexpress SOD to be the most appropriate control because they share a common genetic background with their SOD overexpressing siblings.

4.4. Analysis of Enzyme Activity

7. Any high-intensity light source can be used for development of SOD gels but care must be taken to avoid overdevelopment. For this reason, a fluorescent light box is used. It may be preferable to run the gel at 4°C and all incubations and development can also be carried out at reduced temperatures.
8. Aliquots of leaf extract should be removed before centrifugation to determine chlorophyll content and after centrifugation to determine protein concentration. Although the assay can usually be performed with crude enzyme extracts, if difficulty is encountered, it may be necessary to desalt the extract by G-25 chromatography to remove interfering compounds, such as ascorbate and glutathione.
9. The SOD assay is quite difficult to perform quantitatively. The time of light exposure is critical and must be adjusted in preliminary assays for different light intensities. It is critical that all tubes receive the same light intensity. It is also necessary to run control assays with commercially available SOD (Sigma) to calibrate the assay.
10. One unit of SOD activity is defined as the amount of enzyme required to inhibit the photoreduction of NBT by 50%. Activity is therefore expressed as SOD U/mg protein or U/mg chlorophyll.
11. APX activity is determined by measuring the oxidation of ascorbate at 290 nm. A drop in absorbance of 1 OD U/h in a 1-mL reaction volume equals 21.43 μmol of ascorbate oxidized/h. The specific activity of APX is expressed as μmol ascorbate oxidized/h/mg protein or mg chlorophyll.

4.5. Analysis of Oxidative Stress Tolerance

12. Paraquat resistance can vary with plant age and leaf development; therefore, it is critical to sample leaves of the same developmental stage from plants of the same age.
13. Paraquat solutions ranging from 0.3–2.4 μM have been used for tobacco, but since paraquat sensitivity different species and different varieties varies, it is necessary to carry out preliminary tests on control plants to determine a range of paraquat concentrations that gives a useful range of cellular

damage. It is also important to include leaf disks soaked in solutions without paraquat as a control. The author has attempted to facilitate infiltration of paraquat by placing the leaf disks under vacuum for 5 min prior to the overnight soak, but has not noticed any dramatic improvements using this method.

14. An immersion conductivity cell with a 3-mL capacity should be used. The capacity of some cells can be reduced by filling the extra space with Parafilm. The cell is inverted and Parafilm is used to seal the vent holes. Pour the incubation solutions into the sealed cell and read the conductivity. An inexpensive Orion model 120 has been used for most measurements with very good results.

15. The extent of cellular damage owing to paraquat exposure is calculated as the electrolyte released from paraquat-treated leaf disks as a percent of the electrolyte released from autoclaved leaf disks (by definition, autoclaved leaf disks sustain 100% membrane damage and release all solutes). The comparison of electrolyte leakage of test leak disks with autoclaved leaf disks provides an internal control for each sample and reduces error caused by variations in leaf disk size or thickness. Using these methods the author has found that leaf disks of transgenic SOD overexpressing plants receive significantly less damage than control leaf disks at paraquat concentrations up to 1.2 μM, but at 2.4 μM paraquat the differences were not significant (*12*).

16. The leaf disk assay described here is for analysis of photoinhibition in tobacco leaves. Because the susceptibility to photoinhibition of different plant species varies it will be necessary to adjust the conditions to provide appropriate levels of stress.

17. Susceptibility to photoinhibition changes with leaf maturity and with plant development. In general, younger plants are more susceptible than older plants and younger leaves are more susceptible than older leaves. Leaves acclimated to low light intensity are also more susceptible than those acclimated to high light. It is absolutely critical that leaf samples are collected from comparable leaves of plants of the same developmental stage grown under identical conditions. Generally the first fully expanded leaf should be sampled (usually the fifth leaf from the top of the plant).

18. Pre-equilibration at high light intensity and room temperature helps to ensure that all leaf disks start out under the same conditions and are as physiologically similar as possible. Some leaf disks should be left under these pre-equilibration conditions throughout the experiment for use as nonstressed controls and to ensure that the pre-equilibration conditions do not cause photoinhibition. Be sure that the leaf disks remain fully hydrated throughout the experiment. Leaf disk temperatures should be monitored with a thermocouple thermometer.

19. Maximum photosynthetic rates should be measured under saturating CO_2 and saturating light intensity. For tobacco, a Hansatech O_2 electrode system with a white light source adjusted with neutral density filters to 970 µmol quanta/m^2/s is used. Similar results have been obtained using $^{14}CO_2$ incorporation.

20. Temperature, light intensity, and exposure time needed to achieve the appropriate levels of stress must be determined experimentally for the species and variety being used. For tobacco, 1500 µmol quanta/m^2/s at 3°C is used. If necessary, leaf disks can be placed on a wet ice block to maintain low temperatures.

21. Recovery of photosynthesis after stress treatment can be monitored at 25°C in the O_2 electrode cuvette. In the author's experience, steady-state photosynthetic rates are reached in both SOD overexpressing and control leaf disks within 30 min. Under the conditions described, photosynthesis of leaf disks from mature, vegetative, control plants is reduced by approx 70% in 4 h. Leaf disks from SOD overexpressing plants at the same developmental stage is reduced by 5% or less *(12)*.

References

1. McCord, J. A. and Fridovich, I. (1969) Superoxide dismutase: enzymatic role for erythrocuperin (hemocuperin). *J. Biol. Chem.* **244,** 6049–6055.
2. Cadenas, E. (1989) Biochemistry of oxygen toxicity. *Annu. Rev. Biochem.* **58,** 79–110.
3. Halliwell, B. and Gutteridge, J. M. C. (1989) *Free Radicals in Biology and Medicine*, Clarendon Press, Oxford, UK.
4. Kaiser, W. M. (1979) Reversible inhibition of the Calvin cycle and activation of the oxidative pentose phosphate cycle in isolated chloroplasts by hydrogen peroxide. *Planta* **21,** 377–382.
5. Sandalio, L. M. and Del Rio, L. A. (1988) Intraorganellar distribution of superoxide dismutase in plant peroxisomes. *Plant Physiol.* **88,** 1215–1218.
6. Bowler, C., Van Montagu, M., and Inzé, D. (1992) Superoxide dismutase and stress tolerance. *Annu. Rev. Plant Physiol. Plant Mol. Biol.* **43,** 83–116.
7. Shaalteil, Y. and Gressel, J. (1986) Multienzyme oxygen radical detoxifying system correlated with paraquat resistance in *Conyza bonariensis. Pesticide Biochem. Physiol.* **26,** 22–28.
8. Shaalteil, Y., Chua, N.-H., Gepstein, S., and Gressel, J. (1988) Dominant pleiotropy controls enzymes co-segregating with paraquat resistance in *Conyza bonariensis. Theor. Appl. Genet.* **75,** 850–856.
9. Tepperman, J. M. and Dunsmuir, P. (1990) Transformed plants with elevated levels of chloroplastic SOD are not more resistant to superoxide toxicity. *Plant Mol. Biol.* **14,** 501–511.
10. Pitcher, L. H., Brennan, E., Henley, E., Dunsmuir, P., Tepperman, J. M., and Zilinskas, B. A. (1991) Overproduction of petunia chloroplastic copper/zinc superoxide dismutase does not confer ozone tolerance in transgenic tobacco. *Plant Physiol.* **97,** 452–455

11. Perl, A., Perl-Treves, R., Galili, S., Aviv, D., Shalgi, E., Malkin, S., and Galun, E. (1993) Enhanced oxidative-stress defense in transgenic potato expressing tomato Cu, Zn superoxide dismutases. *Theor. Appl. Genet.* **85,** 568–576

12. Sen Gupta, A., Heinen, J. L., Holaday, A. S., Burke, J. J., and Allen, R. D. (1993) Increased resistance to oxidative stress in transgenic plants that over-express chloroplastic Cu/Zn superoxide dismutase. *Proc. Natl. Acad. Sci. USA* **90,** 1629–1633.

13. Sen Gupta, A., Webb, R. P., Holaday, A. S., and Allen, R. D. (1993) Overexpression of superoxide dismutase protects plants from oxidative stress: induction of ascorbate peroxidase in SOD over-expressing plants. *Plant Physiol.* **103,** 1067–1073.

14. Beyer, W. F. and Fridovich, I. (1987) Assaying for superoxide dismutase activity: some large consequences of minor changes in conditions. *Anal. Biochem.* **161,** 559–566.

CHAPTER 28

Agroinfection

Nigel Grimsley

1. Introduction
1.1. General Information

The term "agroinfection" was first used (1) to describe the use of *Agrobacterium* for the introduction of infectious molecules to plants. This implies infection of the host plant with a molecule, the "infectious agent," generally a virus or viroid, that has the ability to replicate and spread within the plant; the introduction of parts of agents that do not have this potential is thus outside the scope of this chapter. Subsequently, the term "agroinoculation" has also been used by some authors to describe the inoculation step (2). Replication of the agent within the plant often leads to systemic viral or viroidal symptoms that witness T-DNA transfer, independently of T-DNA integration. Nontumorigenic strains of *Agrobacterium* may be used; this provides the additional possibility of regenerating transgenic plants containing all or part of the agent genome integrated in the plant nuclear DNA.

In 1986, several groups independently demonstrated that *Agrobacterium* can be used to introduce the cloned DNA of CaMV, PSTV, and TGMV to plant cells (1,3,4; please refer to Table 1 for a list of abbreviations). Although these molecules are all biologically active when introduced in the form of naked cloned DNA, it was not certain that *Agrobacterium* would provide a suitable route alternative to natural vectors or cloned DNA, because in order to be infectious the self-replicating molecule must first be liberated from the T-DNA in the plant. However, agroinfection turned out to be an efficient means for delivery of these

From: *Methods in Molecular Biology, Vol. 44:* Agrobacterium *Protocols*
Edited by: K. M. A. Gartland and M. R. Davey Humana Press Inc., Totowa, NJ

Table 1

The Uses of Agroinfection with Different Plant Infectious Agents

Abbreviation	Name of agent	Plant	References[a]
ACMV	African cassava mosaic virus	*Nicotiana benthamiana*	6,7 (vir); 8,9 (res)
BCTV	Beet curly top virus	*Nicotiana benthamiana*	10 (vir)
		Lycopersicon esculentum	10 (vir)
		Beta vulgaris	10 (vir)
CaMV	Cauliflower mosiac virus	*Brassica campestris*	1 (inf); 11 (rec,rep); 12 (T-S)
		Brassica napus	13,14 (rec); 15 (agro)
		Brassica rapa	16,17 (agro)
		Raphanus sativus	15 (T-T)
CoYMV	Commelina yellow mottle virus	*Commelina diffusa*	18 (inf)
CMV	Cucumber mosaic virus	*Nicotiana tabacum*	19,20 (res)
DSV	Digitaria streak virus	*Digitaria sanguinalis*	21 (inf)
		Avena sativa	21 (inf)
		Zea mays	21 (inf)
HSV	Hop stunt viroid	*Nicotiana Tabacum*	22 (inf)
MSV	Maize streak virus	*Avena sativa*	23[b](T-T)
		Digitaria sanguinalis	23[b](T-T)
		Hordeum vulgare	23[b](T-T)
		Lolium tementutum	23[b](T-T)
		Panicum milaceum	23[b](T-T)
		Triticum aestivum	23[b](T-T)
		Zea mays	5,24 (inf); 25–30,32 (vir); 17 (agro); 15 (T-S, T-T); 31[a] (T-T); 32,33 (plant); 34,35 (tra)
MiSV	Miscanthus streak virus	*Panicum milaceum*	36 (inf)
		Zea mays	36 (inf)
PSTV	Potato stunt tuber viroid	*Lycopersicon esculentum*	3 (inf); 37 (agro)
PSV	Panicum streak virus	*Zea mays*	38 (inf)

	Rice tungro bacilliform virus	*Oryza sativa*	*39 (inf)*
TMV	Tobacco mosaic virusf	*Nicotiana tabacum*	*40 (inf); 41 (res)*
TGMV	Tomato golden mosaic virus	*Datura stramonium*	*42 (vir)*
		Nicotiana benthamiana	*43 (inf); 44,45 (rep); 42 (vir); 46,47 (vec); 34 (tra)*
		Nicotiana clevelandii	*42 (vir)*
		Nicotiana glutinosa	*42 (vir)*
		Nicotiana tabacum	*43 (inf); 48,49 (rep); 50,51 (vir); 42 (vir); 51,52 (vec); 53 (exp)*
TobRV	Tobacco ringspot virus	*Petunia hybrida*	*4 (inf,rec); 53 (exp)*
		Nicotiana tabacum	*54 (res)*
TYLCV	Tomato yellow leaf curl virus	*Lycopersicon esculentum*	*55 (inf)*
WDV	Wheat dwarf virus	*Aegilops speltoides*	*56 (plant)*
		Hordeum vulgare	*57 (T-T)*
		Triticum aestivum	*58 (inf); 59,60[b] (T-T); 56 (plant)*
		Triticum durum	*56 (plant)*
		Triticum monococcum	*56 (plant)*

[a]Types of work undertaken: (agro) AGRObacterium biology—host range and mutant strains, (exp) transient EXPression, (inf) INFectivity of agent, (plant) PLANT susceptibility/developmental requirement for agroinfection, (rec) RECombination, (rep) REPlication mechanism of agent, (res) plant RESistance, (tra) TRAnsposable element biology, (T-S) T-DNA Structure, (T-T) T-DNA Transfer, (vec) viral expression VECtors, (vir) VIRal biology/mutagenesis.
[b]A variety of *Agrobacterium* strains was studied in these cases.

viruses into plants. In 1987, agroinfection of maize with MSV demonstrated not only that the cloned MSV isolate was infectious (naked DNA was not infectious), but also that T-DNA transfer was possible to a graminaceous plant *(5)*, although *Agrobacterium* does not elicit tumor formation on monocotyledonous plants and its classical host range is limited mainly to dicotyledonous plants. This work opened new avenues of research concerning viral biology and the study of DNA transfer to monocots by *Agrobacterium*. In the following years, the number of laboratories using agroinfection to investigate various aspects of the plant–pathogen interaction has therefore grown. In many of these cases, agroinfection is so far the only way of re-introducing cloned viral DNA back to a host plant.

1.2. Biosafety

Agroinfection involves using a combination of two plant pathogens. The regulations for experimental use of different pathogens varies between different countries, depending on whether or not the pathogen is endemic, and the potential risks, if any, to the local ecosystem and to agriculture. It is therefore important to check that the local safety requirements can be complied with before starting experimentation. A physical containment laboratory is often necessary. In addition, biological safety can be increased by using mutant strains of pathogens that have decreased fitness in natural environments, and keeping infectious agent clones in plasmids that can only be maintained in bacteria by continued selection for a genetic marker (e.g., antibiotic resistance).

1.3. Practical Uses

In this chapter, it is not possible to give detailed protocols concerning all of the procedures used with different virus-bacterium-plant combinations. However, a summary of the reported practical uses of agroinfection, together with references to relevant literature, is presented in Table 1. This table can be used to find the work most closely related to the infectious agents in which you are interested, and provides a starting point for information on the biology of the agents. Several reviews are available concerning DNA transfer by *Agrobacterium* (this volume, and references cited therein). An explanation of the different kinds of usage is given, followed by general methodological principles, and a more specific example.

1.3.1. Storage and Use of Inoculum

Most plant viruses are transmitted by insect vectors. In the laboratory, using insects for inoculation is inconvenient, as it is difficult to maintain the insect cultures, to control the quantity of inoculum, to ensure the use of one particular viral sequence in successive inocula (owing to mutation), and to avoid the danger of cross-contamination (flying inocula!). In those cases where naked DNA is effective as an inoculum, usually about 10 μg of DNA is required for a reproducible infection, necessitating regular preparations of large amounts of plasmid DNA. On the other hand, once an infectious agent is cloned in *Agrobacterium*, it can be conserved indefinitely at −80°C; small amounts of the frozen bacteria are then taken out of stock for growth of the inoculum (*see* Note 1).

1.3.2. Infection of a Plant with a Cloned Viral DNA

In those cases where cloned DNA is infectious (ACMV [6–9], BCTV [10], CaMV [1,11–17], HSV [22], PSTV [37], TMV [40,41], TGMV [4,42–53], TobRV [54]), agroinfection is often used simply as an alternative more efficient way of introducing the viral material. However, cloned viral DNA is not infectious in many cases (CoYMV [18], DSV [21], MiSV [36], MSV [5], PSV [38], RTBV [39], TYLCV [55], WDV [58]), and agroinfection provides the only means of introducing the virus to the plant.

1.3.3. Study of Viral Replication and Recombination

Several properties of the agroinfection system permit investigation of certain aspects of viral replication and recombination. The viral DNA is usually cloned as a tandem dimer (or one-and-a-bit-mer) in *Agrobacterium*. Thus, hybrid molecules containing different viral strains can be introduced. As the T-DNA enters the cell, an infectious molecule must escape, either by recombination or by formation of a replicative intermediate. The nature of this intermediate (the sequence that is amplified and that spreads through the plant) can give information concerning the replicative or recombinational mechanisms involved (4,11,13,14,44,45,48,49).

Plants transgenic for a part of a viral genome or for a defective viral genome can be inoculated with molecules encoding complementing functions or crossed with plants that carry these viral functions, providing the opportunity to observe recombination events that lead to the production of viable viruses. Since the resultant progeny viruses spread systemically in the plant, this provides a sensitive assay for somatic recombination events *(4,13,14)*.

1.3.4. Investigation of Viral Gene Functions

In cases where cloned viral DNA is not infectious, agroinfection permits an in vitro mutational analysis of the viral genome. Integration of mutant or partial viral genomes into the genome of the recipient plant allows dissection of some aspects of viral gene functions, for example, the functions necessary for replication and systemic spread can be distinguished *(6,7,10,25–30,42,50,51)*.

1.3.5. Production of Autonomously Replicating Viral Vectors

The functions required for replication, once defined, can either be used directly to amplify introduced foreign DNA, or be integrated into the host chromosome, together with a sequence of "foreign DNA," which is then replicated to a high copy number in the host plant cells. This system is potentially useful for the expression of large amounts of the gene product of interest in a recipient cell *(46,47,51,52)*.

1.3.6. Transient Expression

Agrobacteria carrying constructs such as those described in Section 1.4. can be infiltrated into leaf discs; as these molecules are then replicated in cells receiving a copy of the T-DNA, high levels of transient expression can be observed *(53)*.

1.3.7. Production of Virus-Resistant Plants

Tandem DNA copies of viral satellite sequences integrated in the genome of a plant remain inactive (nonreplicated) until the plant becomes infected with a virus. Certain viruses provide the satellite replication functions, and as a result the satellite escapes from the genome, and provides a natural means of attenuating the symptoms of the infecting virus *(8,9,19,20,41,54)*.

1.3.8. Investigation of Bacterial Functions Necessary for DNA Transfer to Plants That Are Known to be Hosts for the Virus

The efficiency of DNA transfer from different mutant strains or different isolates of *Agrobacterium* to a plant can be tested. This kind of test has been used most often in cases where tumor formation does not occur in the recipient plant, as, for example, with graminaceous monocotyledonous plants *(15–17,23,31,37,57,59,60)*.

1.3.9. Analysis of T-DNA Intermediates

Since the virus transferred is a part of the T-DNA, the whole T-DNA can itself be replicated under the control of the viral replication functions, providing that the sequences that remain outside of the viral DNA on the T-DNA are short enough to be cloned within the viral replicon. This system has been used to amplify T-DNA intermediates, providing a sensitive way of studying certain kinds of transfer intermediates, before integration into the host genome occurs *(12,15)*.

1.3.10. Study of Transposable Elements

Transposable elements can be introduced to a host plant by agroinfection, by cloning the element into the viral genome. Replication of the virus in the plant facilitates enormously the task of recovering DNA molecules that have been involved in a transposition event, where the element has excised from the virus, permitting the study of many such events *(34,35)*.

1.3.11 Tissue-Specific Susceptibility to Agroinfection

In many cases, particularly if a plant virus is already known to be infectious using cloned DNA as an inoculum, agroinfection can be done by introducing the bacteria to wounded plant tissues in different parts of the plant (*see* ref. *1* for an example). In other cases, particularly when agroinfecting monocotyledonous plants, only certain plant tissues are susceptible to agroinfection *(32,33,56)*. The reason for this specificity for a certain developmental stage is unclear, but may reflect a limitation in the ability of *Agrobacterium* to transfer its T-DNA to certain types of tissues.

1.4. General Approaches to Agroinfection

1.4.1. Cloning the Infectious Agent in Escherichia coli

It is assumed, for the purposes of this chapter, that some information about the genome of the virus is available, as initial characterization of the virus is specialized. RNA molecules must first be converted to cDNA for cloning *(61)*. Each component of multipartite viruses is cloned independently. Genomes of the infectious agents must be cloned with a tandem head-to-tail duplication, because monomer clones are only infectious in certain cases. This limitation arises because in order to escape from the T-DNA either homologous recombination (producing a complete circular infectious molecule) must occur, or production of a replication intermediate occurs (for example, the formation of the long transcript in caulimoviruses).

Often, the sequence of the virus is already known, or at least some restriction mapping data is available. If a clone of a well-characterized viral isolate is available, use this for the first agroinfection attempt. Before attempting further cloning steps, go through the subsequent theoretical steps necessary for transfer of the clone to *Agrobacterium*. Since dimers are not always straightforward to obtain, it may be best to plan construction of a one-and-a-bit-mer; a dimer construction can also be attempted in parallel. One-and-a-bit-mers also have the advantage that one part of the genome is present in only one copy; this facilitates genetic manipulations, such as sequence-directed mutagenesis, that might be planned for later work. However, it is unwise to plan intricate DNA manipulations before the infectivity is proven. If nothing at all is known about the biology of the virus, the safest option is to clone a dimer, by raising the concentration of the insert DNA in the ligation mix. If a simple restriction map of the virus is available, a one-and-a-bit-mer can be made, first by subcloning a fragment of the genome, then by cloning a unit genome in or alongside this.

The viral sequences must be cloned at some stage into a binary vector to permit transfer into *Agrobacterium tumefaciens* by mobilization or by transformation. They can also be moved from an *E. coli* vector to a binary vector, providing that suitable flanking restriction sites are available.

1.4.2. Transfer of the Cloned Agents to A. tumefaciens

Binary vectors are used in preference to integrative vectors because they permit mobilization of a given viral oligomer to different bacterial strains, and they are slowly lost from *Agrobacterium* in the absence of appropriate selection, making them biologically safer. Binary vectors can be transformed or mobilized into *Agrobacterium* (*see* Chapters 29 and 33).

The choice of *Agrobacterium* strain is important; some *A. tumefaciens* nopaline strains, such as C58, generally give high levels of transfer to a variety of hosts, whereas other strains do not transfer (by agroinfection or tumor standards) to certain plants. Some workers have noted that *A. rhizogenes* strain A4 shows a higher level of transfer to wheat than *A. tumefaciens* nopaline strains. A variety of strains have been tested (*23,31,60; see* Table 1). Mobilization of plasmids to exotic strains of *Agrobacterium* is more difficult than using common laboratory strains, because of the lack of suitable selectable markers. Useful tips to overcome difficulties are given in Note 2.

1.4.3. Inoculation of Plants

Inoculation is usually carried out either by rubbing a suspension of bacteria, containing an abrasive, into a leaf, or by injection of a suspension of bacteria into the stem. In the case of graminaceous plants, injection must be done in the region of the meristem *(32)*; only certain regions of plants at different early stages of development are susceptible to transfer (Table 1). Similarly, the choice of plant variety can influence T-DNA transfer.

1.4.4. Observation of Disease Symptoms

Observation for viral or viroidal disease symptoms depends on the virus or viroid being studied. Usually the infection behaves exactly like a natural infection in terms of symptomatology.

2. Materials

A specific example is provided by agroinfection of MSV in which the infectivity of a complete MSV sequence cloned in pBR322 is tested. The techniques necessary for cloning steps are in routine use in many laboratories *(62)*, and will vary depending on different agents, so details are not given. Likewise, more details for transfer of plasmids to *Agrobacterium* can be found in Chapter 23. A detailed procedure for inoculation is presented.

1. Physical containment laboratory: As dictated by local regulations.
2. Sterile equipment: Disposable hypodermic needles (0.45 × 10 mm or 0.40 × 20 mm), Hamilton syringe, magnetic stirrers, pipets, 8.5-cm diameter circular filter papers, Eppendorf tubes, glass bottles, bacterial culture vessels.
3. Sterile solutions: Water, 10 mM magnesium sulfate, bacterial growth media (e.g., Luria-Bertani medium: 10 g/L bacto-tryptone, 5 g/L bacto-yeast, 10 g/L NaCl), and antibiotics (*see* Chapters 16 and 29), 80% glycerol.
4. Plant culture: Culture chamber, compost in pots or trays, maize seeds (e.g., B73, Funk Seeds International, Bloomington, IL).
5. Equipment: Spectrophotometer and disposable cuvets, incubators for growing bacteria, –80°C freezer for bacterial stocks, sterile work station (laminar air flow bench), thimble.
6. Solutions: 1% sodium hypochlorite; 0.05% (w/v) sodium dodecyl sulfate; 70% (v/v) ethanol.

3. Methods
3.1. Cloning MSV

An example overall plan would be:

1. Make a large plasmid preparation of the pBR322 clone.

2. Check for the presence of a few restriction sites that may be useful in the cloning (e.g., *Bam*HI and *Xho*I).
3. Prepare fragments of the genome (by preparative agarose gel electrophoresis) for one-and-a-bit-mer cloning (e.g., *Bam*HI complete genome, *Bam*HI-*Xho*I large DNA fragment, and *Xho*I-*Bam*HI small fragment).
4. Clone the part genomes in a polylinker vector. (For example, clone *Xho*I-*Bam*HI fragments into vector cut with *Sal*I and *Bam*HI. The use of a vector that can be used for preparation of single-stranded DNA will allow sequencing of site-directed mutations that may be desired at a later stage of the project. Check which orientations and vector combinations are likely to be the most useful.) The use of an asymmetrical polylinker in this way leaves a unique site (e.g., *Bam*HI) for further cloning of one complete genome.
5. At this stage, dimer genome constructions can also be attempted (e.g., by using a 5:1 or 10:1 excess of insert (complete genome) DNA in a ligation mix).
6. Make plasmid preparations of the part genome clones and the dimer clones, if the latter are available. Clone into the part genome clones one complete genome from the DNA prepared previously. As there is no blue/white color or marker for screening, the ligation conditions will need to be optimal to avoid too much work in screening mini DNA preps. Keep both kinds of one-and-a-bit-mers (small fragment plus one genome and one genome plus large fragment), as this will permit site-directed mutagenesis on the whole of the genome at a later stage (present as a single copy in either one or the other type of clone).

3.2. Transfer of the Clones to Agrobacterium

1. Prepare plasmid DNA of one-and-a-bit-mers and dimer clones.
2. Using enzymes that do not cut in the virus, but do cut in the polylinker, excise the one-and-a-bit-mers and dimer, and prepare these fragments from an agarose gel. Clone them in a binary vector.
3. Mobilize the clones to *Agrobacterium* C58 using triparental mating and rifampicin plus kanamycin to select for exconjugants. Purify them twice on selective medium by streaking out and stock them at –80°C in glycerol.
4. If desired, the integrity of the viral sequences in *Agrobacterium* can be checked by preparation of DNA and Southern blotting at this stage, but it is recommended that the clones first be tested on maize, as this is easy and the symptoms are clear to see.

3.3. Inoculation
of Maize and Observation for Symptoms

The following procedure has been described *(32)*, and is currently being used with slight modifications *(63)* (*see* Notes 3–9).

3.3.1. Day 1

1. Prepare a set of sterile 9-cm diameter plastic Petri dishes (12–15 kernels will be put in each dish); put three sheets of 8.5-cm diameter filter paper and 10-mL sterile water in each dish.
2. Surface sterilize the maize seeds (e.g., variety B73). Add at least 2 vol of 1% sodium hypochlorite, 0.05% SDS solution to the kernels in a sterile bottle containing a magnetic stirring bar. Stir the seeds for 10–15 min (not longer, viability may decrease).
3. Remove the used bleach solution using a suction device or by decantation.
4. Add 2 vol of sterile water and stir for 10–15 min. Remove the water. Repeat this step once more.
5. Under sterile conditions, put 12–15 kernels in each dish.
6. Put them to incubate in the light or in the dark at 28°C.
7. Streak out on selective medium the *Agrobacterium* strains to be used (from frozen stock cultures; if the frozen stock culture has not been used recently, a prior check for viability is recommended). Put the culture to incubate at 28°C.

3.3.2. Day 3

8. Pick a single colony from the streaked out *Agrobacterium*, and inoculate liquid medium (YEB) containing appropriate antibiotics. Incubate shaking overnight at 28°C.

3.3.3. Day 4

9. The germinated maize seedlings should now be ready to inoculate (shoots about 2-cm long). There are differences in the time required for seedlings to develop to this stage, according to the maize variety being used, but usually 3–5 d is sufficient.
10. Measure the optical density at 600 nm of a tenth dilution of the bacterial culture. Collect the bacteria (a few milliliters) by centrifugation in a microfuge at 14,000g for 3 min. Resuspend the cell pellet in 10 mM magnesium sulfate to a final OD of 1 (this corresponds to about 10^5–10^6 bacteria/microliter for most *Agrobacterium* strains). A portion of the bacteria can be frozen down at this stage, if desired, in case later it is necessary to check the strain in some way.
11. Load the bacterial suspension into a clean Hamilton syringe and inoculate the seedlings with 10–20 μL using a disposable needle. Insert the needle into the meristematic ring of the coleptilar node, then inject the bacteria (Fig. 1); this should be done carefully, since agroinfection frequency drops off rapidly for injections outside of the meristematic region (five- or six-fold reduction 2 mm away from the node [32]). In maize the node can be recognized as a thickening on the shoot. It is convenient to hold the seedling in one hand and the syringe in the other—a thimble can be worn to

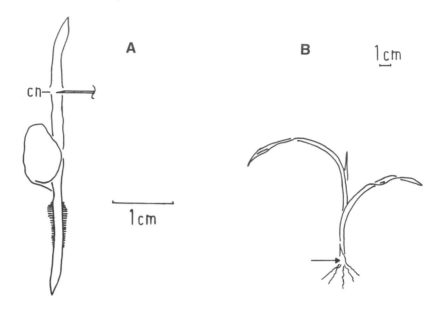

Fig. 1. Position of injection of maize for agroinfection with MSV: (**A**) 3-d-old dark-grown seedling and (**B**) 3-wk-old maize seedling.

    avoid stabbing your finger. When testing a series of bacterial strains the syringe can be sterilized by discarding the disposable needle, flushing out the syringe several times with 70% ethanol, and then flushing it through repeatedly with sterile water.

12. Plant the seedlings in seed trays or pots, and put them in a plant culture chamber. It is advisable to use presterilized compost, but the exact incubation conditions are not critical, because the test will be finished in 3–4 wk using this host–parasite combination. Culture temperature should not exceed 28°C.

3.3.4. Days 6–30

13. Observe the plants regularly for the presence of viral symptoms. Chlorotic spots usually appear 6–7 d after inoculation on the second or third leaf. Most of the plants develop symptoms after 2 wk, and final scoring can be done after 3 or 4 wk.

14. If desired, DNA can be extracted from the plants for confirmation of the presence of MSV or for recloning the systemic virus. Because infected (yellow-striped) maize leaves contain high quantities of MSV circular double-stranded DNA, the choice of plant DNA extraction protocol is not critical (*see* Chapter 15).

4. Notes

4.1. Stability of Infectious Agent Clones

1. Some viruses are difficult to clone because their sequences are poorly tolerated by *E. coli*. This difficulty may be overcome by using a recombination-deficient strain of *E. coli* and lower copy number cloning vectors. Colony hybridizations using part of the viral DNA as a probe can be useful to identify full-length clones. An antibiotic resistance cassette may be cloned in a unique site in a cloned viral genome, to facilitate recloning (of the viral DNA plus the marker) into a part genome length, but the marker must be excised before transfer to *Agrobacterium*. Frozen glycerol stocks of the bacteria must be made as soon as they have been grown up, as the clones will not be stable in stationary phase.

4.2. Agrobacterium *Strains*

2. Mobilization of a constructions to an *Agrobacterium* recipient strain may present a problem if the strain is not commonly used in different laboratories, as there may not be a suitable marker that can be used for selection following triparental mating. One possible solution is to produce antibiotic-resistant strains of *Agrobacterium*. Rifampicin-resistant strains of *Agrobacterium*, for example, can be obtained by plating out high densities of bacteria on selective medium (e.g., 100 µg/mL rifampicin). These can then be used as recipients in matings. Alternatively, try using minimal medium and high levels of selection (e.g., 400 µg/mL neomycin if the binary vector carries Km/Nm resistance) for plating out after triparental mating. Many laboratory strains of *E. coli* will not grow in these conditions, whereas wild-type *Agrobacterium* will grow slowly (L. Otten, personal communication, 1987). *Agrobacterium* can also be transformed (*see* Chapter 33).

4.3. Agroinfection

3. Include negative and positive controls (from a plant–parasite combination shown to work) for comparison.
4. Practice agroinfecting a combination that is known to work, where symptom formation is easily seen. In the case of monocotyledonous plants, the most critical step is the injection of bacteria into the meristematic region. This node is not always easy to identify, depending on the species. Try inoculating different age plants, from 3-d-old seedlings (in the coleoptilar node) to several-wk-old plants (in the base of the plant).
5. Hot lamps near plants in culture rooms may elevate the plant temperature to levels not tolerated by *Agrobacterium* (loss of Ti plasmid above 30°C).

6. It is important to try several different isolates of the viral strain. Cloning of inviable mutant strains is quite common. If possible, re-isolate new viral sequences from an infected plant.

7. Try different varieties of the host plant using *Agrobacterium* strains known to transfer to a broad range of plants before trying many *Agrobacterium* strains.

8. Check the integrity of the viral DNA sequences in *Agrobacterium* by preparation of DNA (preferably from a portion of the bacteria that were frozen down at the time of inoculation) and Southern transfer, using an internal fragment of the virus as a probe.

9. Include 200 μM acetosyringone *(64)* in the suspension of bacteria used for inoculation. This concentration saturates the *vir* gene induction pathway and does not otherwise adversely affect frequency of agroinfection. All plants are likely to induce the bacterial *vir* functions after wounding/inoculation, but this addition may give a difference in T-DNA transfer efficiency in some cases (*see* Chapter 29).

References

1. Grimsley, N., Hohn, B., Hohn, T., and Walden, R. (1986) "Agroinfection"; an alternative route for viral infection of plants by using the Ti-plasmid. *Proc. Natl. Acad. Sci. USA* **83,** 3282–3286.

2. Elmer, J. S., Sunter, G., Gardiner, W. E., Brand, L., Browning, C. K., Bisaro, D. M., and Rogers, S. G. (1988) *Agrobacterium*-mediated inoculation of plants with tomato golden mosaic virus DNAs. *Plant Mol. Biol.* **10,** 225–234.

3. Gardner, R. C., Chanoles, K. R., and Owens, R. A. (1986) Potato spindle viroid infections mediated by the Ti-plasmid of *Agrobacterium tumefaciens. Plant Mol. Biol.* **6,** 221–228.

4. Rogers, S. G., Bisaro, D. M., Horsch, R. B., Fraley, R. T., Hoffmann, N. L., Brand, L., et al. (1986) Tomato golden mosaic virus A component DNA replicates autonomously in transgenic plants. *Cell* **45,** 593–600.

5. Grimsley, N., Hohn, T., Davies, J. W., and Hohn, B. (1987) *Agrobacterium*-mediated delivery of infectious maize streak virus into maize plants. *Nature* **325,** 177–179.

6. Klinkenberg, F. A. and Stanley, J. (1990) Encapsidation and spread of african cassava mosaic virus DNA A in the absence of DNA B when agroinoculated to *Nicotiana benthamiana. J. Gen. Virol.* **71,** 1409–1412.

7. Morris, B., Richardson, K., Eddy, P., Zhan, X. C., Haley, A., and Gardner, R. (1991) Mutagenesis of the AC3 open reading frame of african cassava mosaic virus DNA-A reduces DNA-B replication and ameliorates disease symptoms. *J. Gen. Virol.* **72,** 1205–1213.

8. Stanley, J., Frischmuth, T., and Ellwood, S. (1990) Defective viral DNA ameliorates symptoms of geminivirus infection in transgenic plants. *Proc. Natl. Acad. Sci. USA* **87,** 6291–6295.

9. Frischmuth, T. and Stanley, J. (1991) African cassava mosaic virus-DI DNA interferes with the replication of both genomic components. *Virology* **183,** 539–544.

10. Stenger, D. C., Stevenson, M. C., Hormuzdi, S. G., and Bisaro, D. M. (1992) A number of subgenomic DNAs are produced following agroinoculation of plants with beet curly top virus. *J. Gen. Virol.* **73,** 237–242.

11. Grimsley, N., Hohn, T., and Hohn, B. (1986) Recombination in a plant virus: template-switching in cauliflower mosaic virus. *EMBO J.* **5,** 641–646.

12. Bakkeren, G., Koukolíková-Nicola, Z., Grimsley, N., and Hohn, B. (1989) Recovery of *Agrobacterium tumefaciens* T-DNA molecules from whole plants early after transfer. *Cell* **57,** 847–857.

13. Gal, S., Pisan, B., Hohn, T., Grimsley, N., and Hohn, B. (1991) Genomic homologous recombination *in planta. EMBO J.* **10,** 1571–1578.

14. Gal, S., Pisan, B., Hohn, T., Grimsley, N., and Hohn, B. (1992) Agroinfection of transgenic plants leads to viable cauliflower mosaic virus by intermolecular recombination. *Virology* **187,** 525–533.

15. Grimsley, N. H., Jarchow, E., Oetiker, J., Schlaeppi, M., and Hohn, B. (1991) Agroinfection as a tool for the investigation of plant–pathogen interactions, in *Plant Molecular Biology 2* (Hermann, R. G. and Larkins, B. A., eds.), Plenum, New York, pp. 225–238.

16. Hille, J., Dekker, M., Luttighuis, H., Van Kammen, A., and Zabel, P. (1986) Detection of T-DNA transfer to plant cells by *Agrobacterium tumefaciens* virulence mutants using agroinfection. *Mol. Gen. Genet.* **205,** 411–416.

17. Grimsley, N. H., Hohn, B., Ramos, C., Kado, C., and Rogowsky, P. (1989) DNA transfer from *Agrobacterium* to *Zea mays* or *Brassica* by agroinfection is dependent on bacterial virulence functions. *Mol. Gen. Genet.* **217,** 309–316.

18. Medberry, S. L., Lockhart, B. E. L., and Olszewski, N. E. (1990) Properties of commelina yellow mottle virus's complete DNA sequence, genomic discontinuities and transcript suggest that it is a pararetrovirus. *Nucleic Acids Res.* **18,** 5505–5513.

19. Baulcombe, D. C., Saunders, G. R., Bevan, M. W., Mayo, M. A., and Harrison, B. D. (1986) Expression of biologically active viral satellite RNA from the nuclear genome of transformed plants. *Nature* **321,** 446–448.

20. Harrison, B. D., Mayo, M. A., and Baulcombe, D. C. (1987) Virus resistance in transgenic plants that express cucumber mosaic virus satellite RNA. *Nature* **328,** 799–802.

21. Donson, J., Gunn, H. V., Woolston, C. J., Pinner, M. S., Boulton, M. I., Mullineaux, P. M., and Davies, J. W. (1988) *Agrobacterium*-mediated infectivity of cloned digitaria streak virus DNA. *Virology* **162,** 248–250.

22. Yamaya, J., Yoshioka, M., Sano, T., Shikata, E., and Okada, Y. (1989) Expression of hop stunt viroid from its cDNA in transgenic tobacco plants: identification of tobacco as a host plant. *Mol. Plant–Microbe Interact.* **2,** 169–174.

23. Boulton, M. I., Buchholz, W. G., Marks, M. S., Markham, P. G., and Davies, J. W. (1989) Specificity of *Agrobacterium*-mediated delivery of maize streak virus DNA to members of the Gramineae. *Plant Mol. Biol.* **12,** 31–40.

24. Lazarowitz, S. G. (1988) The molecular characterization of geminiviruses: infectivity and complete nucleotide sequence of the genome of a South African isolate of maize streak virus. *Nucleic Acids Res.* **16,** 229–249.

25. Boulton, M. I., Steinkellner, H., Donson, J., Markham, P. G., King, D. I., and Davies, J. W. (1989) Mutational analysis of the virion sense genes of maize streak virus. *J. Gen. Virol.* **70,** 2309–2323.
26. Lazarowitz, S. G., Pinder, A. J., Damsteegt, V. D., and Rogers, S. G. (1989) Maize streak virus genes essential for systemic spread and symptom development. *EMBO J.* **8,** 1023–1032.
27. Boulton, M. I., King, D. I., Donson, J., and Davies, J. W. (1991) Point substitutions in a promoter-like region and the V1 gene affect the host range and symptoms of maize streak virus. *Virology* **183,** 114–121.
28. Boulton, M. I., King, D. I., Markham, P. G., Pinner, M. S., and Davies, J. W. (1991) Host range and symptoms are determined by specific domains of the maize streak virus genome. *Virology* **181,** 312–318.
29. Shen, W. H. and Hohn, B. (1991) Mutational analysis of the small intergenic region of maize streak virus. *Virology* **183,** 721–730.
30. Schneider, M., Jarchow, E., and Hohn, B. (1992) Mutational analysis of the "conserved region" of maize streak virus suggests its involvement in replication. *Plant Mol. Biol.* **19,** 601–610.
31. Jarchow, E., Grimsley, N. H., and Hohn, B. (1991) virF, the host-range-determining virulence gene of *Agrobacterium-tumefaciens*, affects T-DNA transfer to *Zea mays*. *Proc. Natl. Acad. Sci. USA* **88,** 10,426–10,430.
32. Grimsley, N. H., Ramos, C., Hein, T., and Hohn, B. (1988) Meristematic tissues of maize plants are most susceptible to agroinfection with maize streak virus. *Bio/Technology* **6,** 185–189.
33. Schlaeppi, M. and Hohn, B. (1992) Competence of immature maize embryos for *Agrobacterium*-mediated gene transfer. *Plant Cell* **4,** 7–16.
34. Shen, W. H. and Hohn, B. (1992) Excision of a transposable element from a viral vector introduced into maize plants by agroinfection. *Plant J.* **2,** 35–42.
35. Shen, W. H., Das, S., and Hohn, B. (1992) Mechanism of *Ds*1 excision from the genome of maize streak virus. *Mol. Gen. Genet.* **233,** 388–394.
36. Chatani, M., Matsumoto, Y., Mizuta, H., Ikegami, M., Boulton, M. I., and Davies, J. W. (1991) The nucleotide sequence and genome structure of the geminivirus miscanthus streak virus. *J. Gen. Virol.* **72,** 2325–2331.
37. Gardner, R. and Knauf, V. (1986) Transfer of *Agrobacterium* DNA to plants requires a T-DNA border but not the *vir*E locus. *Science* **231,** 725–727.
38. Briddon, R. W., Lunness, P., Chamberlin, L. C. L., Pinner, M. S., Brundish, H., and Markham, P. G. (1992) The nucleotide sequence of an infectious insect-transmissible clone of the geminivirus panicum streak virus. *J. Gen. Virol.* **73,** 1041–1047.
39. Dasgupta, I., Hull, R., Eastop, S., Poggipollini, C., Blakebrough, M., Boulton, M. I., and Davies, J. W. (1991) Rice tungro bacilliform virus DNA independently infects rice after *Agrobacterium*-mediated transfer. *J. Gen. Virol.* **72,** 1215–1221.
40. Yamaya, J., Yoshioka, M., Meshi, T., Okada, Y., and Ohno, T. (1988) Expression of tobacco mosaic virus RNA in transgenic plants. *Mol. Gen. Genet.* **211,** 520–525.
41. Yamaya, J., Yoshioka, M., Meshi, T., Okada, Y., and Ohno, T. (1992) Cross-protection in transgenic tobacco plants expressing a mild strain of tobacco mosaic virus. *Mol. Gen. Genet.* **215,** 173–175.

42. Von Arnim, A. and Stanley, J. (1992) Determinants of tomato golden mosaic virus symptom development located on DNA-B. *Virology* **186,** 286–293.
43. Hayes, R. J., Coutts, R. H., and Buck, K. W. (1988) Agroinfection of *Nicotiana* spp. with cloned DNA of tomato golden mosaic virus. *J. Gen. Virol.* **69,** 1487–1496.
44. Stenger, D. C., Revington, G. N., Stevenson, M. C., and Bisaro, D. M. (1991) Replicational release of geminivirus genomes from tandemly repeated copies— evidence for rolling-circle replication of a plant viral DNA. *Proc. Natl. Acad. Sci. USA* **88,** 8029–8033.
45. Elmer, J. S., Brand, L., Sunter, G., Gardiner, W. E., Bisaro, D. M., and Rogers, S. G. (1988) Genetic analysis of the tomato golden mosaic virus II. The product of the AL1 coding sequence is required for replication. *Nucleic Acids Res.* **16,** 7043–7060.
46. Hayes, R. J., Coutts, R. H. A., and Buck, K. W. (1989) Stability and expression of bacterial genes in replicating geminivirus vectors in plants. *Nucleic Acids Res.* **17,** 2391–2403.
47. Elmer, S. and Rogers, S. G. (1990) Selection for wild type size derivatives of tomato golden mosaic virus during systemic infection. *Nucleic Acids Res.* **18,** 2001–2006.
48. Hayes, R. J. and Buck, K. W. (1989) Replication of tomato golden mosaic virus DNA B in transgenic plants expressing open reading frames (ORFs) of DNA A: requirement of ORF AL2 for production of single-stranded DNA. *Nucleic Acids Res.* **17,** 10,213–10,222.
49 Hanley-Bowdoin, L., Elmer, J. S., and Rogers, S. G. (1990) Expression of functional replication protein from tomato golden mosaic virus in transgenic tobacco plants. *Proc. Natl. Acad. Sci. USA* **87,** 1446–1450.
50. Gardiner, W. E., Sunter, G., Brand, L., Elmer, J. S., Rogers, S. G., and Bisaro, D. M. (1988) Genetic analysis of tomato golden mosaic virus: the coat protein is not required for systemic spread or symptom development. *EMBO J.* **7,** 899–904.
51. Hanley-Bowdoin, L., Elmer, J. S., and Rogers, S. G. (1989) Functional expression of the leftward open reading frames of the A component of tomato golden mosaic virus in transgenic tobacco plants. *Plant Cell* **1,** 1057–1067.
52. Hayes, R. J., Petty, I. T., Coutts, R. H. A., and Buck, K. W. (1988) Gene amplification and expression in plants by a replicating geminivirus vector. *Nature* **334,** 179–182.
53. Hanley-Bowdoin, L., Elmer, J. S., and Rogers, S. R. (1988) Transient expression of heterologous RNAs using tomato golden mosaic virus. *Nucleic Acids Res.* **16,** 10,511–10,528.
54. Gerlach, W. L., Llewellyn, D., and Haseloff, J. (1987) Construction of a plant disease resistance gene from the satellite RNA of tobacco ringspot virus. *Nature* **328,** 802–805.
55. Kheyrpour, A., Bendahmane, M., Matzeit, V., Accotto, G. P., Crespi, S., and Gronenborn, B. (1991) Tomato yellow leaf curl virus from Sardinia is a whitefly-transmitted monopartite geminivirus. *Nucleic Acids Res.* **19,** 6763–6769.
56. Dale, P. J., Marks, M. S., Brown, M. M., Woolston, C. J., Gunn, H. V., Mullineaux, P. M., et al. (1989) Agroinfection of wheat: inoculation of *in vitro* grown seedlings and embryos. *Plant Sci.* **63,** 237–245.

57. Creissen, G., Smith, C., Francis, R., Reynolds, H., and Mullineaux, P. (1990) *Agrobacterium-* and microprojectile-mediated viral DNA delivery into barley microspore-derived cultures. *Plant Cell Rep.* **8,** 680–683.

58. Woolston, C. J., Barker, R., Gunn, H., Boulton, M. I., and Mullineaux, P. M. (1988) Agroinfection and nucleotide sequence of cloned wheat dwarf virus DNA. *Plant Mol. Biol.* **11,** 35–43.

59. Dale, P. J., Marks, M. S., Woolston, C. J., Gunn, H. V., Mullineaux, P. M., Lewis, D. M., and Chen, D. F. (1989) *Agrobacterium* delivers DNA to wheat, in *Annual report, AFRC Institute of Plant Science Research and John Innes Institute, 1988,* Norwich, UK, pp. 9–10.

60. Marks, M. S., Kemp, J. M., Woolston, C. J., and Dale, P. J. (1989) Agroinfection of wheat: a comparison of *Agrobacterium* strains. *Plant Sci.* **63,** 247–256.

61. Meshi, T., Ishikawa, M., Motoyoshi, F., Semba, K., and Okada, Y. (1986) *In vitro* transcription of infectious RNAs from full-length cDNAs of tobacco mosaic virus. *Proc. Natl. Acad. Sci. USA* **83,** 5043–5047.

62. Sambrook, J., Fritsch, E. F., and Maniatis, T. (1989) *Molecular Cloning: A Laboratory Manual,* 2nd ed., Cold Spring Harbor Laboratory, Cold Spring Harbor, NY.

63. Escudero, J. and Hohn, B. (1994) *The Maize Handbook* (Freeling, M. and Walbot, V., eds.), Springer-Verlag, New York, pp. 599–602.

64. Stachel, S., Messens, E., Van Montagu, M. V., and Zambryski, P. (1985) Identification of the signal molecules produced by wounded plant cells that activate T-DNA transfer in *Agrobacterium tumefaciens. Nature* **318,** 624–629.

CHAPTER 29

T-DNA Transfer to Maize Plants

Wen-Hui Shen, Jesús Escudero, and Barbara Hohn

1. Introduction

Agrobacterium tumefaciens is a soil bacterium that induces, in most dicotyledonous plants, the neoplastic disease called crown gall. These tumors form at the site of wounding. The molecular basis for the tumor formation is the integration into and expression from the plant genome of T-DNA (transferred DNA), a part of the Ti (tumor-inducing) plasmid carried by the bacterium. T-DNA oncogenes specifying the synthesis of auxin and cytokinin are responsible for cell proliferation resulting in tumor growth. Additional T-DNA genes encode enzymes that produce, in the gall, novel amino acid and sugar derivatives, called opines, which are specific growth substrates for the bacterium. Different bacterial strains produce and consume different opines. Thus, strains have been classified accordingly as nopaline, octopine, or other types, depending on their diagnostic opine.

Agrobacterium transformation requires the cis-acting T-DNA border sequences and the trans-acting virulence *(vir)* functions encoded by the Ti plasmid and the chromosome. The *vir* functions and the T-DNA can be located on separate compatible replicons in *Agrobacterium*. This allowed the development of binary vectors. In such systems, the *Agrobacterium* strain contains a wild-type or disarmed Ti plasmid (in the latter plasmid the tumor genes are deleted) that carries the *vir* functions and serves as helper, and a binary vector carrying the T-DNA containing the gene(s) of interest inserted between the T-DNA borders.

From: *Methods in Molecular Biology, Vol. 44:* Agrobacterium *Protocols*
Edited by: K. M. A. Gartland and M. R. Davey Humana Press Inc., Totowa, NJ

Agrobacterium is routinely used to engineer desirable genes into plants. The host range of *Agrobacterium*, as judged by tumor formation, includes most dicots and many gymnosperms. However, the economically important graminaceous plant maize is refractory to tumor induction by *A. tumefaciens*. A sensitive and efficient assay, called agroinfection, has been developed to show T-DNA transfer to maize plants *(1)*. In this assay *(see* Chapter 28), partially or completely duplicated genomes of maize streak virus (MSV) have been inserted between the T-DNA borders of a binary vector. T-DNA transfer was monitored by the appearance of viral symptoms on the plants inoculated with the MSV-containing *Agrobacterium* strain. To study T-DNA transfer from *Agrobacterium* to maize cells in a virus-free system, we used the β-glucuronidase (GUS) gene as a T-DNA expression marker. Histochemical investigation of GUS activity revealed T-DNA transfer to maize shoots *(2)*.

Three factors are essential for detecting T-DNA transfer to maize. First, the marker for T-DNA transfer should be highly specific for expression in the plant. Second, the maize genotype should be competent to accept T-DNA transfer from *Agrobacterium*. Although all maize genotypes tested up to now were agroinfectable, the efficiency of agroinfection could vary tremendously from one genotype to another. Using a maize genotype that is poorly agroinfectable, T-DNA transfer was not detected using the GUS gene as expression marker *(2)*. Third, the knowledge of the type of the *Agrobacterium* strain being used is essential. Although nopaline-type strains of *Agrobacterium* are efficient in agroinfection of maize, octopine-type strains are not *(3)*. The basis for the difference has been shown to be a weak deleterious effect on agroinfection efficiency of the *vir*F gene which is present only on the octopine-type Ti plasmid *(4)* and a major effect of the nopaline specific *vir*A gene product which is, in addition, dominant over the octopine specific *vir*A product *(5)*. Besides these essential factors, an efficient protocol for manipulating maize materials and *Agrobacterium* is desirable. Here we describe a protocol that has been used successfully for demonstrating T-DNA transfer into maize plants.

1.1. Constructs

We have used two GUS constructs (Fig. 1A). In one, the GUS gene is fused to cauliflower mosaic virus open reading frame V *(6)*, the other contains a modified intron of the castor bean catalase gene *(7)* in the

N-terminal part of the GUS coding sequence. Both GUS genes are driven by the 35S promoter and terminated by the nopaline synthase terminator. These GUS genes carried by the binary vector pBIN19 *(8)* did not lead to detectable enzyme activity in *Agrobacterium* cells. However, they were highly expressing in plant cells.

2. Materials

1. Antibiotic stock solutions: Kanamycin (Sigma, St. Louis, MO): 25 mg/mL in water, sterilized by filtration through a 0.22-μm filter. Rifampicin (Sigma): 20 mg/mL in dimethyl sulfoxide (DMSO). These stocks are stored at –20°C. Dilution for use is 1:1000.
2. Acetosyringone (AS; Aldrich Chemie, Steinheim, Germany) stock solution: 80 mg/mL (400 m*M*) in DMSO, stored at –20°C. Dilution for use is 1:2000.
3. YEB medium: 5 g/L Bacto-beef extract, 1 g/L Bacto-yeast extract, 5 g/L peptone, 5 g/L sucrose, 0.493 g/L MgSO$_4$ · 7H$_2$O; sterilized by autoclaving.
4. MS1 medium: MS (Murashige and Skoog) medium *(9)* containing 3% (w/v) sucrose and 1 mg/L thiamine-HCl.
5. Solid MS1 medium: MS1 medium solidified with 1% (w/v) agar.
6. Solid MS2 medium: MS1 medium supplemented with 200 μ*M* AS and solidified with 1% (w/v) agar.
7. 10 m*M* MgSO$_4$ solution, sterilized by autoclaving.
8. Bleach solution: 1.4% (v/v) hypochlorite and 0.05% (w/v) sodium dodecyl sulfate (SDS).
9. Sterile water: Autoclaved distilled water.
10. X-Gluc solution: 0.052% (w/v) 5-bromo-4-chloro-3-indolyl glucuronide (X-Gluc) (Biosynth, Staad, Switzerland) in 100 m*M* phosphate buffer, pH 7.0 in the presence of 0.1% (w/v) sodium azide. The solution was sterilized by filtration through a 0.22-μm filter.
11. Ethanol 70–90% (v/v).
12. Tissue-culture facilities: A laminar flow bench, Petri dishes, Falcon tubes, forceps, scalpels, and a phytotron (25°C, 20,000 lx).

3. Methods

3.1. Preparation of Agrobacterium

Binary vectors containing one of the GUS genes described in Section 1.1. were introduced into the wild-type nopaline strain *A. tumefaciens* C58 (pTiC58) using the electroporation *(10)* *(see* Chapter 33) or the triparental-mating method *(11)*. The resulting *Agrobacterium* strains *(see* Notes 1 and 2) were stored in 15% (v/v) glycerol at –80°C. The bacteria

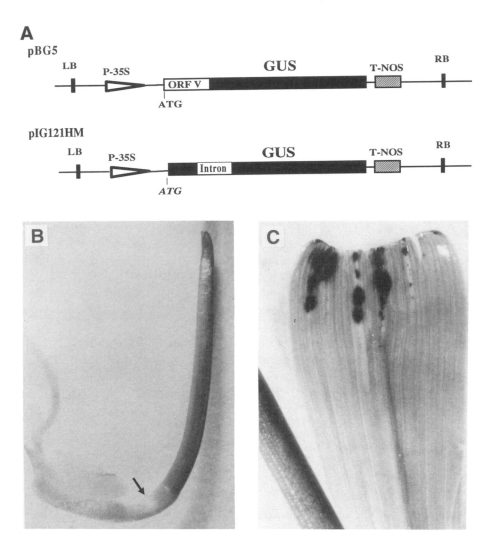

Fig. 1. (A) Schematic representation of the constructs containing the GUS gene in the T-DNA of the binary vector pBIN19. The open reading frame of the original bacterial GUS gene is represented as black boxes. LB: left border, RB: right border, P-35S: the 35S promoter, T-NOS: the nopaline synthase terminator, and ORF V: the sequence coding for the first 29 amino acids of the open reading frame V of the cauliflower mosaic virus. Owing to the fusion with ORF V, the initiation codon (ATG) of the GUS gene in pBG5 is embedded in a sequence of eukaryotic origin. The *ATG* in pIG121HM (identical to pIG221 in 2) is in the original prokaryotic context. Intron: a modified intron of the castor bean catalase gene *(7)*.

were prepared for cocultivation of maize shoots according to the following procedure:

1. Inoculate bacteria from the glycerol stock into 10-mL liquid YEB medium containing rifampicin 20 mg/L (chromosomally located resistance) and kanamycin 25 mg/L (for pBIN19-based, GUS gene-containing binary vectors). The inoculated bacteria are grown at 28°C, with vigorous shaking (about 200 rpm).
2. Subculture the bacteria in 60 mL fresh YEB medium supplemented with the same antibiotics, following a 1:20 dilution of the 2-d-old preculture. The bacteria are grown for another 20 h, reaching a final titer of 1–2×10^9 cells/mL.
3. Harvest the bacterial cells by centrifugation of 50 mL culture in a Sorvall plastic tube (10 min at $15,000g$).
4. Wash the bacterial cells with 25 mL liquid MS1 medium or 10 mM MgSO$_4$.
5. Resuspend the bacterial pellet in 5 mL liquid MS1 medium or 10 mM MgSO$_4$. This results in a final titer of 1–2×10^{10} cells/mL.
6. Add to the bacterial suspension the agrobacterial virulence inducer, AS, to a final concentration of 200 μM (0.5–2 h before cocultivation with maize shoots; *see* Note 3).

3.2. Preparation of Maize Shoots

The maize lines that are frequently used and shown to be highly agroinfectable are Golden Cross Bantam, A188 and B73. Both young seedlings germinated from mature seeds and immature embryos have been shown to be competent for agroinfection *(1,12)*. We have tested shoots excised both from 3-d-old seedlings and from plantlets germinated from immature embryos for T-DNA mediated GUS expression. We found that shoots excised from plantlets germinated from immature embryos give more GUS-positive spots after cocultivation with *Agrobacterium* containing the GUS gene than those from seedlings. Thus, we recommend the use of immature embryo-derived shoots for testing T-DNA transfer by GUS activity assay.

Shoots from plantlets germinated from immature embryos of maize line A188 were prepared as follows:

Fig. 1. *(continued)* (B) A plantlet photographed 5 d after germination of an immature embryo isolated 14 d after pollination. The location for excision of the shoot for cocultivation with *Agrobacterium* is indicated by the arrow. (C) Blue spots on a leaf resulting from cocultivation of the maize shoot with *Agrobacterium* containing pBG5.

1. Harvest immature kernels 14–17 d after pollination.
2. Surface sterilize the kernels by incubating and stirring in 2 vol of bleach solution for 15 min.
3. Remove the used bleach solution and wash the kernels with 2 vol of sterile water for 5 min.
4. Repeat step 3 twice.
5. Isolate embryos from the kernels in a laminar flow bench using sterilized forceps and scalpels.
6. Place the isolated embryos with the scutellar side down onto solid MS1 medium in Petri dishes (25–30 embryos/9-cm diameter dish).
7. Germinate the embryos in a phytotron under a regime of 16-h light (20,000 lx) and 8-h dark, at 25°C (*see* Note 4).
8. Excise the shoots 4–8 d after germination by cutting just below (about 1–3 mm) the coleoptilar node (Fig. 1B) in a laminar flow bench using sterilized forceps and scalpels (*see* Note 5).
9. Keep the excised shoots in liquid MS1 medium until the shoot excision process is finished.

3.3. Cocultivation of Maize Shoots with Agrobacterium

The freshly excised shoots (< 30 min after cutting) are cocultivated with *Agrobacterium* according to the following procedure:

1. Dip the shoots into the *Agrobacterium* suspension in a sterile Falcon tube of size 17 × 100 mm.
2. Subject the tubes to vacuum infiltration (–0.4 to –0.8 atm) for 5 min.
3. Blot the shoots on sterile filter papers to remove excess bacterial suspension and place them onto solid MS2 medium.
4. Culture the shoots with the attached *Agrobacterium* cells in the phytotron under a regime of 16-h light (20,000 lx) and 8-h dark at 25°C.

3.4. GUS Assay

Histochemical assay of GUS activity in maize tissues is performed on shoots according to the following procedure:

1. Collect the shoots 3 d after cocultivation with *Agrobacterium*.
2. Wash the shoots briefly in sterile water to remove the overgrowing bacteria.
3. Blot the shoots dry on sterile filter papers and soak them in X-Gluc solution (20–30 shoots in 5 mL X-Gluc).
4. Subject the soaked shoots to vacuum infiltration for 10 min.
5. Keep the reactions at 37°C for 2 d.
6. Destain for chlorophyll by rinsing with ethanol 70–90% (v/v).

From 1 to >100 blue spots per shoot (Fig. 1C) should be observed on 30–80% of the shoots, depending on the age of immature embryos and the time of germination (*see* Notes 6 and 7; Chapter 17).

4. Notes

1. To be aware of false GUS positives derived from a contaminating micro-organism and/or from an endogenous GUS activity in maize tissues, the *Agrobacterium* strain containing the same GUS gene but devoid of a Ti-plasmid or with mutated virulence gene(s), should be included as a negative control.
2. To test the function of the construct and the virulence of the bacterial preparation, we usually include tobacco leaf disks in the same cocultivation assay. High expression (many blue spots/sectors) of the GUS gene should be observed in tobacco leaf tissues.
3. We found that AS enhances T-DNA mediated GUS expression in maize shoots. Thus, we recommend the maintenance of AS during the cocultivation of maize shoots with *Agrobacterium*, as has been described in Section 3.
4. It is important to germinate the immature embryos and use the germinated shoots. Using immature embryos freshly isolated without germination, cocultivation rarely results in the detection of GUS activity.
5. Extra-wounding with forceps through the surface of the maize shoots just before cocultivation with *Agrobacterium* enhances T-DNA mediated GUS expression. One may include this extra-wounding in the method described.
6. GUS-positive spots were observed mainly on the leaves and the coleoptile. Cocultivation of maize roots or scutellum with *Agrobacterium* containing the GUS gene does not lead to GUS expression *(2)*.
7. We believe that the GUS positive spots are the result of transient expression of the GUS gene on T-DNA transport into maize cells. Integration of T-DNA into the maize genome has not (yet) been observed.

References

1. Grimsley, N., Hohn, T., Davies, J. W., and Hohn, B. (1987) *Agrobacterium*-mediated delivery of infectious maize streak virus into maize plants. *Nature* **325,** 177–179.
2. Shen, W.-H., Escudero, J., Schläppi, M., Ramos, C., Hohn, B., and Koukolíková-Nicola, Z. (1993) T-DNA transfer to maize cells: Histochemical investigation of β-glucuronidase activity in maize tissues. *Proc. Natl. Acad. Sci. USA* **90,** 1488–1492.
3. Boulton, M. I., Buchholz, W. G., Marks, M. S., Markham, P. G., and Davies, J. W. (1989) Specificity of *Agrobacterium*-mediated delivery of maize streak virus DNA to members of the Gramineae. *Plant Mol. Biol.* **12,** 31–40.
4. Jarchow, E., Grimsley, N., and Hohn, B. (1991) *vir*F, the host-range-determining virulence gene of *Agrobacterium tumefaciens*, affects T-DNA transfer to *Zea mays*. *Proc. Natl. Acad. Sci. USA* **88,** 10,426–10,430.

5. Raineri, D. M., Boulton, M. I., Davies, J. W., and Nester, E. W. (1993) *VirA*, the plant-signal receptor, is responsible for the Ti plasmid-specific transfer of DNA to maize by *Agrobacterium*. *Proc. Natl. Acad. Sci. USA* **90,** 3549–3553.

6. Schultze, M., Hohn, T., and Jiricny, J. (1990) The reverse transcriptase gene of cauliflower mosaic virus is translated separately from the capsid gene. *EMBO J.* **9,** 1177–1185.

7. Tanaka, A., Mita, S., Ohta, S., Kyozuka, J., Shimamoto, K., and Nakamura, K. (1991) Enhancement of foreign gene expression by a dicot intron in rice but not in tobacco is correlated with an increased level of mRNA and an efficient splicing of the intron. *Nucleic Acids Res.* **18,** 6767–6770.

8. Bevan, M. (1984) Binary *Agrobacterium* vectors for plant transformation. *Nucleic Acids Res.* **12,** 8711–8721.

9. Murashige, T. and Skoog, F. (1962) A revised medium for rapid growth and bioassays with tobacco tissue cultures. *Physiol. Plant.* **15,** 473–497.

10. Mattanovich, D., Rüker, F., da Câmara Machado, A., Laimer, M., Regner, F., Steinkellner, H., et al. (1989) Efficient transformation of *Agrobacterium* spp. by electroporation. *Nucleic Acids Res.* **17,** 6747.

11. Roger, S. G., Klee, H., Horsch, R. B., and Fraley, R. T. (1988) Use of cointegrating Ti plasmid vectors, in *Plant Molecular Biology Manual* (Gelvin, S. B. and Schilperoort, R. A., eds.), Kluwer Dordrecht, The Netherlands, pp. 1–12.

12. Schläppi, M. and Hohn, B. (1992) Competence of immature maize embryos for *Agrobacterium*-mediated gene transfer. *Plant Cell* **4,** 7–16.

CHAPTER 30

Use of Cosmid Libraries in Plant Transformations

Hong Ma

1. Introduction

Cosmid libraries are important tools for molecular analysis of plant genes because of their useful properties. Cosmids carry the *cos* site from the bacteriophage lambda *(1–4)*; therefore, cosmid libraries have the advantage of being able to be packaged in vitro using commercially available highly efficient packaging extracts, just as lambda libraries can. The high efficiency of in vitro lambda packaging allows a cosmid library with large inserts to be more easily constructed than a simple plasmid library, which relies on transformation of competent *Escherichia coli* cells. In addition, cosmids contain an origin of replication for propagation in *E. coli* and possibly other bacteria. Since cosmids are propagated as plasmids, not as lambda phages, the genes for lambda reproduction are not needed; therefore, a cosmid vector can be much smaller than a lambda vector (Fig. 1). Since lambda phage packages about 37–53 kb of DNA *(5)*, cosmids can have relatively large inserts of more than 40 kb, depending on vector sizes. Another advantage of cosmids is the ease with which to prepare the DNA because plasmid DNA is much more readily isolated than lambda phage DNA. Finally, the fact that cosmids are plasmids allows the recombinant molecules to be transferred from one bacterium to another through conjugation and from *Agrobacterium* to plant cells, as long as the necessary *cis* elements are present. Therefore, cosmid libraries offer a unique combination of very useful features, which make

From: *Methods in Molecular Biology, Vol. 44:* Agrobacterium *Protocols*
Edited by: K. M. A. Gartland and M. R. Davey Humana Press Inc., Totowa, NJ

Lambda Clone

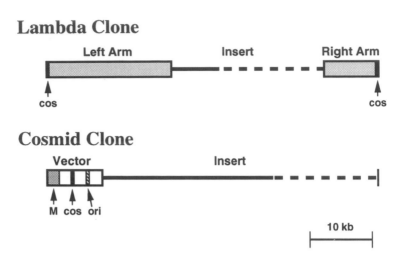

Cosmid Clone

Fig. 1. Comparison of lambda and cosmid clones. The vector for a lambda clone contains all of the genes required for the lytic life cycle of the lambda phage, including genes for phage DNA replication, for the phage structural proteins, and for lysis of *E. coli* cells. These genes are distributed in two regions, called left arm and right arm, usually about 20 and 9 kb in size, respectively. The lambda phage can package 37–53 kb; therefore, the insert sizes range from 8 kb (solid line) to 24 kb (solid and dashed lines). In contrast, cosmid clones propagate as plasmids, so the vector needs only to have a selectable marker (M) and an origin of replication *(ori)*. So a cosmid vector can easily be less than 8 kb, allowing the inserts to be as long as 29 kb (solid line) to 45 kb (solid and dashed lines). Generally, a plasmid vector with a 400-bp fragment containing the *cos* site is a cosmid vector.

them important for many different molecular studies, including those of plant genes.

Agrobacterium-mediated transformation of plant cells is one of the most widely used methods for generating stable transgenic plants in dicotyledonous plants *(6–9)*. In addition, *Agrobacterium*-mediated transformation has been successfully used to introduced DNA into monocotyledonous plants including rice *(10,11)*. DNAs are introduced into plant cells for various studies, from the analysis of gene regulation using reporter genes, to the examination of gene functions using ectopic expression or overexpression. Cosmid libraries that have the T-DNA borders flanking the inserts (therefore are transformation competent) can be used for any experiments that require that large DNAs be introduced into plants via *Agrobacterium*-mediated transformation. A clone from a transformation-competent cosmid library can be used directly for transformation, thus avoiding the necessity of subcloning large pieces of DNA, which is usually difficult. This chapter describes several vectors for cosmid libraries which are also competent for *Agrobacterium*-mediated transformation, and discusses their advantages and shortcomings. In addition, analyses using these cosmid libraries to study plant gene functions are also presented. The detailed procedures for *Agrobacterium*-mediated transformation can be found elsewhere in this volume.

2. Vectors
for Transformation Competent Cosmid Libraries

A number of vectors have been developed that contain both the *cos* site and elements for *Agrobacterium*-mediated transformation. Some of these vectors have both the T-DNA right and left borders flanking both a selectable marker for selection of the transformed plant cells, and a cloning site or polylinker for inserting genomic DNA. Other vectors have only the right border next to the selectable marker and cloning site(s); these vectors allow the entire plasmid to be transferred as T-DNA. Generally the single right border vectors are designed in such a way that the selectable marker for transformed plant cells (e.g., neomycin phosphotransferase, or *npt*, for kanamycin resistance) is transferred after that of the insert DNA. All of the cosmid vectors have a marker for selection in bacteria and origin of replication in *E. coli*. Those with a broad host range origin of replication can propagate in *Agrobacterium* as well as in *E. coli*, and therefore are referred to as binary vectors; those without such broad host range origin of replication can only to be maintained in *Agrobacterium* after integration into a resident Ti plasmid; these are called integrative vectors. Several vectors are described here, and their advantages and shortcomings are discussed.

2.1. The pOCA18 and pLZ03 Vectors

Two cosmid vectors, pOCA18 *(12)* and pLZ03 *(13)*, have been described that are derived from the broad host range plasmid pRK290 *(14)*, which is 20 kb in size and contains the tetracycline *(tet)* resistance gene. pRK290 is derived from the naturally occurring tetracycline-resistant plasmid RK2 (56 kb) by several deletions *(14)*. In addition to the *tet* gene, pRK290 also carries the origin of replication *oriV*, and the genes encoding *trans* factors required for replication, *trf*A and *trf*B. Furthermore, although pRK290 lacks the *trans* factors for transfer during bacterial conjugation, it does have the *cis* element required for the transfer. Therefore, pRK290 DNA can be transferred via conjugation if the *trans* factors are provided from a helper plasmid, such as pRK2013, which needs not to be in the same host as pRK290 *(14)*. Consequently, pRK290 derivatives can be introduced into a number of gram-negative bacteria, including *Agrobacterium*, using a procedure known as triparental mating involving the donor (usually an *E. coli* strain), the recipient, and the helper (an *E. coli* strain with pRK2013, for instance). Olszewski et al. constructed the pOCA18 vector (Fig. 2A) for the purpose of generating cosmid libraries from which clones may be isolated and used for *Agrobacterium*-mediated transformation without further subcloning *(12)*. To construct a shuttle (binary) vector between *E. coli* and *Agrobacterium*, pRK290 was used as the backbone. Several DNA elements were added to pRK290, including the T-DNA left border, a chimeric kanamycin resistance gene *(nos:npt:ocs)* for selection of transgenic plants, the small plasmid pIAN7 that carries a *sup*F gene selectable in appropriate *E. coli* strains, cloning sites for inserting plant DNA, the *cos* site, and T-DNA right border with overdrive sequence (Fig. 2A).

Using pOCA18, Olszewski et al. constructed an *Arabidopsis* genomic library, and tested the efficiency of transferring pOCA18 and a number of clones from *E. coli* to *Agrobacterium (12)*. Efficient transfer was observed for all of the plasmids tested, indicating that pOCA18 and its derivatives can be easily introduced into *Agrobacterium* from *E. coli*. The pOCA18 T-DNA border sequences were shown to be functional by generating transgenic tobacco plants using *Arabidopsis* genomic clones, including some containing a mutant *AHAS* (acetohydroxy acid synthase) gene conferring resistance to the herbicide chlorsulfuron *(15)*. The transgenic shoots with the mutant *AHAS* gene exhibited resistance to chlorsulfuron, indicating that the *Arabidopsis* gene was expressed in the transgenic tobacco shoots. Therefore, clones from a genomic library generated with pOCA18 can be used to test gene function in transgenic plants. In particular, complementation of mutant phenotype with a candidate gene can be achieved using clones from pOCA18-based libraries (*see* Section 3.).

Another vector, which is derived from pOCA18, has been developed by Lazo et al. *(13)*. In addition to the elements present in pOCA13, this new vector, pLZ03 (Fig. 2B), contains a second selectable marker, *gat*

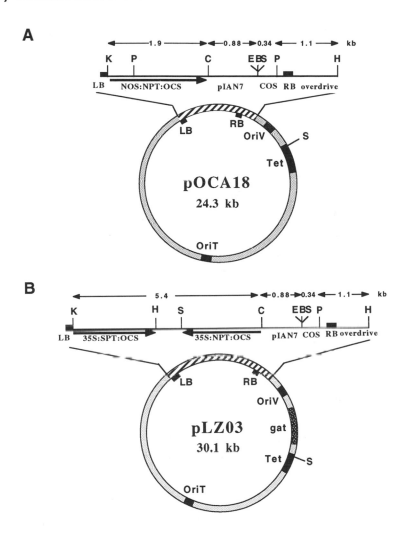

Fig. 2. Maps of pOCA18 (**A**) and pLZ03 (**B**). The pOCA18 map was redrawn from Olszewski et al. *(12)* by permission of Oxford University Press. The pLZ03 map was redrawn from Lazo et al. *(13)* with permission. The *nos:npt:ocs* and 35S:*npt:ocs* chimeric genes confer kanamycin resistance to plant cells, and the 35S:*spt:ocs* chimeric gene confers streptomycin resistance to plant cells. Genes and DNA elements: *cos*, the lambda *cos* sequence; *gat*, gentamicin acetyltransferase gene; LB, left border; *ori*T, origin of conjugal transfer between bacterial strains; *ori*V, broad host range origin of replication; pIAN7, contains a ColE1 origin of replication and a bacterial *sup*F tRNA gene; RB, right border; *tet*, tetracycline resistance. Restriction endonuclease sites: B, *Bam*HI; C, *Cla*I; E, *Eco*RI; H, *Hin*dIII; K, *Kpn*I; P, *Pst*I; S, *Sal*I.

(encoding gentamicin acetyltransferase), placed adjacent to the *tet* gene. The *gat* marker was added to allow a much tighter selection in *Agrobacterium*. In addition, the pLZ03 T-DNA region has also been modified to include two chimeric 35S:*npt:ocs* and 35S:*spt:ocs* genes, which contain the cauliflower mosaic virus 35S promoter (35S), coding regions for *npt* and *spt*, respectively, and the octopine synthase gene 3' nontranscribed region. Therefore, in addition to the selectable kanamycin resistance marker, pLZ03 has a scorable marker for streptomycin resistance for plant cells. The large size (30.1 kb) of pLZ03 allows neither the vector-only monomer nor dimer to be packaged into the lambda phage head; therefore, during cosmid library construction, using pLZ03 can minimize the production of vector-only phages, and increase the frequency of recombinant phages. However, the large sizes of pOCA18 and pLZ03 make manipulations with these vectors more difficult than smaller vectors, such as the pCIT vectors described in Section 2.2.

Since transfer *en masse* of cosmid libraries from *E. coli* to *Agrobacterium* often reduces the complexity of the library, it is desirable to construct cosmid libraries directly into *Agrobacterium*. For this purpose, an *Agrobacterium* strain, AGL0, has been generated that expresses the lambda phage receptor LamB *(13)*. Furthermore, an AGL0 derivative, AGLI, was constructed that is deficient in general recombination, thus suitable for stably harboring genomic libraries. A cosmid library was constructed by ligating *Arabidopsis* genomic DNA into pLZ03, and introduced into AGLI, to generate more than 20,000 recombinant colonies. These clones, with an average insert size of 15–20 kb, represent about 3–4 equivalents of the *Arabidopsis* haploid genome.

2.2. The pCIT30 and pCIT101 Through pCIT104 Vectors

It is often useful to have cosmid libraries containing relatively large (30–40 kb) segments of genomic DNAs. For example, when a cosmid library is used in a chromosome walk experiment, larger inserts allow the same distance to be covered with fewer clones. Similarly, when multiple clones from a given region are to be tested for complementation following transformation, larger inserts mean fewer clones need be tested. In addition, when all of the necessary regulatory elements of a gene have not been defined yet, larger inserts increase the chances of retaining the elements in one clone. Several smaller vectors have been constructed *(16)*: pCIT30 and pCIT101-104 (Fig. 3), which can take up to 46 kb of insert DNA. The vectors pCIT30, pCIT101, and pCIT102 were derived from pMON200 *(17,18)*. The elements derived from pMON200 include the *StrR/SpR* gene for selection in bacteria, the scorable marker *nos* (encoding nopaline synthase), the T-DNA right border, the pBR origin of replication, and Ti homology (the left inside homology, or LIH). The vectors can be transferred from *E. coli* to *Agrobacterium* using triparental mating with a helper strain (e.g., an *E. coli* strain with pRK2013). Because pMON200 lacks the broad host range origin of replication (*ori*V), it or its derivatives can only be propagated after a recombination to form a cointegrate with a resident Ti plasmid in *Agrobacterium (18)*. The recombination can occur either between the LIH on the pCIT plasmids and a disarmed octopine-type Ti plasmid (e.g., pTiB6S3-SE; *17,18*), or between the pBR sequence in the pCIT plasmids and any disarmed Ti plasmids with an introduced pBR homology (e.g., pGV3850, *19*). These vectors are therefore integrative vectors. For selection of transformed plant cells, pCIT30 carries a chimeric *Hph* gene conferring hygromycin resistance, whereas pCIT101 and pCIT102 contain a chimeric *nos:npt:nos* gene (with both 5' and 3' sequences from the *nos* gene) for kanamycin resistance. These vectors are about 10 kb; therefore, the insert sizes can range from 27–43 kb, considerably larger than those in pOCA18 or pLZ03.

Two other vectors (pCIT103 and pCIT104), which can take even larger inserts, have been constructed from the pMON721 vector *(16)* by inserting a *cos* site-containing fragment and cloning sites. The pMON721 vector is only 6.6 kb, and the derivatives pCIT103 and 104 are only 7.0 kb; therefore, the genomic inserts can be as long as 46 kb. Since pMON721 (and pCIT103 and pCIT104) contain the broad host range origin of replication *ori*V from pRK2 *(20,21)*, these vectors are binary in the *Agrobacterium* host that contains a functional *trf*A gene, such as those that carry the pMP90RK plasmid *(22)*. The *trf*A gene encodes a replication factor that interacts with the *ori*V *(20,21)*. In the absence of *trf*A, the pCIT103 and pCIT104 plasmids are incapable of replication and must be maintained as a cointegrate in *Agrobacterium* hosts containing either the LIH or an introduced pBR region on the resident Ti plasmid, similar to the pMON200 derivatives.

Several cosmid libraries were constructed using pCIT30 as the vector. One of these contained about 6×10^5 independent clones carrying

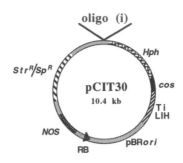

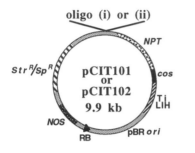

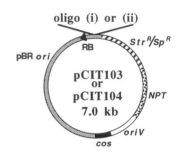

Oligo (i) sequence

pCIT30 &
pCIT103b

        SfiI   T7 Promoter   BglII   SP6 Promoter   NotI
gacGGCCccccccGGCCtaatacgactcactatAGATCTatagtgtcacctaaatGCGGCCGCgtc

pCIT101 &
pCIT103a

       NotI   SP6 Promoter   BglII   T7 Promoter   SfiI
gacGCGGCCGCatttaggtgacactatAGATCTatagtgagtcgtattaGGCCggggggGGCCgtc

Oligo (ii) sequence

pCIT102

   HindIII XbaI EcoRI XhoI HindIII EcoRI XbaI
gacAAGCTTCTAGAATTCCTCGAGAAGCTTGAATTCTAGAgtc

pCIT104

   XbaI EcoRI HindIII XhoI EcoRI XbaI HindIII
gacTCTAGAATTCAAGCTTCTCGAGGAATTCTAGAAGCTTgtc

Fig. 3. Maps of pCIT30 and pCIT101-104. These maps were from Ma et al. *(16)*. The pCIT30 vector contains most of pMON200, including the bacterial selectable marker streptomycin/spectinomycin resistance gene (*Str*R/*Sp*R), the scorable marker *nos* for plant cells, the ColE1 origin of replication (pBRori), the T-DNA right border, and the Ti left inside homology (Ti LIH) *(17,18)*. The chimeric *nos:npt:nos* gene on pMON200 was replaced by the *Hph* gene (encoding hygromycin phosphotransferase) from pMON754. In addition, A 403-bp *Hinc*II

genomic sequences of an average length of about 35 kb from *Arabidopsis thaliana* (Landsberg *erecta*). This library was used to isolate the genomic sequences of several *Arabidopsis* genes *(23,24)*, including the floral homeotic gene *AG (25)*, which is required for the organ identity of the floral reproductive organs, and for the determinate growth of the floral meristem *(26)*. The identity of the isolated *AG* gene was confirmed by introducing a candidate cosmid into plant cells homozygous for an *ag* mutation using *Agrobacterium*-mediated transformation, and observing functional complementation *(25)*. Another pCIT30-based genomic library was used to clone the *ETR1* gene by chromosome walking *(27)*. The *etr1-1* mutant, which is insensitive to the hormone ethylene and defines the *ETR1* gene, is dominant over the wild-type *(28)*. Therefore, a cosmid library was constructed using the genomic DNA from the mutant and the pCIT30 vector. DNA sequences from the candidate *etr1-1* mutant clones were shown to confer enthylene insensitivity when introduced into the wild-type background by *Agrobacterium*-mediated transformation *(27)*.

The small sizes of the pCIT vectors allow recombinant molecules with two or more copies of the vector to be packaged into a single-phage head.

Fig. 3. *(continued)* fragment carrying the *cos* site *(2,4)* was inserted at the *Hind*III site of pMON200 after it was filled-in using the Klenow DNA polymerase I large fragment. An oligo [oligo (i)] with the cloning *Bgl*II site and flanking elements was inserted into a unique *Stu*I site. The *Sfi*I, *Bgl*II, and *Not*I sites are unique. The T7 and SP6 promoters flanking the cloning site allow the synthesis of RNA probes using in vitro transcription. The orientation of the arrowhead at the RB is such that it leads the transferred DNA (T-DNA). Therefore, the *nos* gene is transferred first in pCIT30.

The pCIT101 and pCIT102 vectors were derived from pMON200 by inserting the *Hinc*II *cos* fragment, followed by the insertion of oligos (i) or (ii), respectively, into the *Stu*I site. The pCIT103 and pCIT104 vectors were derived from pMON721, which carries the broad host range origin of replication (*ori*V), by inserting the *cos* site containing *Sal*I fragment into a unique *Sal*I site, followed by the ligation of oligo (i) or (ii), respectively, with the vector digested by *Sma*I, which removes the original pMON721 polylinker. The restriction sites in oligo (ii) flanking the *Xho*I cloning allow easy release of an end fragment, which can be labeled using the random priming method. These vectors all contain the streptomycin/spectinomycin resistance gene (*Str*R/*Sp*R), the *nos:npt:nos* gene *(npt)*, the *cos* site, the ColE1 origin of replication (pBRori), and the right border (RB).

Furthermore, two or more short genomic DNA fragments may be ligated to the same vector to produce a chimeric clone. Such chimeric clones can be very troublesome in a chromosome walk experiment. Therefore, genomic DNA fragments need to be properly size selected before ligation to the vector DNA to minimize the generation of phages with more than one copy of the vector. In addition, vector–vector or insert–insert ligation can be prevented by partially filling in the ends of the digested vector and insert DNA with the complement of two of the four single-stranded nucleotides, such that the vector end no longer matches with itself, nor the insert end with itself, but the vector end matches the insert end. For example, for pCIT30, pCIT101, and pCIT103, the cloning site is *Bgl*II, which after digestion results in a 3'-CTAG-5' overhang; partial filling-in with nucleotides G and A leads to a 3'-AG-5' two base overhang. If the genomic DNA is digested with *Sal*I or *Xho*I (both produce 3'-AGCT-5' overhangs), and partially filled-in with T and C, then the remaining 3'-CT-5' will be complementary with the 3'-AG-5' of the vector. The relatively infrequent distribution of *Sal*I and *Xho*I sites makes the pCIT30, pCIT101, and pCIT103 vectors not very well suited for this partial filling-in procedure. On the other hand, the pCIT102 and pCIT104 vectors contain an *Xho*I site as the cloning site, so that, after being digested with *Xho*I and partially filled-in with T and C, these vectors can be ligated with genomic DNAs that are partially digested with *Sau*3AI (which produces the same overhang as *Bgl*II, 3'-CTAG-5') and partial filled-in with G and A. Because *Sau*3AI sites occur much more frequently, partial digests of genomic DNA will represent the genome more evenly than *Sal*I or *Xho*I digests.

3. Uses of Transformation-Competent Cosmid Libraries

3.1. Testing Candidate Genes by Functional Complementation

When a candidate gene which corresponds to a mutation has been isolated, it is desirable, even necessary, to confirm the identity of the gene by functional complementation. This can be achieved by re-introducing the cloned gene into the mutant plant background, using *Agrobacterium*-mediated transformation if the plant under study is receptive to such a method. The transgenic plants can then be analyzed for the restoration of the wild-type function. The complementation test is particularly important when other means of confirmation are scarce, such as when there are few mutant alleles that could be analyzed to provide a confirmation of the identity of the gene, or when the mutation is owing to a T-DNA insertion for which reversion analysis similar to those done with transposon insertional mutants is not possible. As discussed earlier, for example, the identity of the *Arabidopsis* floral homeotic gene *AG* was confirmed by functional complementation using a candidate genomic cosmid clone *(25)*. Another example is the *GA1* gene, which is involved in gibberellin biosynthesis. *GA1* was isolated by genomic subtraction using a deletion mutant; the identity of the gene was confirmed by transforming the *ga1* mutant with a cosmid clone carrying the wild-type candidate *GA1* gene *(29)*. In cases where the gene being tested is relatively well defined, subcloning the genomic *(30,31)* or cDNA *(32)* clones of the gene into a T-DNA vector also allows one to functionally test the gene. However, the subcloning relies on availability of appropriate restriction sites, whereas the use of cosmid clones does not.

3.2. Isolating Genes
Among a Small Number of Clones

In plants studied genetically, many genes are defined only by mutations that do not lead to polymorphisms of the genomic DNAs that are detectable by restriction digests, polymerase chain reaction, and electrophoresis. For plants with little dispersed repetitive DNAs, such as *A. thaliana*, chromosome walking is a viable means of isolating these genetically defined genes *(33,34)*. During a chromosome walk, successive overlapping genomic DNAs are isolated starting from a previously identified DNA marker that has been mapped closed to the gene of interest. Technologies, such as the yeast artificial chromosome (YAC; *35*), the P1 phage *(36)*, and bacterial artificial chromosome (BAC; *37*), are often used during the early stages of a chromosome walk to cover a relatively long region in as few steps as possible. With the information on the map position of the gene, the progress of the walk is usually monitored by genetic mapping using a set of high resolution recombinants in the region of interest, and using newly isolated clones which uncover DNA polymorphisms. As the walk proceeds toward the gene, the number of recombinants decreases between the isolated DNA (polymorphic marker) and the genetic locus (phenotypic marker). Further, it is necessary to walk across the gene of interest; this is indicated by detecting recombinant(s) between the DNA and the gene, but on the side of the gene opposite to the side where the walk originally started. Due to the low resolution of the genetic map, often when the region of interest is completely covered by isolated clones, the location of the gene may still be uncertain over a region of up to several hundred kilobases. The intermediate range of the insert sizes (20–45 kb) of the cosmids have made them an indispensable component of chromosome walk efforts *(38)*. When large clones have been isolated from YAC or other libraries with 300–1000-kb inserts, cosmids can serve as the next tier of clones with a higher resolution. To localize the gene of interest onto particular cosmid(s), it is usually necessary to transform a number of the cosmids into the mutant background for complementation tests. With organisms for which large scale genome projects are being carried out, physical mapping will eventually produce overlapping set(s) of clones covering most, if not all, of the genome. Then, it is possible to localize a gene to a region of several hundred kilobases when the genetic position of the gene is determined. At that point, it is not necessary to carry out chromosome walk experiments. However, it will still be necessary to test individual cosmid clones for their ability to complement the mutation. For this purpose, cosmids that contain the genomic inserts as part of the T-DNA can be used to avoid additional subcloning before transformation.

The relatively large inserts of cosmid clones make the number of clones needed to cover a particular region small. When sufficient numbers of overlapping cosmid clones were used to transform mutant plant cells, a positive result by the relevant functional assay would identify those that contain the gene. Analysis of the insert DNAs can then determine the region where the gene resides. Further analysis can then follow to identify transcripts, to isolate cDNA clones, and to determine DNA sequences. This application of a transformation competent cosmid library is illustrated by the recent isolation of the *Arabidopsis AXR1* gene *(39)*, which affects a wide range of cellular responses to the phytohormone auxin *(40)*. The *AXR1* gene was localized to a region of about 100 kb on chromosome 1 using a combination of RFLP mapping and chromosome walking experiments with both YAC and cosmid clones *(39)*. Several cosmids which correspond to the region of the *AXR1* gene were isolated from a pOCA18-based cosmid library *(12)* and introduced into an *axr1* mutant by *Agrobacterium*-mediated transformation, two overlapping cosmids were able to restore the wild-type phenotypes *(39)*. Since many genes are now being cloned by chromosome walking, transformation competent cosmid libraries are going to be important tools for these efforts.

3.3. Isolation of Genes by Functional Complementation

Although there are a number of methods for introducing DNAs into plant cells, including electroporation, PEG precipitation, and particle bombardment, the most widely and successfully used method for transformation of dicot cells is the one mediated by *Agrobacterium*. Nowadays it is routine to introduce several constructs into plant cells to generate a number of transgenic plants per construct by this method. In addition to the well-established tissue-culture procedures for transforming cells from various parts of the plant, methods for nontissue-culture transformations have also been developed *(41,42)*. As the efficiency of transformation continues to be improved *(42)*, it is likely that in the near future, the frequency of transformation will increase to a level that allows the method to be used in shotgun transformation of genomic libraries. When the transformation frequency is high enough to generate at least 10^4–10^5 transformants with a reasonable amount of effort, this method can then be used to isolate genes based on functional complementation of mutations, as is true for microorganisms such as *E. coli* and yeast. For this purpose, cosmid libraries of plant genomic DNA established in an *Agrobacterium* host are very useful *(13)*, although a cosmid library with a sufficient number of clones established in *E. coli* can also be used *(12)*.

3.4. Testing Gene Function in Heterologous Plants

It is often desirable to introduce a gene that has a particular property from the plant in which the gene was identified into another plant. For example, herbicide resistance genes have been identified in *A. thaliana* *(15,43)*, and some of them have been cloned and shown to be functional in other plants *(12,44,45)*. It is of great agricultural importance to transfer these genes to crop plants to confer herbicide resistance. Another example is disease resistance genes, which have been studied in many plants, including *Arabidopsis (46,47)*. The newly identified disease resistance genes can be transferred to other plants to test for function using cosmid libraries constructed from the DNA of the disease-resistant plants. The relatively large inserts of cosmids increase the chances of retaining in a single insert all of the necessary regulatory elements flanking a gene. Again, the isolation of a clone that is already competent for transformation eliminates the need to subclone the gene. Finally, if there is a selection for the gene of interest, especially at the tissue-culture level, then the transfer may be done by transformation with *Agrobacteria* carrying a cosmid library made from genomic DNA of the donor plant, followed by selection for the gene of interest.

4. Conclusion

Cosmid libraries that are competent for *Agrobacterium*-mediated transformation of plant cells are useful for a number of studies of plant genes. Several vectors have been described that can be used to construct transformation-competent cosmid libraries. The pOCA18 and pLZ03 vectors allow inserts of 20–30 kb, and are suitable for testing candidate genes. However, the large sizes of the vectors make the DNA manipulations more difficult. The pCIT vectors are much smaller, particularly the pCIT103 and pCIT104 vectors, which are only 7 kb, allowing easier manipulations. The smaller sizes of the pCIT vectors do require that the genomic DNAs be size selected. The larger inserts of the pCIT-based libraries make them more suitable in a chromosome walk project. In addition, the pCIT vector contain only the right border of the T-DNA, which allows the transfer of the entire plasmid. These vectors should be useful tools for the studies of plants using transformation. Cosmid libraries made using these and other similar vectors may be used to test candidate genes by functional complementation, and to isolate genes using chromosome walk procedures. As transformation efficiency increases, cosmid libraries may be introduced into plant cells *en masse* to isolate the gene of interest by functional complementation, and to transfer selectable genes from one plant to another without first cloning the gene.

Acknowledgments

The author thanks Kevan Gartland for comments. The work in the author's laboratory was supported by grants from the National Science Foundation (DCB 9004567 and DCB 9105260) and funds from the Robertson Foundation.

References

1. Emmons, S. W. (1974) Bacteriophage lambda derivatives carrying two copies of the cohesive end site. *J. Mol. Biol.* **83,** 511–525.
2. Hohn, B. (1983) DNA sequences necessary for packaging of bacteriophage λ DNA. *Proc. Natl. Acad. Sci. USA* **80,** 7456–7460.

3. Hohn, B. and Collins, J. (1980) A small cosmid for efficient cloning of large DNA fragments. *Gene* **11,** 291–298.
4. Meyerowitz, E. M., Guild, G. M., Prestidge, L. S., and Hogness, D. S. (1980) A new high-capacity cosmid vector and its use. *Gene* **11,** 271–282.
5. Feiss, M., Fisher, R. A., Crayton, M. A., and Egner, C. (1977) Packaging of the bacteriophage chromosome: the chromosome length. *Virology* **77,** 281–293.
6. Nester, E. W., Gordon, M. P., Amasino, R. M., and Yanofsky, M. F. (1984) Crown gall: a molecular and physiological analysis. *Annu. Rev. Plant Physiol.* **35,** 387–413.
7. Klee, H. J., Horsch, R. B., and Rogers, S. G. (1987) *Agrobacterium* mediated plant transformation and its further applications to plant biology. *Annu. Rev. Plant Physiol.* **38,** 467–486.
8. Binns, A. N. and Thomashow, M. F. (1988) Cell biology of *Agrobacterium* infection and transformation of plants. *Annu. Rev. Microbiol.* **42,** 575–606.
9. Zambryski, P., Tempe, J., and Schell, J. (1989) Transfer and function of T-DNA genes from *Agrobacterium* Ti and Ri plasmids in plants. *Cell* **56,** 193–201.
10. Raineri, D. M., Bottino, P., Gordon, M. P., and Nester, E. W. (1990) *Agrobacterium*-mediated transformation of rice (*Oryza sativa* L.). *Bio/Technology* **8,** 33–38.
11. Hooykaas, P. J. J. and Schilperroort, R. A. (1992) Detection of monocot transformation via *Agrobacterium tumefaciuns. Methods Enzymol.* **216,** 305–313.
12. Olszewski, N. E., Martin, F. B., and Ausubel, F. M. (1988) Specialized binary vector for plant transformation: expression of the *Arabidopsis thaliana* AHAS gene in *Nicotiana tobacum. Nucleic Acids Res.* **16,** 10,765–10,781.
13. Lazo, G. R., Stein, P. A., and Ludwig, R. A. (1991) A DNA transformation-competent *Arabidopsis* genomic library in *Agrobacterium. Bio/Technology* **9,** 963–967.
14. Ditta, G., Stanfield, S., Corbin, D., and Helinski, D. R. (1980) Broad host range DNA cloning system for Gram-negative bacteria: construction of a gene bank of *Rhizobium meliloti. Proc. Natl. Acad. Sci. USA* **77,** 7347–7351.
15. Haughn, G. W. and Somerville, C. (1986) Sulfonylurea-resistant mutations of *Arabidopsis thaliana. Mol. Gen. Genet.* **204,** 430–434.
16. Ma, H., Yanofsky, M. F., Klee, H. J., Bowman, J. L., and Meyerowitz, E. M. (1992) Vectors for plant transformation and cosmid libraries. *Gene* **117,** 161–167.
17. Fraley, R. T., Rogers, S. G., Horsch, R. B., Eichholtz, D. A., Flick, J. S., Fink, C. L., et al. (1985) The SEV system: a new disarmed Ti plasmid vector system for plant transformation. *Bio/Technology* **3,** 629–635.
18. Rogers, S. G., Horsch, R. B., and Fraley, R. T. (1986) Gene transfer in plants: production of transformed plants using Ti plasmid vectors. *Methods Enzymol.* **118,** 627–640.
19. Zambryski, P., Joos, H., Genetello, C., Leemans, J., Van Montagu, M., and Schell, J. (1983) Ti plasmid vector for the introduction of DNA into plant cells without alteration of their normal regeneration capacity. *EMBO J.* **2,** 2143–2150.
20. Figurski, D. H. and Helinski, D. R. (1979) Replication of an origin-containing derivative of plasmid RK2 dependent on a plasmid function in trans. *Proc. Natl. Acad. Sci. USA* **76,** 1648–1652.
21. Thomas, C. M. (1981) Complementation analysis of replication of maintenance functions of broad host range plasmids RK2 and RP1. *Plasmid* **5,** 277–291.

22. Koncz, C. and Schell, J. (1986) The promoter of T_L-DNA gene 5 controls the tissue-specific expression of chimaeric gene carried by a novel type of *Agrobacterium* binary vector. *Mol. Gen. Genet.* **204**, 383–396.

23. Ma, H., Yanofsky, M. F., and Meyerowitz, E. M. (1990) Molecular cloning and characterization of *GPA1*, a G protein α subunit gene from *Arabidopsis thaliana*. *Proc. Natl. Acad. Sci. USA* **87**, 3821–3825.

24. Ma, H., Yanofsky, M. F., and Meyerowitz, E. M. (1991) *AGL1-AGL6*, an *Arabidopsis* gene family with similarity to floral homeotic and transcription factor genes. *Genes Dev.* **5**, 484–495.

25. Yanofsky, M. F., Ma, H., Bowman, J. L., Drews, G. N., Feldmann, K. A., and Meyerowitz, E. M. (1990) The protein encoded by the *Arabidopsis* homeotic gene *agamous* resembles transcription factors. *Nature* **346**, 35–39.

26. Bowman, J. L., Smyth, D. R., and Meyerowitz, E. M. (1991) Genetic interactions among floral homeotic genes of *Arabidopsis*. *Development* **112**, 1–20.

27. Chang, C., Kwok, S. F., Bleecker, A. B., and Meyerowitz, E. M. (1993) *Arabidopsis* ethylene-response gene *ETR1*: similarity of product to two-component regulators. *Science* **262**, 539–544.

28. Bleecker, A. B., Estelle, M. A., Somerville, C., and Kende, H. (1988) A dominant mutation confers insensitivity to ehtylene in *Arabidopsis thaliana*. *Science* **241**, 1086–1089.

29. Sun, T.-P., Goodman, H. M., and Ausubel, F. M. (1992) Cloning the Arabidopsis *GA1* locus by genomic subtraction. *Plant Cell* **4**, 119–128.

30. Herman, P. L. and Marks, M. D. (1989) Trichome development in *Arabidopsis thaliana*. II. Isolation and complementation of the *GLABROUS1* gene. *Plant Cell* **1**, 1051–1055.

31. Roe, J. L., Rivin, C. J., Sessions, R. A., Feldmann, K. A., and Zambryski, P. C. (1993) The *TOUSLED* gene in Arabidopsis thaliana encodes a protein kinase homologue that is required for leaf and flower development. *Cell* **75**, 939–950.

32. Arondel, V., Lemieux, B., Hwang, I., Gibson, S., Goodman, H. M., and Somerville, C. R. (1992) Map-based cloning of a gene controlling omega-3 fatty acid desaturation in *Arabidopsis*. *Science* **258**, 1353–1355.

33. Meyerowitz, E. M. and Pruitt, R. E. (1985) *Arabidopsis thaliana* and plant molecular genetics. *Science* **229**, 1214–1218.

34. Meyerowitz, E. M. (1989) Arabidopsis, a useful weed. *Cell* **56**, 263–269.

35. Burke, D. T., Carle, G. F., and Olson, M. V. (1987) Cloning of large segments of exogenous DNA into yeast by means of artificial chromosome vectors. *Science* **236**, 806–812.

36. Pierce, J. C. and Sternberg, N. L. (1992) Using bacteriophage P1 system to clone high molecular weight genomic DNA. *Methods Enzymol.* **216**, 549–574.

37. Shizuya, H., Birren, B., Kim, U.-J., Mancino, V., Slepak, T., Tachiiri, Y., and Simon, M. (1992) Cloning and stable maintenance of 300-kilobase-pair fragments of human DNA in *Escherichia coli* using an F-factor-based vector. *Proc. Natl. Acad. Sci. USA* **89**, 8794–8797.

38. Evans, G. A., Snider, K., and Hermanson, G. G. (1992) Use of cosmids and arrayed clone libraries for genome analysis. *Methods Enzymol.* **216**, 530–548.

39. Leyser, H. M. O., Lincoln, C. A., Timpte, C., Lammer, D., Turner, J., and Estelle, M. (1993) *Arabidopsis* auxin-resistance gene *AXR1* encodes a protein related to ubiquitin-activating enzyme E1. *Nature* **364,** 161–164.

40. Lincoln, C., Britton, J. H., and Estelle, M. (1990) Growth and development of the *axr1* mutants of *Arabidopsis*. *Plant Cell* **2,** 1071–1080.

41. Feldmann, K. A. and Marks, M. D. (1987) *Agrobacterium*-mediated transformation of germinating seeds of *Arabidopsis thaliana*: a nontissue culture approach. *Mol. Gen. Genet.* **208,** 1–9.

42. Bechtold, N., Ellis, J., and Pelletier, G. (1993) *In planta Agrobacterium* mediated gene transfer by infiltration of adult *Arabidopsis thaliana* plants. *Life Sci. (C. R. Acad. Sci. Paris)*, **316,** in press.

43. Kreps, J. A. and Town, C. D. (1992) Isolation and characterization of a mutant of *Arabidopsis thaliana* resistant to α-methyltryptophan. *Plant Physiol.* **99,** 269–275.

44. Haughn, G. W., Smith, J., Mazur, B., and Somerville, C. (1988) Transformation with a mutant *Arabidopsis* acetolactate synthase gene renders tobacco resistant to sulfonylurea herbicides. *Mol. Gen. Genet.* **211,** 266–271.

45. Li, Z., Hayashimoto, A., and Murai, N. (1992) A sulfonylurea herbicide resistance gene from *Arabidopsis thaliana* as a new selectable marker for production of fertile transgenic rice plants. *Plant Physiol.* **100,** 662–668.

46. Debener, T., Lehnackers, H., Arnold, M., and Dangl, J. (1991) Identification and molecular mapping of a single *Arabidopsis thaliana* locus determining resistance to a phytopathogenic *Pseudomonas syringae* isolate. *Plant J.* **1,** 289–302.

47. Kunkel, B. N., Bent, A. F., Dahlbeck, D., Innes, R. W., and Staskawicz, B. J. (1993) *RPS2*, an Arabidopsis disease resistance locus specifying recognition of *Pseudomonas syringae* strains expressing the avirulence gene *avrRpt2*. *Plant Cell* **5,** 865–875.

CHAPTER 31

Regulation
of *Agrobacterium* Gene Manipulation

Frank Dewhurst

1. Introduction

The regulation of biotechnology within the member states of the European Community* has developed very significantly over the last 3 yr, and a comprehensive body of legislation, guidance notes, and codes of practice have appeared in the United Kingdom in particular. The recent incident at Birmingham University, where UK Health and Safety Executive inspectors halted work on a genetic manipulation project, highlights the need for scientists to be aware of legislation, to understand its implications, and to comply adequately with its requirements.

In this chapter the nature of the legal environment within which biotechnology operates is explained briefly, the significance of European law (which in this context means European Community/European Union law) outlined, the United States system for regulation of biotechnology commented on, and finally the European approach, as illustrated by UK national law, discussed. The legislation will of necessity not be subjected to detailed examination, as the UK guidance notes on regulation and control of deliberate release of genetically modified organisms *(1)* and the guide to the contained use regulations *(2)* alone run to 130 pages of information.

The purpose of this chapter is to broadly outline the main features of the regulation, with particular reference to *Agrobacterium*, and to indicate where more detailed and further information can be obtained.

*In this chapter, the terms European Community (EC), European Economic Community (EEC), and European Union (EU) are used interchangeably.

From: *Methods in Molecular Biology, Vol. 44:* Agrobacterium *Protocols*
Edited by: K. M. A. Gartland and M. R. Davey Humana Press Inc., Totowa, NJ

2. Law and Science

The law may be divided into Public Law, which is concerned with the rights and duties of citizens within the state, and Private Law, which concerns rights and duties of individual citizens (or organizations) with respect to other citizens, and so on. In the Private Law domain, contract and tort create obligations between persons. Tort is the duty we owe to others not to negligently injure them, damage their property, or interfere with their enjoyment of their property. The remedies in private law are to constrain the defendant from causing further harm and to compensate their victims by payment of damages. These damages may be very substantial indeed, and, in a case involving many deaths and extensive damage to health or property, can run into hundreds of millions of dollars.

In the Public Law domain, administrative law governs the way in which public bodies, such as regulatory authorities, discharge their functions and is relevant with regard to appeals against regulatory decisions, potential procedural abuses, or abuse of regulatory powers. The criminal law is the Public Law area of most direct concern to the laboratory scientist, and health and safety and environmental protection legislation form part of the criminal law. Sanctions against those who break the criminal law are normally punishment in the form of fines or imprisonment. In the UK, both the Health and Safety at Work etc. Act 1974 and the Environmental Protection Act 1990 (Part VI covering genetically modified organisms [GMOs]), which are the major basis for regulation of contained use and deliberate release of GMOs, include both substantial fines and imprisonment as sanctions. Imprisonment is not impossible but most unlikely in the light of past experience. Fines that may seem modest for a company take on a different aspect when an individual scientist has to pay them.

There are also sanctions available that do have a substantial impact on scientific organizations. The law frequently involves the granting of an authorization to carry out work subject to certain conditions. Serious breaches of the law can lead to withdrawal of the authorization, imposition of new conditions, and, in certain cases, cast doubt on the suitability of an individual for authorization purposes *(3)*. Regulatory enforcement powers also include rights to enter premises, seize, make safe, or destroy GMOs in cases where "it is a cause of imminent danger of damage to the environment" *(4)*. Prohibition notices may be issued by the Secretary of State to stop a person's activities if "he is of the opinion" the activities "would involve a risk of causing damage to the environment" (Environmental Protection Act 1990—EPA—s110). Inspectors under the Health and Safety at Work etc. Act 1974 (HASAWA) have similar powers to destroy or render harmless a cause of imminent danger (HASAWA s25) and may issue Prohibition Notices (having immediate effect) or improvement notices (HASAWA ss21 and 22).

In criminal law "intention" is often a requirement for an act to be criminal. With regulations designed to prevent harm to health or the environment this would often be inappropriate and in consequence there is usually "strict liability." This means that simple commission of the act constitutes an offense. Ignorance of the law's requirements or the consequences of an action being unintentional will not be a defense.

It is thus important to familiarize oneself with the appropriate national law. In regulatory offenses it is usually a defense to be able to show that you have exercised "all reasonable care" and taken "all reasonable precautions." These, in UK law, have "been said to relate to those risks of harm reasonably foreseeable when a prudent and competent person applies his mind seriously to the situation" *(5)*. The various guidance notes, codes of practice, and standards issued by regulatory authorities are not legally binding as government legislation is (in the UK Acts of Parliament, Regulations, and Orders). They are of evidential value, however, in showing what a prudent and competent person would be expected to do.

The importance of proper record keeping, documentation of procedures, and so on, for scientists working with GMOs cannot be overstressed, both as proof of "serious application of the mind to the situation" and as a basis for constructive relationships with regulators and insurers.

The interpretation of the wording of legislation is essentially common sense, bearing in mind that in the case of GMOs, it is aimed at a technically trained audience with some knowledge of the subject. The UK used to be unusual in the extent to which the law was interpreted on the narrow meaning of the words. With health and safety and environmental legislation now being based on European Community law it is interpreted in the European "purposive" fashion.

Legislation is examined in the light of the objectives, provisions, system of related legislation, and the spirit of the treaty on which Community law is based. Courts in all Community member states have followed this approach. In the UK in *Pickstone v Freemans plc (6)*, the highest national court held that it was permissible to give a purposive construction to regulations to make them comply with European Community law, and in *Litster v Forth Dry Dock and Engineering Co. (7)*, a lower court was held to have been in error for adopting a literal approach rather than the proper purposive (or "teleological" approach). In *Marleasing (8)*, it has been held that in applying Member State national law, a national court must interpret it, as far as possible, in the light of the wording and purpose of any relevant EEC directive, whether the new law originated before or after the adoption of the directive.

In the interpretation of legislation it is important to read any definition section it contains. Courts also tend to follow previous decisions in order to achieve certainty as to what the law actually means. Commentaries are frequently published on more important pieces of legislation that will refer to such previous decisions or "precedents." The text by Stephen Tromans *(3)* on the UK Environmental Protection Act 1990 is an example. It is worth noting that in UK legislation, the words "shall" and "shall not" impose an absolute obligation to do or not do the act or thing in question and it is not permissible to argue practicability, or even impossibility as a defense. Within the UK, EPA s1 *(3)* states "'Pollution of the environment' means pollution of the environment due to the release (into any environmental medium) from any process of substances which are capable of causing harm to man or any other living organism supported by the environment." The use of the word capable should be noted, as this means that there is no requirement for *proof* of actual harm but simply the *potential* to cause harm.

The UK, US, and many former Commonwealth States have what is termed "common law jurisdictions," in contrast to the "civil law jurisdictions" found in the rest of the world. In many civil law jurisdictions, a basic distinction is drawn between criminal and administrative sanctions. For example, in Germany "ordnungswidrigkeiten" attract administrative rather than criminal financial penalties—"Bussgeld fines." There are a more limited range of "real" criminal offenses that include a number involving "serious environmental damage or pollution and endangerment of life and limb" that are enforced by police and state prosecution authorities. Heine briefly reviewed national differences in environmental protection and criminal law *(9)*.

In the regulatory area it is important to realize that the relationship between regulatory enforcement inspectors and potential offenders, i.e., in our context, anyone working or planning to work with GMOs, is not a simple "cops and robbers" relationship. The role of the inspectorate is not to catch and punish offenders but to prevent offenses in the first place. They are thus very approachable and willing to discuss problems and advise on compliance and good practice. During inspections they tend to adopt a constructive approach, particularly if they perceive a client is trying to comply with the law. They do, however, have real powers and a duty to use them. In the Birmingham case *(10)*, after a visit to the laboratories of the University Medical School, an inspector issued a prohibition order to stop work that was considered dangerous and not being carried out using proper procedures. It is of interest that preliminary reports indicate overcrowding in the laboratory was a significant factor in the inspector's decision.

If harm is caused to human health or to the environment, then the incident can serve as the basis both for a criminal action and a tort action for compensation. It will be obvious that a conviction for the criminal offense makes the pursuit of a claim for compensation a great deal easier. The financial implications of claims for damages can be very high. Insurance may be used to meet some of these costs. Insurers, however, make regulatory style demands for information, impose conditions with regard to working practices, and make inspections. Recent trends to deregulate activities in some countries (including UK proposals in this area) will probably, in the end, tend to replace the state regulatory system with unofficial regulation by insurers who may well prove less cooperative and flexible than the old officials.

3. European Community Law

The Member States of the European Community (EC) are all covered by community Law. Membership of the community has created a new type of legal order that must not be confused with the results of signing a simple treaty or with a federal state. Essentially, the Member States have ceded some of their law-making powers to the institutions of the Community and accepted the European Court of Justice (ECJ) as the final arbiter on the interpretation of Community Law. The ECJ also resolves disputes between Member States and Community institutions.

The fundamental laws of the Community and its aims and objectives are contained in the 1959 Treaty of Rome as amended by the 1986 Single European Act (SEA) and the terms of the recent Maastricht Treaty. The SEA introduced a new environmental section into the Treaty (Title VII).

It simplified the introduction of legislation in the health and safety and environmental areas by the replacement of a requirement for unanimity among Member States with a qualified majority only being needed to pass most legislation. A qualified majority means the voting system is weighted to prevent dominance by unrepresentative minority groupings of large or small states.

Articles 130R and 130T indicate the Community approach.

Article 130R

1. Action by the Community relating to the environment shall have the following objectives:
    (i)   to preserve, protect and improve the quality of the environment;
    (ii) to contribute towards protecting human health;
    (iii) to ensure a prudent and rational utilization of natural resources.
2. Action by the Community relating to the environment shall be based on the principles that preventive action should be taken, that environmental damage should as a priority be rectified at source, and that the polluter should pay. Environmental protection requirements should be a component of the Community's other policies.
3. In preparing its action relating to the environment, the Community shall take account of:
    (i)   available scientific and technical data;
    (ii) environmental conditions in the various regions of the Community;
    (iii) the potential benefits and costs of action or of lack of action;
    (iv) the economic and social development of the Community as a whole and the balanced development of its regions.
4. The Community shall take action relating to the environment to the extent to which the objectives referred to in paragraph I can be attained better at Community level than at the level of the individual Member States. Without prejudice to certain measures of a community nature, the Member States shall finance and implement the other measures.

Article 130T

The protective measures adopted in common pursuant to Article 130S shall not prevent any Member States from maintaining or introducing more stringent protective measures compatible with this Treaty.*

The Community is empowered under Art. 189 of the Treaty to make Regulations, issue Directives, take Decisions, make Recommendations, or deliver Opinions.

*Title VII is now Title XVI of the Post Maastricht Treaty with some change of wording, but not of emphasis.

A Regulation shall have general application. It shall be binding in its entirety and directly applicable in all Member States.

A Directive shall be binding, as to the result to be achieved, upon each Member State to which it is addressed, but shall leave to the national authorities the choice of form and methods.

A Decision shall be binding in its entirety upon those to whom it is addressed.

Recommendations and Opinions shall have no binding force.

Direct applicability in the case of a Regulation means that it automatically becomes part of Member State law at the commencement date specified in the Regulation. In the case of a Directive, each Member State must incorporate it in a suitable form that meets the aims and objectives of the Community Legislation. The exact form of the incorporation is left to each Member State and in the case of GMOs the Contained Use Directive was incorporated as UK national regulations under HASAWA, whereas the Deliberate Release Directive was incorporated under the provisions of the EPA 1990 (an Act of Parliament) and by national regulations subsequently introduced under that act. Both the EPA and HASAWA are examples of enabling acts that confer powers to introduce secondary national legislation such as regulations without need for the full legislative procedure required for an act.

All Directives have a date by which they should be implemented by Member States, and pressure can be brought by the Community on Member States that drag their feet, to ensure implementation. In many cases Directives can give rights to individuals which they can enforce in their national courts if the Member State has failed to implement a Directive either in part or totally. This is termed "Direct Effect." Community law takes precedence over national law within the Member States and the supremacy of Community law has always been accepted by all Member States. In the Factortame cases *(11)*, for example, the UK courts gave unqualified acceptance to EC principles for Direct Effect and Supremacy of EC law.

In some circumstances Member States are allowed to introduce and apply legislation that is stricter than that applied by the Community itself, and acts as a barrier to trade. This can be justified in terms of the "Mandatory Requirement" for Member States to protect the health and life of humans, animals, or plants. In the Danish Bottles case *(12)* the ECJ established that protection of the environment was an example of a "Mandatory Requirement." The wording of Article 130T should be noted in this context.

The Maastricht Pact speaks of subsidiarity, which means only taking community level action when the objectives of legislation are attained better at the Community level than at the individual state level. This is not a new concept and was included in Article 130 R(4). It has, particularly in the context of a move toward deregulation, prompted thoughts of national legislation weaker than Community legislation being acceptable. In a recent UK case, however, the subsidiarity principle was used to justify stricter, not more lenient, national legislation *(13)*.

It is a feature of both national and Community law that draft legislation and guidance notes are produced and circulated for comment from interested parties. Scientists should avail themselves of the opportunity to make an input into legislation, and they provide a basis for forward planning in terms of likely implications for future equipment, facilities, training, and finance. The *Official Journal of the European Community* (OJ) is the source for information on Community legislative activity as illustrated for the GMO Directives:

Directives 90/219 and 90/220 on genetically modified organisms

Commission proposal	16/05/1988 OJ 1988, No. C198/19
Parliament's opinion	24/05/1989 OJ 1989, No C158/122
ECOSOC's (Economic	08/11/1988 OJ 1989, No C23/45
and Social Committee	14/03/1990 OJ 1990, No. C96/87
of the European	
Community) opinion	
Commission amendments	23/08/1989 OJ 1989, No. C246/5 and 6
Council Directives	23/04/1990 OJ 1990, No. L117/1 and 15

4. Planned European Legislation

There are currently a number of proposals under discussion and measures being drafted by the EC which are concerned mostly with the development of the commercial exploitation of biotechnology. These cover such things as biotechnology patents, plant varieties, pesticides, medicinal products, new plants, and transgenic animals (including environmental risk assessment, seeds, silage organisms, animal feedstuffs, and the transport of GMOs) *(14)*. There are also proposals at both national and Community level to introduce more streamlined procedures. It will probably be some time before any of these become law. Commission discussion document XI/506/94-Rev I outlines proposals for streamlined procedure and directives 94/51 and 94/15 amend/adapt the contained use and deliberate release directives (90/219 and 90/220).

The European Parliament voted in October 1993 to demand the update of food labeling rules to require the descriptions "contains genetically modified organisms" or "produced with gene technology" where appropriate. Again, this requirement is unlikely to become law in the near future.

5. Standardization in Biotechnology

The European Commission recognized the necessity for common standards applicable throughout the European Union and together with the European Free Trade Association (EFTA) mandated the Comité Européen de Normalisation (CEN), the European Committee for standardization to develop such standards covering almost the whole of biotechnology. A technical committee of CEN (TC 233) was set up with representation from the EFTA and EU states and four working groups are expected to produce the majority of the standards by 1995 and complete their work by July 1997. The terms of reference speak of developing "standards in the field of biotechnology including safety." The CEN standards will replace any national standards in the 18 Member States and will be pub lished in English, French, and German. The American Society for Testing and Materials has also recently held a workshop sponsored by the US Department of Commerce Technology Administration National Institute of Standards and Technology.

It will be neccssary to take account of these new standards, which are expected to consider the magnitude of hazards and facilitate compliance auditing assessment. Kirsop *(15)* recently discussed these standards more fully and listed them together with implementation dates and the names of project leaders. A summary of the projects is given in Table 1.

The legal status of the standards is uncertain as yet with the UK wishing to use them for guidance only, but Germany reported wishing to incorporate them into regulations. They will have contractual importance in equipment specification.

6. United States Legislation

It is often said in Europe that US legislation is characterized by a "can do" approach. It is certainly the case that in 1990, over 60% of all field tests worldwide (mostly involving transgenic plants) took place in the US. It cannot be said, however, that American legislators have been quite so "can do," and a well-thought-out comprehensive system of legislation designed to meet the needs of biotechnology on the lines of the European Community legislation does not exist. Vice President Gore *(16)* recog-

Table 1
Biotechnology Standards Under Development
by the European Committee for Standardization (CEN)

Group I. Projects covering laboratories for research, development, and analysis
Ten projects dealing with categorization of microbiology laboratories and definition of equipment needed; guidelines for containment of experimental plants and animals; reporting on hosts and vectors used to construct group I organisms; classification work studies on microorganisms (relating to Directive 90/679 EEC); Good Laboratory Practice codes, and methods for waste handling, inactivation, and testing.

Group II. Projects covering large-scale processes and production work
Seven projects concerned with plant building and equipment provision in relation to degree of hazard; strain conservation; fermentation and extraction procedures; waste, raw material, and energy supply control; and personnel, including training.

Group III. Projects covering modified organisms for application in the environment
Thirteen projects dealing with insert characterization, detection, and identification (including sampling) methods for GMOs; inserts and free DNA; diagnostic kit quality control methods; and assessment methods for microorganisms and products.

Group IV. Projects covering equipment standards
Three projects dealing with standard testing procedures for cleanability, sterilization, and leak tightness together with 21 setting performance criteria for equipment, ranging from autoclaves through sampling methods and shaft seals to valves and tubes.

nized the harmful effects this can have for the development of the biotechnology industry in the US with, in one case, a local Monterey County, CA, ordinance halting a field trial. The US federal regulatory process is balkanized, with responsibility being divided between six federal agencies under the mosaic of federal statutes termed the "Coordinated Framework" *(17)*. The agencies are the US Department of Agriculture (USDA), Environmental Protection Agency (EPA), Food and Drug Administration (FDA), National Institutes of Health (NIH), National Science Foundation (NSF), and Occupational Safety and Health Administration (OSHA). The Coordinated Framework is, in fact, an interagency cooperative agreement and agency jurisdiction is based primarily on the type and function of the GMO and the intended product use. Preexisting legislation is used for regulating GMOs so that the EPA used the Federal Insecticide, Fungicide, and Rodenticide Act (FIFRA) to grant an experimental use permit (EUP) for a field trial of cottonseed containing the delta-endotoxin gene *(18)*. This calls for an "unreasonable adverse effects" test of was the risk posed an "unreasonable risk" when balanced against "the economic, social and environmental costs and benefits for the use." The EPA also uses the Toxic Substances Control Act (TSCA) to regulate deliberate releases regarding GMOs as chemical substances and the Premarketing Notification (PMN) and Significant New Use Rule (SNUR) requirements. Once the manufacturer has proved product efficacy, the EPA must determine if an unreasonable risk to the environment is posed.

The USDA Animal and Plant Health Inspection Service (APHIS) has published regulations setting out field testing permit procedures using a "no significant impact" assessment to classify the Calgene FLAVR SAVR tomato as not a regulated article on the grounds of the use of a disabled vector *(18)*. The FDA is considering the same tomato under the Food, Drug and Cosmetics Act as a foodstuff and the kanamycin-resistance marker gene and its protein, utilized in producing the tomato, under the FDA food additive regulations *(18)*.

In 1992 the Office of Science and Technology Policy published a new comprehensive policy on "Federal Oversight" with risk determination based on comparison with past, safe releases of similar unmodified organisms *(19)*. The US regulation has been criticized for laxity, leading to low levels of public protection, lack of account of the "dread factor" leading to lack of public confidence and increased public opposition and, finally, the creation of a confused and complex situation for industry. A detailed review of the position by P. Mostow, a biotechnologist turned lawyer *(20)*, is well worth study, as is the review by Saperstein *(21)*. APHIS recently introduced a rule to reduce the regulatory burden of field testing *(22)* with the emphasis on the product rather than the process by which it is produced. Goldhammer *(23)* recently reviewed the regulation of agricultural biotechnology and notes other potential quasilegal restraints on the industry, with mainstream orthodox Jewish groups accepting microbially produced calf chymosin as *kosher pareve* for use in cheesemaking under religious laws (an interesting example of the product over process approach).

In the US, state-level legislation exists, and that from North Carolina is more integrated than the national federal legislation. The North Carolina legislation has, however, been subject to the same criticism as the federal legislation *(21)*.

The importance of public confidence in the approval processes for biotechnology cannot be overstressed, as shown, for example, by a proposal to plant 10,000 genetically modified petunias in an open field near Cologne, Germany resulted in 16,000 objections (all of which would have to be considered at a public hearing) and the Dutch "eco-saboteurs" group "Fiery Virus," who caused about $2 million worth of damage to experimental plant fields. Lack of public confidence can spread to insurers, with 18% of all biotechnology companies rating the cost of product liability insurance as the most important problem facing their firms *(24)*.

Interesting discussions of the relationship between biotechnology and public perceptions and of North American and UK rules and regulations on releases of GMOs are to be found in BCPC Monograph No. 55 *(25)*.

7. UK GMO Legislation

The Environmental Protection Act 1990, part VI, sections 106–127, brought in new provisions imposing controls over activities involving GMOs for the purpose of "preventing or minimizing any damage to the environment which may arise from the escape or release from human control of genetically modified organisms." Section 106 covers the meaning of GMOs and related expressions whereas s107 deals with "damage to the environment," "control," and related expressions. The term "capable of causing harm" is used in the definition of harm to the environment, and this may be construed as covering GMOs acting individually, collectively, or through their ability to produce descendants. With regard to "control," insertion of a gene ensuring death on exposure to sunlight, so that the GMO would not survive in the open, or the adoption of other means to ensure the organisms or their descendants would be harmless if they did enter the environment, would constitute "control."

Section 108 requires a risk assessment before import, acquisition, release, or marketing of any GMO, whereas s109 creates duties for:

(a) those importing or acquiring GMOs to identify risks of damage to the environment and prohibits any such activity if, despite any precautions, such risks cannot be avoided;
(b) those keeping GMOs to identify past damage, the risks of damage from continuing to keep them despite any additional precautions which can be taken and to cease keeping them if such risks cannot be prevented;
(c) those proposing to release or market to identify risks of environmental damage from release or marketing and prohibiting such activities if such risks cannot be prevented.

There is a duty on all those keeping or releasing GMOs to use Best Available Techniques Not Entailing Excessive Costs (BATNEEC).

BATNEEC covers not only technology but the way it is applied, including training, layout of premises, and working procedures. Excessive cost will not take financial resources of those involved into account—if you cannot afford the necessary precautions you should not do that work! Guidance has been published on the meaning of BATNEEC *(26).* In s119 the onus of proof as to the use of BATNEEC is placed on the defendant and failure to produce compliance records is admissible as evidence. Again, the importance of records must be stressed.

There are powers to issue prohibition notices and general provisions, including rights of entry and inspection, powers to require information and to deal with imminent danger to the environment.

Section 111 allows the Secretary of state to prescribe activities that may then only be undertaken with a consent and in accordance with any conditions attached to that consent. The UK Deliberate Release Regulations have been made under these provisions and anything requiring a consent under s111 does not come under ss108 and 109. Section 112 covers limitations and conditions (which may be both implied and express) and subsection 7 provides a general duty for those keeping, releasing, or marketing GMOs to take reasonable steps to stay informed on development of techniques to avoid environmental harm and to inform the Secretary of State when better techniques than those specified in the conditions become available. This last provision takes into account that scientists working with GMOs will include those at the forefront of development of control and containment techniques in their field.

Sections 122 and 123 deal with public registers of information, confidentiality, and exclusion of material from registers, whereas ss125 and 126 deal with delegation of functions (with particular regard to the Health and Safety Inspectorate) and joint exercise of certain functions with the Ministry of Agriculture, Food, and Fisheries

7.1. Contained Use Regulations

The European Community Directive 90/219/EEC on the contained use of genetically modified microorganisms (8.5.90 OJ No. L 117/1) was implemented in the UK by the Genetically Modified Organisms (Contained Use) Regulations 1992 (SI 3217) and the Genetically Modified Organisms (Contained Use) Regulations 1993 (SI 15). These regulations replaced the Genetic Manipulation Regulations 1989, although there are some transitional provisions for work authorized under the older regulations. The regulations have been made under the Health and Safety at Work etc. Act 1974, but are explicitly stated as being for the protection of the environment as well as for protecting the health of workers and other persons. The regulations are administered by the Health and Safety Executive.

The Contained Use Regulations were made without reference to the EPA 1990 to overcome definition problems and go beyond the Contained Use Directive, which is restricted to microorganisms. The UK regulation covers microorganisms but extends to plants and animals also. The activities covered include holding modified animals in animal houses or farm animals with adequate fencing, as well as fermenters, growth rooms, and glasshouses. The guide to the regulations *(2)* points out that for plants, physical barriers alone may not sufficiently limit contact with the environment. Some designs of glasshouse may be inadequate to contain pollen.

"Genetic modification" is defined and means altering the genetic material in an organism in a way that does not occur naturally by mating or natural recombination or both. Schedule 1* gives examples to illustrate this definition. Techniques constituting genetic modification include:

(a) recombinant DNA techniques consisting of the formation of new combinations of genetic material by the insertion of nucleic acid molecules, produced by whatever means outside the cell, into any virus, bacterial plasmid or other vector system so as to allow their incorporation into a host organism in which they do not occur naturally but in which they are capable of continued propagation;

(b) techniques involving the direct introduction into an organism of heritable material prepared outside the organism including micro-injection, macro-injection and micro-encapsulation; and

(c) cell fusion (including protoplast fusion) or hybridization techniques where live cells with new combinations of heritable genetic material are formed through the fusion of two or more cells by means of methods that do not occur naturally.

In vitro fertilization, polyploidy induction, conjugation, transduction, transformation, or other natural processes are not considered to result in genetic modification (for the purposes of these regulations) if they do not involve the use of recombinant-DNA molecules or GMOs. Techniques to which the regulations do not apply (provided they do not involve the use of GMOs as recipient or parental organisms) are mutagenesis, construction, and use of somatic hybridoma cells (i.e., for monoclonal antibody pro-

---

*Guidance on Regulations documents in the UK generally include the regulations themselves.

duction) and cell fusion (including protoplast fusion) of plant cells where the resulting organisms can also be produced by traditional breeding methods. The regulations also do not apply to self-cloning of non-pathogenic naturally occurring microorganisms that fulfill the criteria for group I for recipient microorganisms and certain other nonpathogens.

Activities with GMOs are subdivided into two types. Type A are essentially small scale, for teaching, research, development, or for nonindustrial or noncommercial purposes, so that culture volumes and organism numbers are such that the necessary containment system reflects good microbiological practice and good occupational safety and hygiene. Standard laboratory decontamination techniques should easily render organisms inactive. Type B operations means any activity involving the genetic modification of an organism other than a type A operation. The EC Contained Use Directive suggests a volume of 10 L for a type A operation. Type B operations will be mainly under industrial conditions and frequently repetitive.

The guidance notes indicate that the term "micro-organism" includes viruses and viroids, the uncharacterized agent responsible for transmissible spongiform encephalopathy, cell cultures, and tissue cultures, including those from plants, animals, and humans. It does not cover naked DNA or naked plasmids. The term "organism" covers, in addition, all multicellular organisms not defined as microorganisms, including plants and animals.

The regulations classify microorganisms into group I and group II organisms applying criteria set out in schedule 2 of the regulations. These criteria relate to recipient or parental organism, vectors/inserts, and GMOs. Part 2 of schedule 1 sets out the guidelines published in European Commission Decision No. 91/448/EEC for the further interpretation of the criteria.

The guidance notes *(2)* state:

> Group I organisms may not be based upon recipients with pathogenic characteristics or those which may harm the environment. *As a general guide*, the following recipient micro-organisms will not qualify for Group I status and are therefore automatically considered to be Group II:
> micro-organisms listed in hazard groups 2, 3 or 4 in the Advisory Committee on Dangerous Pathogens (ACDP) Guidance document *Categorisation of pathogens according to hazard and categories of containment,* 1990, 2nd Ed.;
> organisms listed in the Plant Health (Great Britain) Order 1987, or any other plant pathogen or pest;

Agrobacterium strains that retain all of the plant interactive genes of the wild type plasmid;

pathogens of animals, poultry, fish, or bees controlled by Agriculture Departments (listed in Appendix N of the above ACDP publication);

any other micro-organism capable of harming the environment by any means including the production of metabolites.

Disabled strains of pathogenic species such as *E. coli* K12 may satisfy the host/vector criteria for Group I status set out in Schedule 2.

Agrobacterium tumefaciens strains which carry modified plasmids deficient in all tumorigenic genes may satisfy the host/vector criteria for Group I as long as the plasmid has not also been altered to increase pathogenicity or alter the host range.

The Ministry of Agriculture, Fisheries, and Food/SOAFD license requirements under the Plant Health legislation for genetic modifications involving plant pests or pathogens such as *Agrobacterium* species must be met irrespective of group I or group II status.

Schedule 6 of the guidance notes lays down the containment measures called for in type B operations using group II microorganisms applying three levels of provision: B2, B3, and B4 set out in ACGM/HSE/DOE Note 6. To work with group II microorganisms in type A operations, four physical containment levels are laid down (*see* ACGM/HSE/DOE Note 8).

The Advisory Committee on Genetic Manipulation (ACGM) in conjunction with the Department of the Environment (DOE) and the Health and Safety Executive (HSE) has produced a series of guidance notes listed in Table 2. These guidance notes are designed to indicate to the scientist how to comply with the regulations and protect health and environment during work with GMOs.

The Contained Use Regulations require an assessment of human and environment risk (listing the parameters in Schedule 3), notification of first use of premises for genetic modification work, and notification of the activity itself. After categorization of the work based on risks to health and safety and of environmental safety, type of activity, and the nature of the organism, the activity may commence only after express authorization in some cases. In other cases it may start after a stipulated period unless objected to by the HSE, but in cases involving least risk notification is replaced by a requirement to keep records and provide information annually.

The regulations also require establishment of a local genetic modification safety committee to advise on risk assessment, lay down requirements for standards of safety and containment, require notification of

Table 2
Current Titles in the ACGM Guidance Note Series

ACGM/HES/DOE Note 1.	(2nd Revision)—Guidance on construction of recombinants containing potentially oncogenic nucleic acid sequence.
ACGM/HSE/DOE Note 2.	Disabled host/vector systems (now incorporated in Note 7).
ACGM/HSE/DOE Note 3.	The intentional introduction of genetically manipulated organisms into the environment.
ACGM/HSE/DOE Note 4.	Guidelines for the health surveillance of those involved in genetic manipulation at laboratory and large-scale.
ACGM/HSE/DOE Note 5.	Guidance on the contained use of eukaryotic viral vectors in genetic modification.
ACGM/HSE/DOE Note 6.	Guidelines for the large-scale use of genetically manipulated organisms.
ACGM/HSE/DOE Note 7.	Guidelines for the risk assessment of operations involving the contained use of genetically modified microorganisms.
ACGM/HSE/DOE Note 8.	Laboratory containment facilities for genetic manipulation.
ACGM/HSE/DOE Note 9.	Guidelines on work with transgenic animals.
ACGM/HSE/DOE Note 10.	Guidelines on work involving the genetic manipulation of plants and plant pests.
ACCM/HSE/DOE Note 11.	Genetic manipulation safety committees.

accidents (giving an agreed format), and, where appropriate, require the drawing up of emergency plans. Again, reflecting its EPA 1990 and European Directive origins, there is a requirement for disclosure and registers of information and for fees for notification (the polluter pays). The regulations, like all HASAWA 1974-based regulations, can provide a basis for civil actions for damages.

7.2. Deliberate Release Regulations

The UK Deliberate Release Regulations implement Directive 90/220/EEC on the Deliberate Release into the Environment of Genetically Modified Organisms (8.5.90 OJ No. L117/15), which is supplemented with a Handbook on Implementation *(27)* (as is the Contained Use Directive 90/219/EEC). It should be noted the Community Directives are available in all the Community Official Languages.

The Directive has been implemented under the EPA 1990 in the Genetically Modified Organism (Deliberate Release) Regulations 1992

(SI 3280) as amended by the Genetically Modified Organism (Deliberate Release) Regulations 1993 (SI 152). The DOE have published a comprehensive Guidance Note *(1)* on these regulations. A consultation paper on proposals for amending the Genetically Modified Organisms (Deliberate Release) Regulations 1992 appeared in December 1994.

The regulations, being based on similar Directives and the EPA 1990, share many features with the Contained Use Regulations, particularly with regard to definitions, assessment, information registers, and enforcement powers. However, whereas the Contained Use Regulations list *examples* of artificial techniques of genetic modification, those listed in the Deliberate Release Regulations are the *only* prescribed techniques for the purposes of Part VI of the EPA 1990. This means that consent requirements can differ under the two sets of UK Regulations. The Deliberate Release Regulations Guidance Note draws attention to the Member State authorities agreed interpretation of "traditional breeding methods":

> Traditional breeding means practices which use one or more of a number of methods (e.g., physical and/or chemical means, control of physiological processes) which can lead to successful crosses between plants of the some botanical family.

This means cross family protoplast fusion, for example, *Trifolium repens* (Leguminosae) and *Triticum aestivum* (Gramineae), fall within the scope of control, although fusion of *Brassica rapa* with *Raphanus sativa* would be exempted as, although their genera differ, they are both Cruciferaceae. It should also be noted that although a general purpose of the EPA 1990 is to prevent harm to the environment, there is within s4 of the contained release regulations a strictly limited provision to harm some natural organisms through pesticidal or toxic waste disposal activity of GMOs and their products. This is allowed under section 107 *(8)* of the EPA 1990, but a net benefit to the environment should result.

Consents are required for releasing organisms, applications for consent must be advertised in local newspapers, and specified persons and authorities notified. There is also a general consent condition to inform the Secretary of State of any new information on risks and the effects of releases (Section 9.). Release consents are also of limited duration, although this may extend over several years and the question of a consent covering several releases or a single release must be considered.

Table 3
Release or Marketing Consent Application Format for GMOs

Part A. Schedule 1 Consent to Release or Market
Part I. General Information
  This incudes name, qualifications, and training of scientists and other relevant persons.
Part II. Information Relating to Organisms
  Characteristics of donor, parental, and recipient organisms (13 subsections).
  Characteristics of the vector (4 subsections).
  Characteristics of the modified organism (5 subsections).
  Characteristics of the GMO (9 subsections with 15 other subsections under
  characteristics in relation to human health).
Part III. Conditions of Release
  The release (11 subsections).
  The environment both on site and in the wider environment (12 subsections)
  including migratory species (note the nccd to be aware of wildlife and habitat
  protection legislation), planned developments or changes in land use, relationship
  to drinking water, supply, and so on.
Part IV. The Organism and the Environment
  Characteristics affecting survival (3 subsections).
  Interactions with the environment (7 subsections).
  Potential environmental impact (9 subsections including known or predicted
  involvement in biogeochemical processes).
Part V. Monitoring, Control, Waste Treatment, and Emergency Plans
  Monitoring techniques (4 subheadings).
  Control of release (3 subheadings).
  Waste treatment (4 subheadings including type and amount expected).

Marketing consents are always needed, except for "approved products" and where transitional provisions apply. Applications are needed to cover both an initial marketing and a subsequent marketing for a different purpose (including for a different class of user).

A fees and charges scheme and guidance notes have been published *(28)* and the fees reflect the amount of work required to assess the application. In this context it is worth noting that an application to release genetically modified plants modified only by introduction of one or more of GUS (β-glucuronidase) *Lac ZY* (β-galactosidase) *XyE* (catechol dioxygenase), *Luc* (Luciferase), and *Lux* (bacterial luminescence genes) have a $775 application fee as against the usual $2,700.

An application format with some guidance notes has been published *(29)* and its nature is outlined in Table 3. These reveal the type of information required when applying for consents.

7.3. Oncogenes

The work halted at Birmingham University involved oncogenes or tumor-causing genes being inserted into a virus. *Agrobacterium tumefaciens* and *rhizogenes* carry the Ti (tumor-inducing) or Ri (root-inducing) plasmids and are often referred to as "oncogenic" because of their ability to produce neoplastic growth in plant tissue, termed crown gall tumors or hairy roots, respectively. The HSE have recently published a Specialist Inspector Report on laboratory work with oncogenes *(30)*. In the glossary to this report the term oncogene is used to describe a gene that is capable of progressing cells through a stage in the multistage process of cancer. *"c-onc"* is an oncogene that is present in a cell, these having been isolated from human, animal, and avian cells, and *"v-onc"* is an oncogene that is carried by a virus and is usually a homolog of a cellular oncogene. The Advisory Committee on Genetic Modification (ACGM) has defined oncogenic sequences as DNA sequences that induce tumors in experimental animals or that cause transformation of cells in vitro, leading to an escape from normal growth control, immortalization of cells, or induce anchorage independent growth, e.g., *Harvey Ras, met, myc,* and *abl* oncogene.

The normal Ti or Ri plasmids do not contain genes that induce animal tumors or transform animal cells by coding for enzymes that synthesize auxin or cytokinin and thus cause proliferation of undifferentiated plant (but not animal) tissue. In normal *Agrobacterium* work, it is unlikely that any animal oncogenic sequences will be used. The report should be consulted if any such work is planned and the appropriate ACGM/HSE/DOE Guidance Notes (Notes 1 and 7) consulted and any resultant risk assessment reviewed by a local genetic modification safety committee.

It is worth noting:

1. That ACGM/HSE/DOE Note 1 recommends 24 m^3 for each worker in a laboratory working with oncogenic sequences.
2. There is some evidence from HSE-sponsored research to indicate that oncogenic DNA can induce tumors when applied to abraded skin in mice *(31)*.

7.4. Animal Work

It is not very likely that most scientists working with *Agrobacterium* would use live animals but any such work (including keeping transgenic animals) would, in the UK, come under the Animals (Scientific Procedures) Act 1986, which implements Directive 86/609/EEC. There would almost certainly be a requirement for appropriate licences and certificates and a need for designated premises. A guidance note and code of practice exist *(32,33)*, which should be consulted if any such work is envisaged. The UK Home Office Inspectorate are always happy to answer queries and must be contacted before any such work commences.

8. Other Relevant UK Legislation*

Although the deliberate release and contained use regulations provide a comprehensive integrated framework for biotechnology research and development, they do not replace other legislation that can be relevant to work that involves GMOs. That legislation includes controls on plant pathogens under the Plant Health (Great Britain) Order 1987 and on the introduction of novel species under the Wildlife and Countryside Act 1981. These pieces of legislation must be consulted when work is planned with *Agrobacterium*. It may also be that the Medicines Acts 1968 and 1971 and the Food and Enrironmental Protection Act 1985 are relevant for work involving human and veterinary medicines and pesticides, respectively.

If the application is for consent to market, additional general information is required which includes a description of the person expected to use the product and "additional relevant information" covering emergency measures for escape or misuse, storage, handling, packaging (including its appropriateness to avoid escape), and information for production and import in the EC. Parts A2 to A6 must then be filled in, however, for marketing only parts A2 to A5 of the application.

A2. Details of data or results from previous releases. (Appropriate cross-referencing to answers given for the purpose of Part A1 is acceptable.)
A3. Details of previous applications for release. (Appropriate cross-referencing to answers given for the purpose of Part A1 is acceptable.)
A4. Risk evaluation statement.
A5. Assessment of confidentiality of information provided.
A6. Statement on whether information about GMO and purposes of the release have been published.

Detailed guidance on each of the points listed in Schedule 1 can be found in Chapter 3 of the deliberate release guide *(1)*. It should be noted that although it may be sufficient to state that a particular point is not applicable, you may be asked to provide further information or justify your response to part A.

*A draft of new guidelines for working with alien plant pathogens appeared in December 1994.

Part B of the application covers the information about the application to be included in the public register. The Freedom of Access to Environmental Information Directive 90/313/EEC (23.6.90 OJ 1990 No. L158 56) requires Member States to ensure access to information relating to the environment held by public bodies, but in Article 3 *(2)* states "Member States *may* provide for a request for such information to be refused where it affects commercial and industrial confidentiality, including intellectual property." The UK Environmental Information Regulations 1992 (SI No. 3240) implement this Directive nationally and s4 *(2)* (e) states that such information is capable of being treated as confidential within the terms of the UK regulation.

Part C of the application covers the information to be sent to the Commission of the European Communities when applying for consent to release or market a GMO.

References

1. *The Regulation and Control of the Deliberate Release of Genetically Modified Organisms.* (1993) DOE/ACRE Guidance Note 1. Department of Environment, London.
2. *A Guide to the Genetically Modified Organisms (Contained Use) Regulations 1993.* (1993) Health and Safety Executive, HMSO, London.
3. Tromans, S. (1993) *The Environmental Protection Act 1990 Text and Commentary*, 2nd ed. Sweet and Maxwell, London, p. 43–78.
4. *Idem*, p. 43–269.
5. *Idem*, p. 43–272.
6. *Pickstone v Freemans plc* (1988) 2 AER 803.
7. *Litster v Forth Dry Dock and Engineering Co.* (1989) 2WLR 634.
8. *Marleasing SA v La Comercial International De Alimentation SA* (Case C-106/89). (1992) 1 C.M.L.R. 305.
9. Lomas, O. (ed) (1991) *Frontiers of Environmental Law*, Chancery Lane Publishing, London, p. 75.
10. *Independent* Newspaper, Friday, 4th Feb. 1994.
11. *R v SS for Transport ex p. Factortame* (No. 2) (1991), 1AER 70 and Factortame (No. 3) (1991) 3 AER 759.
12. *Commission v Denmark* (1989) 54 CMLR 619.
13. *R v London Boroughs Transport Committee, ex p. Freight Transport Association and others* (1992) Env. L. R. 62.
14. Hodgson, J. (1992) Europe, Maastricht and Biotechnology. *Biotechnology* **10,** 1421–1426.
15. Kirsop, B. (1993) European standardization in biotechnology. *TIBTECH.* **11,** 375–378.
16. Gore, A. (1991) Planning a new biotechnology policy. *Harvard J. Law Technol.* **5,** 19–30.

17. The Office of Science and Technology Policy's Coordinated Framework for Regulation of Biotechnology (1986) 51 *Federal Register* 23, 302.
18. Parr, K. D. (1993) Developments in agricultural biotechnology. *William Mitchell Law Rev.* **19**, 457–480.
19. Exercise of Federal Oversight within Scope of Statutory Authority: Planned Introduction of Biotechnology Products into the Environment (1992) 57 *Federal Register* 6753.
20. Mostow, P. (1992) Reassessing the scope of federal biotechnology oversight. *Pace Environ. Law Rev.* **10**, 227–273.
21. Saperstein, R. (1991) The monkey's paw: regulating the deliberate environmental release of genetically engineered organisms. *Washington Law Rev.* **66**, 247–265.
22. Genetically Engineered Organisms and Products; Notification Procedures for the Introduction of Certain Regulated Articles and Petition for Nonregulated Status (1993) 58 *Federal Register* 17, 044.
23. Goldhammer, A. (1993) The regulations of agricultural biotechnology: an industrial perspective. *Food Drug J.* **48**, 501–510.
24. Stovsky, M. D. (1991) Product liability barriers to the commercialization of biotechnology: improving the competitiveness of the US biotechnology industry. *High Technol. Law J.* **6**, 363–381.
25. *Opportunities for Molecular Biology in Crop Production* (BCPC Monograph No. 55). (1993) British Crop Protection Council, Farnham, pp. 285–304.
26. *Integrated Pollution Control: A Practical Guide.* (1993) DOE, London.
27. *Handbook for the Implementation of Directive 99/220/EEC on the Deliberate Release of Genetically Modified Organisms to the Environment.* (1992) Commission of the European Communities, X1/322/92-EN.
28. *The Genetically Modified Organisms (Deliberate Releases) Fees and Charges Scheme 1993 and Guidance Notes.* (1993) DOE, London.
29. *Format for Application for Consent to Release or Market Genetically Modified Organisms (GMOs).* (1993) DOE, London.
30. Bosworth, D. A. and South, D. S. (1993) *Laboratory Work with Chemical Carcinogens and Oncogenes.* Health and Safety Executive Specialist Inspector Reports No. 41, London.
31. Burns, P. A., Jack, A., Neilson, F., Haddow, S., and Balmain, A. (1991) Transformation of mouse skin endothelial cells *in vivo* by direct application of plasmid DNA encoding the human *T24 H ras* oncogene. *Oncogene* **6**, 1973–1978.
32. *Guidance on the operation of the animals (Scientific Procedures) Act 1986.* (1990) HMSO, London.
33. *Code of practice for the housing and care of animals used in scientific procedures.* (1989) HMSO, London.

CHAPTER 32

Transgenic Oilseed Rape

How to Assess the Risk of Outcrossing to Wild Relatives

Alain Baranger, Marie-Claire Kerlan, Anne-Marie Chèvre, Frédérique Eber, Patrick Vallée, and Michel Renard

1. Introduction

The increasing concern about growing transgenic plants in commercial fields makes it necessary to assess the risks linked to such crops. In the case of plants transformed via *Agrobacterium*, different kinds of risks can be considered.

Some risks are linked to the engineering itself, depending on the DNA fragment transferred (size, presence or absence of a bacterial replicative origin, and so on) or to the possible change of the target genome (suppression of a gene, pleiotropic effects, synthesis of a new molecule, expression of a virus previously inactive). This type of risk is usually considered low: Indeed, the selection of the best transformants, their molecular description (copy number, integrity of the border sequence), and field trials prior to commercialization ensure reasonably good safety as far as this type of risk is concerned *(1)*.

Some risks are linked to the new character conferred. It is essential to estimate what can be the result of introducing a new biological pathway in plants that are to be visited by pollinators looking for pollen or nectar, or to be eaten by various consumers (cattle, humans, and so on). For each gene transferred into a crop, toxicological studies must therefore be considered *(2)*. The type of gene transferred and the selection pressure to which it can be exposed are also of crucial importance concerning the

From: *Methods in Molecular Biology, Vol. 44:* Agrobacterium *Protocols*
Edited by: K. M. A. Gartland and M. R. Davey Humana Press Inc., Totowa, NJ

potential spreading of the gene within a species. Some new functions, such as resistance to antibiotics or modifications of the seed quality, can be regarded as being "neutral," for they do not confer any advantage to the transformed plant whatever the ecosystem may be. The fate of resistance or tolerance genes to herbicides, insects, viruses, fungi, or environmental stresses can, on the contrary, differ depending on whether they are or are not exposed to a selection pressure. The maintenance of an herbicide resistance gene will, for instance, be very different out of the field or in the field (or areas sprayed with the herbicide). Finally, a character such as apomixy, which confers the capacity of vegetative multiplication, seems to give the transformed plant an advantage that would lead to the rapid spreading of the gene within populations or species *(3)*.

It is obvious that the destiny of the transgene within a species also depends on the floral biology of this species and on the way it disseminates (pollen, seeds, tubers). The ability of the crop to become a weed itself will depend on the pollen transfer potential over various distances, the survival of the seeds in the soil, and the competitiveness of the crop as a weed *(4)*.

Finally, the last type of risk that must be taken into account is linked to the uncontrolled spreading of the transgene through horizontal or vertical transfers. Although it still seems unlikely today, transfer of genetic information from plants to microorganisms through infection or symbiosis cannot be totally excluded *(5)*. The occurrence of a transfer through sexual hybridization seems more likely, for the presence of wild relatives in the field has been recorded for many cultivated crops. In some cases, the consequences of such crosses for the release of transgenic crops have been studied; for instance, such results are reported for foxtail millets *(6)*.

We have therefore focused our work on this type of risk. A transgenic oilseed rape resistant to an herbicide appeared to be an adequate model to study this risk for the following reasons:

1. The numerous wild relatives, carrying different genomic structures, which are found as weeds in rapeseed cultivated areas; and
2. The properties of an herbicide resistance gene, in particular the economical and ecological impacts its transfer to weeds can have, the easy detection through generations, and finally its susceptibility to selection pressure, applied or not.

The potential risk that an herbicide resistance gene, transferred into rapeseed via *Agrobacterium*, would be transferred to related wild species by cross-hybridization, can be considered according to three points:

1. To what extent is it possible to cross-hybridize rapeseed, either transgenic or not, with related wild species?
2. What are the features of the F1 hybrids obtained?
3. Is the maintenance of the transgene likely?

The authors describe some of the ways they have tried to answer these questions.

2. Materials

A Canadian spring rapeseed variety (*Brassica napus* L., AACC, *n* = 19), named "Westar," homozygous for the *bar* gene conferring resistance to phosphinothricin (commercial name BASTA), was used for the crosses under controlled conditions. Seeds were kindly provided by Plant Genetic Systems (Gent, Belgium).

A cytoplasmic male sterile spring rapeseed variety, "Brutor" *(7)*, was used in the field experiments. This variety was not transgenic.

The wild species considered were hoary mustard (*Brassica adpressa* Boiss. Moench or *Hirshfeldia incana* L. Lagrèze-Fossat, AdAd, *n* = 7) and wild radish (*Raphanus raphanistrum* L., RrRr, *n* = 9). They are common weeds occuring in rapeseed fields, and were locally collected.

3. Methods

3.1. Production of F1 Hybrids Between Rapeseed and Wild Species

Hybridization experiments were carried out either in the greenhouse or in the field.

3.1.1. Under Controlled Conditions

Reciprocal crosses between the transgenic Westar and the two wild species were performed manually in the greenhouse. Ovaries were excised 4 to 6 days after pollination and cultured on a E12 medium *(8)* in a growth chamber (22–23°C d/15°C night; 16 h photoperiod). Seedlings were placed at 4°C for 10 d and then transferred to a MS medium *(9)*, containing 20 g/L of sucrose. Plantlets were finally transferred to the greenhouse at the 4–6 leaf stage.

The number of pollinated flowers and the subsequent recovery of embryos after in vitro rescue are shown on Table 1. Results vary significantly depending on the female parent: More *B. napus* × *B. adpressa* embryos were obtained than for the reciprocal cross. The opposite result was observed for *R. raphanistrum*, contradicting the idea that embryo

Table 1
Production of F1 Interspecific Hybrids

Hybrid structure	Under controlled conditions		Under natural conditions		
	No. flowers	Embryos/ 100 flowers	No. flowers	Seeds/ 100 flowers	No. seeds/m²
B. napus × *R. raphanistrum*	414	1.2	infinite	9.6	3734
R. raphanistrum × *B. napus*	888	2.8	—	—	—
B. napus × *B. adpressa*	326	11.6	infinite	1.9	512
B. adpressa × *B. napus*	1418	2.5	—	—	—

yield is higher when the parent with the higher ploidy level is used as female *(10)*.

3.1.2. Production of F1 Hybrids by Using Cytoplasmic Male Sterility

From these results, we tried to get closer to natural conditions and to maximize the hybridization potential by preventing self- and cross-pollination in rapeseed.

In order to increase the pollination pressure from the wild species, male sterile rapeseed was sown with each wild species on isolated trials. Each trial was composed of alternate plots (*B. napus*/wild species), every plot consisting of five rows 2-m long. One trial involved *B. adpressa* with 25 replications (250 m²). The second one, involving *R. raphanistrum*, had six replications (60 m²). Numbers of flowers, pods, and seed production/m² were assessed on the female parent *(B. napus)*.

Pods harvested on the rapeseed parent were small and buckled, and contained two seed-size groups: Seeds showing a higher diameter than 2 mm were not hybrids. The second group, containing seeds of approx 1.6 mm in diameter, were interspecific hybrids.

The main results given in Table 1 only take into account this second group. The number of hybrid seeds/m² and the number of seeds/100 open pollinated flowers were much higher for the *B. napus* × *R. raphanistrum* trial than for the B. napus × *B. adpressa* one *(11)*.

3.2. Main Features of the F1 Hybrids

Whatever their origin (in vitro culture or field experiments), the F1 hybrid characterization followed the same steps:

1. Isozyme analysis was performed for four systems (phosphoglucoisomerase, phosphoglucomutase, 6 phosphogluconate dehydrogenase, and leucine aminopeptidase), according to the protocols given by Arüs et al. *(12)*.
2. Chromosome counts were made on root-tip dividing cells according to the technique described by Jahier et al. *(13)*.
3. Meiotic behaviors and chromosome pairing were observed on pollen mother cells at the Metaphase I (MI) stage *(13)*.
4. Pollen fertility was estimated by the percentage of mature pollen grains stained by acetocarmine.
5. When the parental rapeseed was transgenic (controlled conditions), resistance to Basta was followed in the F1 hybrids using polymerase chain reaction (PCR) amplification of the *bar* gene by specific primers (5'GAAACCCACGTCATGCCAGTTCC3' and 5'GCACCATCGTCAA CCACTACATCG3', kindly provided by Plant Genetic Systems), in correlation with a nondestructive foliar resistance test.

3.2.1. Results

The hybrid state of the embryos produced in vitro and of the seeds harvested in the field was established using allozymes showing different electrophoretic mobilities in rapeseed and in the wild species. Six loci were used for the hybrids with *B. adpressa* and four for the hybrids with *R. raphanistrum (11)*. All the embryos developed from in vitro culture appeared to be interspecific hybrids. For both field trials the larger seeds were determined to be *B. napus*, while the smaller were interspecific hybrids showing bands from both parental species. Morphology of the interspecific F1 hybrids was very close to rapeseed. Hairiness and flower size were intermediate.

Cytological studies showed that the hybrids obtained had the expected triploid structure (ACX), except for one amphiploid plant obtained on the *R. raphanistrum* field trial (AACCRrRr), probably resulting from unreduced gametes (Table 2).

The analysis of chromosome pairing in the meiosis of the triploid hybrids produced under controlled conditions showed that pairing was much higher in the hybrids with *R. raphanistrum* than with the *B. adpressa* ones, whatever the female parent (Table 2). The same tendency was observed in the hybrids produced in the field.

Haploid rapeseed controls were studied to quantify chromosome pairing between the A and C genomes, which can be very different according to the cultivar considered (7.45 for Westar, 12.56 for Brutor). The hybrids

Table 2
Cytological Characterization of F1 Interspecific Hybrids

	Mitotic analyses			Meiotic analyses			
Hybrid structure	No. plants	2n	Genomic structure	No. plants	2n	Mean number of paired chromosomes/cell	Range of male fertility
				Produced under controlled conditions			
B. napus × R. raphanistrum	3	28	ACRr	3	28	13	0
R. raphanistrum × B. napus	8	28	RrAC	8	28	13.92	0–10%
B. napus × B. adpressa	6	26	ACAd	4	26	7.72	0
B. adpressa × B. napus	11	26	AdAC	8	26	7.62	0
Haploid B. napus control	4	19	AC	4	19	7.45	—
				Produced under natural conditions			
B. napus × R. raphanistrum	188	28	ACRr	7	28	15.29	0–30%
	1	56	AACCRrRr	1	56	48.0	a
R. raphanistrum × B. napus	—	—	—	—	—	—	—
B. napus × B. adpressa	84	26	ACAd	2	26	6.68	0%
				6	26	13.02	0–30%
B. adpressa × B. napus	—	—	—	—	—	—	—
Haploid B. napus control	2	19	AC	2	19	12.56	—

[a]Data not obtained.

produced with *B. adpressa* showed, whatever their origin, a mean number of paired chromosomes per cell similar to the corresponding haploid rapeseed except for two plants. These latter showed an inhibition of chromosome pairing probably owing to the Ad genome. Pairing between AC and Rr genomes in the ACRr and RrAC hybrids was much higher than pairing in the AC haploids, indicating that recombination may occur between the genomes of *B. napus* and *R. raphanistrum* (Table 2).

Male fertility of the hybrids was usually low. However, some of the triploid hybrids reached up to 30% fertility, which is noteworthy for the hybrids produced in the field knowing that they carry a cytoplasm inducing male sterility in rapeseed.

The hybrids produced under controlled conditions were expected to carry the *bar* gene. This was checked on five hybrids with *R. raphanistrum* and six with *B. adpressa*. All were Basta resistant and gave a positive signal by PCR amplification *(14)*. Since it is possible to produce interspecific transgenic hybrids, it is important to assess the likely fate of these hybrids.

Table 3
Production of a Progeny from F1 Interspecific Hybrids

Hybrid structure	Under controlled conditions			Under natural conditions		
	No. plants	No. flowers	Embryos/ 100 flowers	No. plants	No. flowers	Seeds/ 100 flowers
B. napus × *R. raphanistrum*	3	480	0.6	281	infinite	0.05
R. raphanistrum × *B. napus*	8	277	0.6	—	—	—
B. napus × *B. adpressa*	4	384	0.3	158	infinite	0.22
B. adpressa × *B. napus*	8	589	0	—	—	—

3.3. Likelihood of Maintenance of the Transgene in the F1 Progeny

Experiments were carried out either in the greenhouse or in the field in order to produce back-crosses with the wild species.

3.3.1. Under Controlled Conditions

Methods were the same as the ones described for the F1 hybrid production. The F1 hybrids were always used as female parents. The number of pollinated flowers and embryos/100 pollinated flowers are detailed in Table 3. They were very much lower than the results obtained in the F1 hybrid production (Table 1). No progeny could be obtained from the *B. adpressa* × *B. napus* hybrids. On the whole, the embryo yield was higher for the hybrids with *R. raphanistrum* than for the ones with *B. adpressa*. Attempts to get progeny by selfing the hybrids gave no embryos at all.

3.3.2. Production of F1 Hybrid Progeny in the Field

A 20-m² plot of the interspecific F1 hybrids produced with cytoplasmic male sterility (CMS) was sown in an isolated field. In each trial the hybrids were surrounded by their corresponding parental weed population. The main results can be found in Table 3. Numbers of seeds/100 pollinated flowers were again much lower than for the F1 hybrid production. Very few F1 hybrid plants produced pods (18.3% for the *B. napus* × *B. adpressa* hybrids, and 8.5% for the *B. napus* × *R. raphanistrum* hybrids), and these pods contained low amounts of seeds (11).

3.3.3. Main Features of the Progeny

Chromosome counts, meiotic behavior, and fertility of the progeny were determined following the same steps as for the F1 hybrids. Results are available for the trial involving *R. raphanistrum*. Of the nine plants observed (one could not be analyzed further because it was totally sterile), three showed the expected structure AACCRrRr, $2n = 37$ (with one showing a 56% pollen fertility) and five were amphiploids AACCRrRr, $2n = 56$ (two showing 65 and 67% pollen fertility). The analysis of the seeds obtained in the *B. adpressa* trial is in progress.

4. Notes

4.1. Risk Assessment Procedures

The procedures might be very different, depending on the presence or not of wild relatives in the cultivated area, and the potential hybridization between the crop and the wild species. It seems that for a crop such as maize, which has no contact with relatives in our European countries, the spreading of the gene is most unlikely. Likewise, potato does not cross-hybridize with wild species under natural conditions. The problem however could occur when growing transgenic lines at the diversification sites of these crops, i.e., South America for potato and Central America for maize. For these two crops, risk assessment in Europe has to focus on other issues, such as dispersal within the species *(15)*. In contrast, some crops can freely cross-hybridize with wild species flowering at the same period. It is the case for beet, for instance, where interspecific hybrids can be produced even from two fertile parents *(16)*. Similar effects have been observed in rapeseed grown in Northern Europe, where *Brassica rapa* often occurs as a wild species. Spontaneous *B. napus* × *B. rapa* hybrids have been described *(17)*. When spontaneous infertility of the crop and the wild species is not obvious, our study shows that some features, such as the availability of a CMS, can be successfully used for risk assessment. The model shown is restricted to the hybridization process using *B. napus* as a female. The availability of a male sterility for each parental species would be helpful. As it is not the case for the wild species we studied, we plan to take advantage of the self-incompatibility shown by *R. raphanistrum* to produce seeds from the reciprocal cross under isolated cages containing one plant of each species.

4.2. Hybridization Frequencies

The production of F1 hybrids is possible and easier than BC1 production in vitro as well as in the field. There seems to be a bottleneck at the BC1 level, due to the poor fertility in the interspecific F1 hybrids.

Embryo rescue techniques are interesting for getting F1 hybrid or BC1 plants but not relevant to quantitative evaluation of their production. Indeed, there are differences between the species in the number of embryos or seeds/100 pollinated flowers according to the technique used.

This could be owing to significant differences between the techniques used. Intraspecific variability of the wild species is better exploited in the field than in the greenhouse, because the number of flowers is not limited and pollination occurs at different stages after anthesis. On the other hand, when using in vitro techniques, pollination before anthesis may break down incompatibility systems, and embryo development is favored. These last arguments could explain the better results obtained for *B. adpressa* under controlled conditions.

The effect of the rapeseed genotype is still difficult to assess. We are now trying to extend the experiments to different rapeseed male sterile and transgenic cultivars, grown in similar field design for the production of F1 hybrids with *R. raphanistrum*.

4.3. Features of the Hybrids

Viability of the hybrids was satisfactory. When the parental rapeseed was transgenic, the hybrids also carried the transgene and expressed it. Until now, no regulatory effect of the interspecific background on transgene expression was observed.

The risk of introgression of a transgene into the genome of a wild species is closely linked to the occurrence of meiotic recombination during chromosome pairing in the meiosis of the F1 hybrids. Triploid hybrids (ACX) are more likely to give rise to recombination than amphiploid hybrids (AACCXX), but it seems recombination events also depend on regulation effects varying from one species to the other. Pairing is sometimes inhibited in hybrids with *B. adpressa*, while it is favored in hybrids with *R. raphanistrum*. Therefore, recombination of a chromosome fragment carrying a transgene into *R. raphanistrum* genome looks more likely, and this result is supported by the successful introduction under selection pressure of a fertility restorer gene from *R. sativus* to *B. napus (11,18)*.

Recombination in the meiosis of amphiploid hybrids AACCXX, so that a transgene would be transferred to the wild species genome, seems unlikely. However, the high fertility they show raises the possibility of them becoming a new species as a whole, in which a transgene would be maintained easily. The same question can be pointed out for the BC 1 with the AACCRrRr structure showing increased male fertilities, or even for the subsequent generations. After all, synthetic rapeseed AACC, after some generations of selfing, may become a fertile stable cultivar *(19)*.

4.4. Features of the Progeny

The issues of the maintenance under selection pressure of the chromosome carrying the transgene, the introgression by recombination or the possible occurrence of unreduced gametes come again at every generation. Recombination will again depend on regulatory systems, on the insertion site of the transgene in the rapeseed genome, and on the possible loss of the transgene at one generation. Unreduced gametes are more likely to be observed when the fertility of the mother plant is low.

4.5. Fitness

The next issue to be considered is the fitness of transgenic F1 hybrids and progeny (successive backcrosses or selfings) in various ecosystems. Their competitiveness will very much depend on the features described in Section 4.3. and on selection pressure, so that one can consider that case by case experiments are needed according to the nature of the transgene and the features of the crop.

4.6. Safety Regulations

National committees, such as CGB (Commission du Génie Biomoléculaire) in France, the Advisory Committee on Release to the Environment in the UK, or the Provisional Committee on Genetic Modification in the Netherlands, tend to set up regulations and confinement regimes for the experimental use of transgenic crops aiming at risk assessment, and for the commercial release of such crops. The EEC 90/220 directive draws together future restraints in the use of transgenic crops throughout Europe.

Acknowledgment

This work was supported by the European Economic Community-Biotechnology Research for Innovation, Development and Growth in Europe (EEC-BRIDGE) program.

References

1. Dale, P. J., Irwin, J. A., and Scheffler, J. A. (1993) The experimental and commercial release of transgenic crop plants. *Plant Breeding* **111,** 1–22.
2. Kessler, D. A., Taylor, M. R., Maryanski, J. H., Flamm, E. L., and Kahl, L. S. (1992) The safety of foods developed by biotechnology. *Science* **256,** 1747–1750.

3. Arnould, J., Gouyon, P. H., Lavigne, C., and Reboud, X. (1993) OGM: une théorie pour les risques [GMO: a theory to evaluate the risks]. *Biofutur* **124,** 45–50.

4. Crawley, M. J., Hails, R. S., Rees, M., Kohn, D., and Buxton, J. (1993) Ecology of transgenic oilseed rape in natural habitats. *Nature* **363,** 620–623.

5. Bryngelsson, T., Gustaffson, M., Green, B., and Lind, C. (1988) Uptake of host DNA by the parasitic fungus *Plasmodiophora brassicae. Physiol. Mol. Plant Path.* **33,** 163–171.

6. Till-Bottraud, I., Reboud, X., Brabant, P., Lefranc, M., Rherissi, B., Vedel, F., and Darmency, H. (1992) Outcrossing and hybridization in wild and cultivated foxtail millets: consequences for the release of transgenic crops. *Theor. Appl. Genet.* **83,** 940–946.

7. Pelletier, G., Primard, C., Vedel, F., Chétrit, P., Remy, R., Rousselle, P., and Renard, M. (1983) Intergeneric cytoplasmic hybridization in *Cruciferea* by protoplast fusion. *Mol. Gen. Genet.* **191,** 244–250.

8. Delourme, R., Eber, F., and Chèvre, A. M. (1989) Intergeneric hybridization of *Diplotaxis erucoides* with *Brassica napus.* I-Cytogenetic analysis of F1 and BC1 progeny. *Euphytica* **41,** 123–128.

9. Murashige, T. and Skoog, F. (1962) A revised medium for rapid growth of bioassay with tobacco tissue culture. *Physiol. Plant* **15,** 473–497.

10. Kerlan, M. C., Chèvre, A. M., Eber, F., Baranger, A., and Renard, M. (1992) Risk assessment of outcrossing of transgenic rapeseed to related species: I. Interspecific hybrid production under optimal conditions with emphasis on pollination and fertilization. *Euphytica* **62,** 145–153.

11. Eber, F., Chèvre, A. M., Baranger, A., Vallée, P., Tanguy, X., and Renard, M. (1994) Spontaneous hybridization between a male sterile oilseed rape and two weeds. *Theor. Appl. Genet.* **88,** 362–368.

12. Arüs, P., Chèvre, A. M., Delourme, R., Eber, F., Kerlan, M. C., Margalé, E., and Quiros, C. F. (1991) Isozyme nomenclature for eight enzyme systems in three *Brassica* species, in *Proceedings of the GCIRC Eighth International Rapeseed Congress,* vol. 4, Saskatoon, Canada, pp. 1061–1066.

13. Jahier, J., Chèvre, A. M., Tanguy, A. M., and Eber, E. (1989) Extraction of disomic addition lines of *B. napus-B. nigra. Genome* **32,** 408–413.

14. Kerlan, M. C., Chèvre, A. M., and Eber, E. (1993) Interspecific hybrids between a transgenic rapeseed (*Brassica napus* L.) and related species: cytogenetical characterization and detection of the transgene. *Genome* **36,** 1099–1106.

15. Tynan, J. L., Williams, M. K., and Conner, A. J. (1990) Low frequency of pollen dispersal from a field trial of transgenic potatoes. *J. Genet. Breed* **44,** 303–306.

16. Boudry, P., Mörchen, M., Saumitou-Laprade, P., Vernet, P., and Van Dijk, H. (1993) The origin and evolution of weed beets: consequences for the breeding and release of herbicide resistant transgenic sugar beets. *Theor. Appl. Genet.* **87,** 471–478.

17. Kapteijns, A. J. A. M. (1993) Risk assessment of genetically modified crops. Potential of four arable crops to hybridize with the wild flora. *Euphytica* **66,** 145–149.

18. Pellan-Delourme, R. and Renard, M. (1988) Cytoplasmic male sterility in rapeseed (*Brassica napus* L.): female fertility of restored rapeseed with "Ogura" and cybrid cytoplasm. *Genome* **30,** 234–238.

19. Prakash, S. and Hinata, K. (1980) Taxonomy, cytogenetics and origin of crop *Brassicas,* a review. *Opera Bot.* **55,** 1–57.

CHAPTER 33

Electroporation Protocols
for *Agrobacterium*

Garry D. Main, Stephen Reynolds,
and Jill S. Gartland

1. Introduction

Agrobacterium spp. are widely used to transform new genes into plant tissue. Transferring useful genes into bacteria in order to transform them into the target plant is an essential part of the process, which can be achieved by electroporation.

Electroporation is a technique involving exposing cells to an electric field for a short duration, in order to facilitate the cross membrane transfer of material. This involves the transient production of pores or holes in the cell membrane. In order to gain an understanding of the cellular events caused by electroporation the following questions need to be answered:

1. How does a cell interact with the applied electric field?
2. What causes the formation of pores?
3. What happens when the cell membrane is disrupted?
4. How does the cell recover?

The interaction between the cell and the electric field is electrostatic. The bacterial cell membrane is composed of a phospholipid bilayer. This is a charged structure. The phosphate groups on the outer layers provide a hydrophilic region, the lipids in the center are hydrophobic. This provides a good permeation barrier for the cells.

When a cell suspended in a physiological salt solution ($\approx 0.15M$ KCl or NaCl) is exposed to an electric field within the electroporation appara-

From: *Methods in Molecular Biology, Vol. 44: Agrobacterium Protocols*
Edited by: K. M. A. Gartland and M. R. Davey Copyright © 1995 Humana Press Inc., Totowa, NJ

tus, there is a rapid redistribution of cations, e.g., Ca^{2+} and Mg^{2+}, in the vicinity of the plasma membrane. The result of this is a transmembrane potential. The composition of the membrane and the charge across it will determine the number and size of the pores formed *(1)*.

The maximum transmembrane potential that may be generated is given by the following equation:

$$\Delta\psi = 1.5rE_a$$

where $\Delta\psi$ is the maximum membrane potential, r is the radius of the cell, and E_a is the direct current strength of the field applied. This assumes that the duration of the field was greater than the time required for it to rise to its maximum level. The rise time is a constant imposed by the discharging system, generally a capacitor bank.

The application of the electric field can cause rapid isomerization of the—CH_2—CH_2—bonds within the phospholipid molecules in the membrane. If the field strength is of a sufficient magnitude the disruption will spread to the surrounding molecules, causing the formation of a pore. The normal state of the membrane is an equilibrium between the molecules in the solid and fluid phase. The electrical pulse will cause more molecules to enter the fluid phase aiding in the formation of a pore *(1)*.

During the existence of the pore, molecules such as plasmids pass into the cell. The size of plasmids entering depends on the cell types and the field strengths. Plasmid number is affected by size and concentration and the duration of the pore. The longer the pore exists, the greater the risk of losing cell contents to the surrounding media. Most of the material that may be lost during the pore's short existence can be replaced, but during longer time periods cell death is risked. Extended poration times may allow the transfer of larger material, but if the cell does not survive there will not be much to be gained.

During a short duration pulse, the cell membrane will suffer disruption, but when the field is removed, the transmembrane potential will drop back to normal, allowing the molecules within the membrane to realign to their normal state, thus closing the pore.

The field strength required to create pores is dependent on the bacteria being used. Even between strains of *Agrobacterium* there are differences. Most of the current protocols use field strengths of approx 12 kV/cm *(2–6)*.

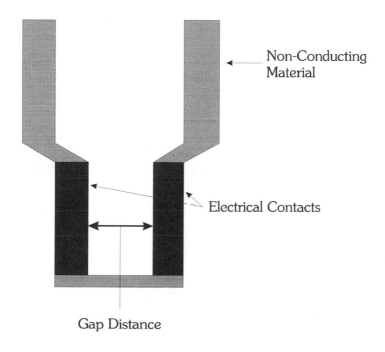

Fig. 1. Electroporation cuvet.

2. Materials

The equipment required for electroporation is minimal and many people make their own equipment. Alternatively, a commercial apparatus is available, e.g., Gene Pulser from BioRad (Richmond, CA) *(7)*. The requirements are as follows:

1. Electroporation cuvets: These are available in a variety of sizes. As shown in Fig. 1, the cuvet resembles a normal cuvet, with the addition of metal contacts on two sides at the base. These extend through the cuvet, but are not linked. The gap distance is the space between these two metal plates. During electroporation, the cells are suspended in the media in this gap (*see* Note 1).
2. Safety equipment: Appropriate safety equipment is vital for this type of experimentation. It may only consist of a box over the cuvet and electrodes while firing, but some operator protection from the voltage must exist. The voltage ranges in use vary from 220–3000V, which are potentially very damaging (*see* Note 2).

   Figure 2 shows a typical cuvet holder. The cuvet, shown in Fig. 1, is mounted in a sliding mechanism. The contacts to complete the circuit and

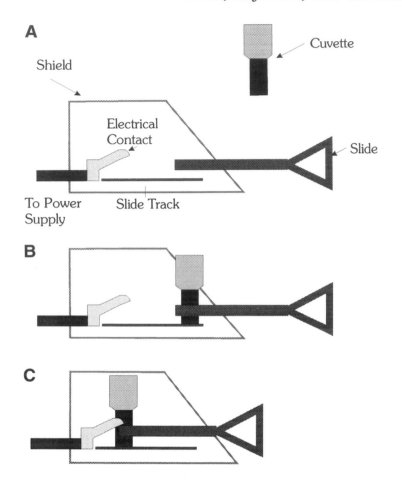

Fig. 2. Cuvet holder. **(A)** Safety setup prior to loading. **(B)** Cuvet loaded into slide. **(C)** Slide used to place cuvet within the safety shield. The cuvet is now part of the electrical circuit, and the operator is protected.

   link the cuvet to the power supply are within a shield, when the cuvet cell is pushed forward into the apparatus it completes the circuit (*see* Note 3).
 3. Power supply: This may range from a constant voltage power supply either alternating current (ac) or direct current (dc) to a microprocessor controlled stepped voltage supply (*see* Note 4).
      The most commonly used low-cost commercial system is based on the electrical discharge of a capacitor bank. The capacitors are charged via a set of resistors to the required voltage (*see* Note 5). A trigger system is used to discharge the capacitor bank through the cuvet (*see* Note 6).

4. Plasmid DNA: A suitable plasmid is required for introduction into the *Agrobacterium*. The ideal plasmid should have a gene of interest, a selectable marker, and be capable of replication in both *Agrobacterium* and *E. coli* for cloning purposes (*see* Note 7).

5. *Agrobacterium* growth media: A typical medium for *Agrobacterium* is YMB. Some strains may have different requirements. YMB broth (1 L): 0.4 g Yeast extract, 10 g mannitol, 0.1 g NaCl, 0.2 g $MgSO_4 \cdot 7H_2O$, 0.5 g K_2HPO_4, pH 7.0–7.2.

6. Cell storage: The prepared *Agrobacterium* cells may be stored at –70°C in a 10% (v/v) glycerol solution until required. Details of the cell preparation are given in Section 3.1.

7. Recovery medium (SOC): Rapid transfer of the electroporated *Agrobacterium* cells to growth media improves the transformation efficiency by increasing the recovery rate of the cells. The use of SOC reduces losses from cell death (*see* Note 8). SOC medium: 2% (w/v) bacto tryptone, 0.5% (w/v) bacto yeast extract, 10 mM NaCl, 2.5 mM KCl, 10 mM $MgCl_2$, 10 mM $MgSO_4$, 20 mM glucose.

8. Selective medium: After the transformation has been undertaken, some method by which transformed cells can be identified from nontransformants is required. Usually the plasmid that has been inserted will contain a selectable marker. The plating out of the cells, after recovery, onto the selective media will thus allow the identification of transformants. If the cells are first spread onto both selective and nonselective growth media, the transformation efficiency can be obtained (*see* Note 16).

3. Methods
3.1. Cell Preparation

1. Prepare a 10-mL YMB culture of the *Agrobacterium* strain.
2. Grow cells at 25°C and shake vigorously until OD_{600} reaches 0.5–0.7 (*see* Note 9).
3. Place the flask on ice for 15–30 min prior to harvesting.
4. Harvest by centrifugation at 4000 rpm for 15 min.
5. Remove as much supernatant as possible with a sterile pipet.
6. Resuspend the cell pellet in 7 mL sterile-distilled water. Centrifuge as in step 4.
7. Repeat step 6 (*see* Notes 10 and 11).
8. Resuspend in a final vol of 100 μL in 10% (v/v) sterile glycerol in an Eppendorf tube.
9. Store the cells at –70°C until required.

3.2. Electroporation

1. Gently thaw the cells and place on ice.
2. In a prechilled Eppendorf tube, mix 40 μL of cell suspension with 2 μL of DNA, typically 2 ng (*see* Note 12).

3. Mix well and stand for 2 min on ice.
4. Transfer the mixture of cells and DNA to a prechilled cuvet (2 mm gap distance).
5. Set the cuvet into the apparatus, ensuring all safety features are in place.
6. Adjust the field strength settings to the desired level.
7. Pulse at the desired settings, e.g., 12.5 kV/cm (*see* Note 13).
8. Switch off all power and ensure apparatus is fully discharged.
9. Immediately remove the cuvet and add 1 mL of prechilled (on ice) SOC recovery medium (*see* Note 14).
10. Resuspend the cells with a sterile pipet.
11. Transfer the cells to a sterile Eppendorf tube.
12. Incubate at 28°C for 1–1.5 h (*see* Note 15).
13. Plate onto selective and nonselective media (*see* Note 16).
14. Incubate at 28°C for 2–3 days.

4. Notes

1. The gap distance may vary depending on the dimensions and capacity of the cuvet. The standard cuvet size is 1.5 mL, with a gap distance of 2 mm.
2. Only suitably trained personnel should be allowed to operate this device.
3. This ensures that it would be quite difficult for any person to accidentally place his or her fingers in the circuit. The voltages and field strengths used for electroporation are similar to those used in hospital defibrilators and will have a similar effect on any person tampering with the system.
4. Most electroporation setups, however, make use of dedicated power supplies. There are three types of pulse that have been used for transformation, namely exponential decay, square wave, and ac pulse. The exponential decay is used in the method described here. This is the easiest form of pulse to generate, because most electric companies supply electrical power in this manner.

   The square wave is a modification of this, the supplied sine wave being smoothed out via an electronic circuit to produce a stepped voltage. The square wave and sine waves allow repeated cycling of the field in the cuvet. The field direction changes gradually in the sine wave and abruptly in the square wave to the opposite orientation, as shown in Fig. 3. The time taken to switch the field direction can be controlled. An attempt to create multiple pores within the cell membrane and allow as much material as possible to enter the cell must be balanced by the increased risk of cell damage.
5. Altering the voltage alters the field strength, as explained earlier.
6. The rise to maximal field strength can be altered by the use of ballast resistors to slow the discharge and hence the rise rate.
7. The DNA to be inserted must be suspended in a low ionic strength buffer such as TE. High salt concentrations increases the conductivity within the

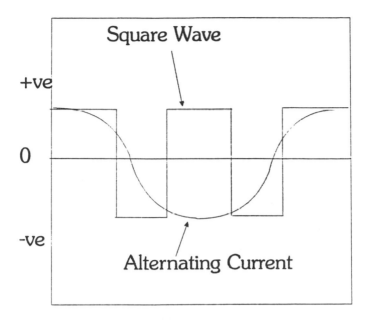

Fig. 3. Illustration of square and sine waves.

cuvet, and hence leads to a decrease in the available maximum field strength. It may also lead to a feature known as arcing, flashes of light within the cuvet, indicating that the pulse has discharged directly across the cuvet, with little or no electric field being produced, lowering the efficiency of transformation. TE Buffer: 10 mM Tris-HCl, pH 8.0, 1 mM EDTA.

8. Immediately after electroporation, the cells should be transferred to growth media, SOC, or YMB, and incubated at 28°C for 1–1.5 h.

9. As electroporation works best with cells that are growing rapidly, the timing of the harvesting is crucial to the success of the transformation. The optimal optical density may vary depending on the *Agrobacterium* strain used and it may be necessary to assess previously the bacterial growth curve.

10. Washing the cells during the harvesting stage helps to remove any surface material, bacterial slimes, and so on, that may be on the cells. The removal of this material increases the transformation efficiency.

11. The presence of any contaminating material in the reaction mix may lead to arcing within the cuvet. This may lead to damage to the cuvet and apparatus. Therefore, great care must be taken during the washing stages to remove all contaminating material.

12. The ratio of cells to DNA may vary, 100 µL cells to 2 ng of DNA is one such ratio. The amounts required vary between *Agrobacterium* species and

also with the size of plasmids which are in use. The length of pulse and field strength used also varies among samples *(2,3,5)*.

13. The field strength to be used has to be selected by experimentation; 12.5 kV/cm is a useful starting point *(2,4)*.
14. The rapid addition of cold SOC after the pulse is crucial to successful electroporation. This increases the recovery rate of the cells.
15. Shaking the tubes gently during the recovery period may increase the transformation efficiency.
16. The electroporated cells should then be plated out onto selective media for analysis and further work. After 48–72 h, those cells that have the plasmid should have grown into discrete colonies on the plates. Positive and negative controls should be used, of course. If the cell suspension is split prior to plating, so that a sample is plated onto selective and nonselective media, it should be possible to determine the efficiency of the transformation.

Efficiency calculation example: 2750V, 2 ng DNA, 40 µL *Agrobacterium,* 10-µL culture aliquots plated on selective and nonselective media:

$$\text{Effciency } \% = (\text{colonies on selective media/colonies on nonselective media}) \times 100 = 12/350 = 3.42\%$$

References

1. Chernomordik, L. V. (1992) Electropores in lipid bilayers and cell membranes, in *Electroporation Guide* (Chang et al., eds.), Academic, London.
2. Joersbo, M. and Brunstedt, J. (1991) Electroporation: mechanism and transient expression. *Physiol. Plant.* **81,** 256–264.
3. Mersereau, M., Pazour, G. J., and Das, A. (1990) Efficient transformation of *Agrobacterium tumefaciens* by electroporation. *Gene* **90,** 149–151.
4. Mozo, T. and Hooykass, P. J. J. (1991) Electroporation of megaplasmids into *Agrobacterium. Plant Mol. Biol.* **16,** 917,918.
5. Mattaanovich, D., Rüker, F., Da Câmara Machado, A., Laimer, M., Regner, F., Steinkellner, H., et al. (1989) Efficient transformation of *Agrobacterium spp*, by electroporation. *Nucleic Acids Res.* **17,** 6747.
6. Wen-Jun, S. and Forde, B. G. (1989) Efficient transformation of *Agrobacterium spp.* by high voltage electroporation. *Nucleic Acids Res.* **17,** 8385.
7. Gerand, S. (1993) Electropulsation: important parameters to consider. *Internat. Biotechnol. Lab.* **11,** 22.

Index

A

AB medium, 24
Abscisic acid analysis, 259, 260
Acetosyringone, 24, 35, 37–45, 61,
    124
ACGM Guidance Note Series, 385
Agrobacterium strains, 90, 91, 155
Agroinfection, tissue-specific
    specificity, 331
Agroinfection, 325–342, 344–350
Analysis transformants, 101–119
Antisense RNA, 264–266
Arabidopsis thaliana, 135–147
Arachis hypogea, 87–100
Ascorbate peroxidase assay, 317
AT minimal medium, 38
att, 15
Attractants, 30
ATx2 stock solution, 38
Augmentin, 143
Autonomously replicating viral
    vectors, 330
Azocoupling GUS histochemistry,
    190, 191

B

5-Bromo-4-chloro-3-indolyl
    beta-D-glucuronide, *see*
    X-glucuronide
B5 vitamin stock, 63
bar gene primers, 397
Bare root solution culture, 299
BATNEEC, 380, 381

Benedict's reagent, 10
Beta-galactosidase, 22
Beta-galactosidase assay, 25, 26
Beta-glucuronidase, *see* GUS
Binary vectors, 16, 47–58
Biosafety, 328, 402
Biovars, 9, 10
Blackberries, 129–133
Blackcurrants, 129–133
Blindwell assays, 32
Brassica napus, 71–78, 79–85
Broadleaved trees, 149–165

C

Cadmium,
    field studies, 300, 301
    partitioning, 295–308
    quantitation, 299, 300
Callus culture (CC) medium, 73
Capillary assays, 31
cDNA constucts, 263, 264
Chemical fusion, 178
Chemotaxis, 29–36
Chemotaxis medium, 30
Chloramphenicol acetyltransferase,
    53
*chv*A, B, 15
Cocultivation, 74
Cocultivation protoplasts, 287–289
Contained use regulations, 381–385
Cosmid libraries, 351–367
Cosmid libraries, vectors, 353–360
CTAB DNA isolation, 105, 106 108,
    109

Cu/Zn SODs, 309–323
Cuvet holder, 408
Cytokinin analysis, 251–257
Cytokinin-N-glucosidase, 214

D

Deliberate release regulations,
    385–388
Deuterated tracers, 248
Diazomethane preparation, 246, 247
Diethylaminoethyl (DEAE)-Sephadex
    A25 preparation, 246
Differential acid production assay,
    10, 11
Direct current strength, 406
Disarmed vectors, 47, 48
Disarming, 16

E

Ecotypes, 135
Electrical safety, 407, 410
Electrofusion, 178, 179
Electroporation, 152, 153 264–266
    269, 270, 405–412
Electroporation cuvet, 407
Enhancer tag vector, 783–285
ENZ 1/3/5 enzyme solution, 72
European Community law, 369,
    373–377

F

F1 hybrid production, 395–399
Feasibility studies, 154–156
Federal oversight, 379
Feeder plates, 79
Fluorometric GUS analysis,
    195–199
Fragaria, 129–133
Freeze-drying, 2, 3
Freezing, 3
Functional complementation, 363

G

Gap distance, 10
Gas chromatography-coupled mass
    spectrometry, *see* GC-MS
Gas chromatography-electron capture
    detection, *see* GC-ECD
GC-ECD, 259
GC-MS, 258–260
Gel filtration medium, 297
Gene manipulation, safety, 328,
    369–391
Genomic blots, 106, 111–115
Glycine max, 101–119
Growth conditions, 3
Growth regulators, preparation, 73
GUS, 16, 53, 89, 153
    assay buffer, 196
    extraction buffer, 196

H

^{2}H-ABA preparation, 248
Hairy root disease, 207
High hormones (HH) medium, 123
High performance liquid
    chromatography, *see* HPLC
Histochemical GUS analysis,
    185–193, 348, 349
Hornibrook medium, 2
HPLC-diode array, 251, 252
HPLC-fluorescence, 257, 258
HPLC-MS, 253–257
Hybridity, confirmation, 180, 181
Hydroxycinnamoyl Amide Analysis
    226, 230–232
Hypervirulent plasmid pTOK47, 61

I

IAA analysis, 237–244 257–259
    extraction, 240, 241
    derivatization, 241–242
*iaa*H, 15
*iaa*M, 15

IAM measurement, 242, 243
Immunodetection, 275–277
Indigogenic GUS histochemistry, 189, 190
Indole-3-acetamide, *see* IAM
Indole-3-acetic acid, *see* IAA
Indole-beta-glucosidase, 214, 215
Inducers, 24–27
Inheritance kanamycin resistance, 83
Inheritance test, 139, 140
In situ/localization, 186–188
Intrinsic fluorescence, 199
Ion-exchange paper, 203
ipt, 15

K

3-Ketolactose production, 10
3-Ketolactose test, 11
Kanamycin resistance, 74, 75, 82

L

Lactose agar, 10
Lambda clones, 352, 353
LB medium, 24
Leaf and stem regeneration (LSR-I) medium, 123
Leaf-disc transformation, 59–70, 266, 267, 270, 271
Lettuce transformation, 63–69
Localization accumulated Cd, MT, 301, 302
Localization artifacts 188, 189
Luciferase, 53
Lycopersicon, 167–183

M

4-Methyl-umbelliferyl-beta-D-glucuronide, 196
Maize inoculation, 334–336
Maize shoots, cocultivation, 348
Maize streak virus, 333, 334

Maximum transmembrane potential, 406
MCP labeling, 33
Metallothioncin (MT), 295–297
Minimal medium (MM), 2
Motility, 30
    assay, 10, 11
    medium, 11
Mouse MT-I gene, 63
MS rooting medium, 63
MS selection medium, 63
MS suspension medium, 63
MS104 medium, 63
MS30 liquid medium, 123
MSR medium, 63
MUG, 196
Murashige and Skoog media, 170
Mutant analysis, 289, 290

N

Neomycin phosphotransferase, *see* *npt*II
Northern blotting, 267, 268, 271–274
*npt*II, 201–206
    extraction buffer, 203
    phosphate buffer, 203
    reaction buffer, 203
    saturation solution 203
NPTII assay, 96, 97, 103, 105, 108
Nurse culture plates, 65
Nutrient glucose agar (NGA), 10

O

Oil-seed rape protoplasts, 71–78
Oil-seed rape stem disks, 79–85
Opine detection, 94–96
Outcrossing risks, oilseed rape, 393–403

P

PCR, 111, 175, 176
PCR analysis transformants, 104, 106, 110, 111

Peanut, 87–100
Pectate medium PGA, 11
Pectolytic activity, 11
pGV3850, 72
Photoinhibition analysis, 318, 319
Photosynthesis,. manipulating,
    263–280
Phytohormone analysis, combined
    extraction, 248–251
Plant infectious agents, 326, 327
Polyamine analysis, 224, 225,
    227–230
Polygalacturonic acid (PGA) test, 10
Polymerase chain reaction (PCR), 9
Potato, 121–128
Potato dextrose agar (PDA) calcium
    carbonate medium, 10
Potato shoot cultures, initiation, 124
Promoter analysis, 55, 56
Protein concentration, determination,
    204
Protoplast culture (PC) medium, 73
Protoplast fusion, 172, 177–180
Protoplast isolation, 74, 171, 177
*psc*A, 15

Q

Quantifying phytohormones,
    237–244, 245–262

R

Radioactive tracers, 247
Radioimmunoassay 252, 253
Raspberries, 129–133
Recognition avirulent mutants, 38, 41
Reporter genes, 17
Revival lyophilized bacteria, 4
Rhizobium minimal medium (RMM),
    2
Ri plasmids, 207–222
Ribes, 199–133
Rise time, 406

Risk assessment, 380–391, 400
rol genes,
    functions, 217
    morphological development,
        208–212
    root induction 212, 213
Root-forming selective medium, 297
Rubus, 129–133
Ruthenium red, 12, 15

S

Shoot forming selective medium, 297
Shoot regeneration (SR) medium, 73
Sine wave, 410, 411
Skimmed milk, 5
SOC recovery medium, 409
SOD,
    assay, polyacrylamide gel, 313,
        316, 317
    assay, spectrophotometric 313,
        314, 317
    plant stress protection, 311
    oxidative stress tolerance, 311,
        312
Sodium hypochlorite solution, 72
Soft fruit multiplication medium, 131
Solanum tuberosum, 121–128
Somatic hybrids, 167–183
Soybean, 101–119
Square wave, 410, 411
Standardization, 377, 378 8
Stilbene synthase, 71, 79
Stop buffer, 196
Storage, 1
Strawberries, 129–133
Superoxide-dismutase, *see* SOD
Surface sterilant, 15

T

T-DNA structure, 16, 17
T-DNA tagging, 281–294
Testing gene function, 363

Thidiazuron, 133
Tobacco protoplast isolation,
    285–287
Tomato, 167–183
Transfection protoplasts, 290, 291
Transformation,
    cotyledons, 138
    E. coli, 270
    efficiency, 144, 145
    internodes, 160, 161
    root explants, 138, 139
    leaf disks, 158, 159
    process, 21
    trees, 150
Transgene maintenance, 399, 400
Transmembrane potential, 406, 407
Triparental mating, 91, 93
Tryptone-yeast extract medium (TY), 2

U

UK legislation, 380, 381, 389, 390
United States legislation, 377–380

V

Vancomycin, 143
vir genes 19, 29, 37
Viral replication, 329, 330
Virulence, 15–27
Virulence complementation analysis,
    44

W

W13M salt solution, 72
Western blotting, 269, 274, 275

X

X-glucuronide, 53, 54, 187, 188

Y

Yeast mannitol broth medium
    (YMB), 2, 24
YEB medium, 75, 80
YEP medium, 24